Encyclopaedia Britannica
11th Edition, “Crocoite” to “Cuba”

By
Various

PUBLISHED BY: 2024 by BTB Publishing

ISBN: 978-1-63652-286-9

ENCYCLOPAEDIA BRITANNICA

11TH EDITION, "CROCOITE" TO "CUBA"

VARIOUS

TABLE OF CONTENTS

CROCOITE

A mineral consisting of lead chromate, $PbCrO_4$, and crystallizing in the monoclinic system. It is sometimes used as a paint, being identical in composition with the artificial product chrome-yellow; it is the only chromate of any importance found in nature. It was discovered at Berezovsk near Ekaterinburg in the Urals in 1766; and named crocoise by F. S. Beudant in 1832, from the Greek κϱόκος, saffron, in allusion to its colour, a name first altered to crocoisite and afterwards to crocoite. It is found as well-developed crystals of a bright hyacinth-red colour, which are translucent and have an adamantine to vitreous lustre. On exposure to light much of the translucency and brilliancy is lost. The streak is orange-yellow; hardness 2½-3; specific gravity 6.0. In the Urals the crystals are found in quartz-veins traversing granite or gneiss: other localities which have yielded good crystallized specimens are Congonhas do Campo near Ouro Preto in Brazil, Luzon in the Philippines, and Umtali in Mashonaland. Gold is often found associated with this mineral. Crystals far surpassing in beauty any previously known have been found in the Adelaide Mine at Dundas, Tasmania; they are long slender prisms, 3 or 4 in. in length, with a brilliant lustre and colour.

Associated with crocoite at Berezovsk are the closely allied minerals phoenicochroite and vauquelinite. The former is a basic lead Chromate, $Pb_3Cr_2O_9$, and the latter a lead and copper phosphate-chromate, $2(Pb, Cu)CrO_4.(Pb, Cu)_3(PO_4)_2$. Vauquelinite forms brown or green monoclinic crystals, and was named after L. N. Vauquelin, who in 1797 discovered (simultaneously with and independently of M. H. Klaproth) the element chromium in crocoite.

(L. J. S.)

CROCUS

A botanical genus of the natural order Iridaceae, containing about 70 species, natives of Europe, North Africa, and temperate Asia, and especially developed in the dry country of south-eastern Europe and western and central Asia. The plants are admirably adapted for climates in which a season favourable to growth alternates with a hot or dry season; during the latter they remain dormant beneath the ground in the form of a short thickened stem protected by the scaly remains of the bases of last season's leaves (known botanically as a "corm"). At the beginning of the new season of growth, new flower- and leaf-bearing shoots are developed from the corm at the expense of the food-stuff stored within it. New corms are produced at the end of the season, and by these the plant is multiplied.

These crocuses of the flower garden are mostly horticultural varieties of *C. vernus*, *C. versicolor* and *C. aureus* (Dutch crocus), the two former yielding the white, purple and striped, and the latter the yellow varieties. The crocus succeeds in any fairly good garden soil, and is usually planted near the edges of beds or borders in the flower garden, or in broadish patches at intervals along the mixed borders. The corms should be planted 3 in. below the surface, and as they become crowded they should be taken up and replanted with a refreshment of the soil, at least every five or six years. Crocuses have also a pleasing effect when dotted about on the lawns and grassy banks of the pleasure ground.

Some of the best of the varieties are:—*Purple*: David Rizzio, Sir J. Franklin, purpureus grandiflorus. *Striped*: Albion, La Majestueuse, Sir Walter Scott, Cloth of *Silver*, Mme Mina. *White*: Caroline Chisholm, Mont Blanc. *Yellow*: Large Dutch.

The species of crocus are not very readily obtainable, but those who make a specialty of hardy bulbs ought certainly to search them out and grow them. They require the same culture as the more familiar garden varieties; but, as some of them are apt to suffer from excess of moisture, it is advisable to plant them in prepared soil in a raised pit, where they are brought nearer to the eye, and where they can be sheltered when necessary by glazed sashes, which, however, should not be closed except when the plants are at rest, or during inclement weather in order to protect the blossoms, especially in the case of winter flowering species. The autumn blooming kinds include many plants of very great beauty. The following species are recommended:—

Spring flowering:—*Yellow*: *C. aureus*, *aureus* var. *sulphureus*, *chrysanthus*, *Olivieri*, *Korolkowi*, *Balansae*, *ancyrensis*, *Susianus*, *stellaris*. *Lilac*: *C. Imperati*, *Sieberi*, *etruscus*, *vernus*, *Tomasinianus*, *banaticus*. *White*: *C. biflorus* and vars., *candidus*, *vernus* vars. *Striped*: *C. versicolor*, *reticulatus*.

Autumn flowering:—*Yellow*: *C. Scharojani*. *Lilac*: *C. asluricus*, *cancellatus* var., *cilicicus*, *byzantinus* (*iridiflorus*), *longiflorus*, *medius*, *nudiflorus*, *pulchellus*, *Salzmanni*, *sativus* vars. speciosus, zonatus. White: caspius, cancellatus, hadrialicus, *marathonisius*.

Winter flowering:—*C. hyemaeis*, *laevigatus*, *vitellinus*.

CROESUS

Last king of Lydia, of the Mermnad dynasty, (560-546 B.C.), succeeded his father Alyattes after a war with his half-brother. He completed the conquest of Ionia by capturing Ephesus, Miletus and other places, and extended the Lydian empire as far as the Halys. His wealth, due to trade, was proverbial, and he used part of it in securing alliances with the Greek states whose fleets might supplement his own army. Various legends were told about him by the Greeks, one of the most famous being that of Solon's visit to him with the lesson it conveyed of the divine nemesis which waits upon overmuch prosperity (Hdt. i. 29 seq.; but see SOLON). After the overthrow of the Median empire (549 B.C.) Croesus found himself confronted by the rising power of Cyrus, and along with Nabonidos of Babylon took measures to resist it. A coalition was formed between the Lydian and Babylonian kings, Egypt promised troops and Sparta its fleet. But the coalition was defeated by the rapid movements of Cyrus and the treachery of Eurybatus of Ephesus, who fled to Persia with the gold that had been entrusted to him, and betrayed the plans of the confederates. Fortified with the Delphic oracles Croesus marched to the frontier of his empire, but after some initial successes fortune turned against him and he was forced to retreat to Sardis. Here he was followed by Cyrus who took the city by storm. We may gather from the recently discovered poem of Bacchylides (iii. 23-62) that he hoped to escape his conqueror by burning himself with his wealth on a funeral pyre, like Saracus, the last king of Assyria, but that he fell into the hands of Cyrus before he could effect his purpose.[1] A different version of the story is given (from Lydian sources) by Herodotus (followed by Xenophon), who makes Cyrus condemn his prisoner to be burnt alive, a mode of death hardly consistent with the Persian reverence for fire. Apollo, however, came to the rescue of his pious worshipper, and the name of Solon uttered by Croesus resulted in his deliverance. According to Ctesias, who uses Persian sources, and says nothing of the attempt to burn Croesus, he subsequently became attached to the court of Cyrus and received the governorship of Barene in Media. Fragments of columns from the temple of Attemis now in the British Museum have upon them a dedication by Croesus in Greek.

See R. Schubert, *De Croeso et Solone fabula* (1868); M. G. Radet, *La Lydie et le monde grec au temps des Mermnades* (1892-1893); A. S. Murray, *Journ. Hell. Studies*, x. pp. 1-10 (1889); for the supposition that Croesus did actually perish on his own pyre see G. B. Grundy, *Great Persian War*, p. 28; Grote, *Hist. of Greece* (ed. 1907), p. 104. Cf. CYRUS; LYDIA.

---

1 This is probably a Greek legend (cf. the Attic vase of about 500 B.C. in *Journ. of Hell. Stud.*, 1898, p. 268).

CROFT, SIR HERBERT

Bart. (1751-1816), English author, was born at Dunster Park, Berkshire, on the 1st of November 1751, son of Herbert Croft (see below) of Stifford, Essex. He matriculated at University College, Oxford, in March 1771, and was subsequently entered at Lincoln's Inn. He was called to the bar, but in 1782 returned to Oxford with a view to preparing for holy orders. In 1786 he received the vicarage of Prittlewell, Essex, but he remained at Oxford for some years accumulating materials for a proposed English dictionary. He was twice married, and on the day after his second wedding day he was imprisoned at Exeter for debt. He then retired to Hamburg, and two years later his library was sold. He had succeeded in 1797 to the title, but not to the estates, of a distant cousin, Sir John Croft, the fourth baronet. He returned to England in 1800, but went abroad once more in 1802. He lived near Amiens at a house owned by Lady Mary Hamilton, said to have been a daughter of the earl of Leven and Melville. Later he removed to Paris, where he died on the 26th of April 1816. In some of his numerous literary enterprises he had the help of Charles Nodier. Croft wrote the Life of Edward Young inserted in Johnson's *Lives of the Poets*. In 1780 he published *Love and Madness, a Story too true, in a series of letters between Parties whose names could perhaps be mentioned were they less known or less lamented*. This book, which passed through seven editions, narrates the passion of a clergyman named James Hackman for Martha Ray, mistress of the earl of Sandwich, who was shot by her lover as she was leaving Covent Garden in 1779 (see the Case and Memoirs of the late Rev. Mr James Hackman, 1779). *Love and Madness* has permanent interest because Croft inserted, among other miscellaneous matter, information about Thomas Chatterton gained from letters which he obtained from the poet's sister, Mrs Newton, under false pretences, and used without payment. Robert Southey, when about to publish an edition of

Chatterton's works for the benefit of his family, published (November 1799) details of Croft's proceedings in the *Monthly Review*. To this attack Croft wrote a reply addressed to John Nichols in the *Gentleman's Magazine*, and afterwards printed separately as *Chatterton and Love and Madness* ... (1800). This tract evades the main accusation, and contains much abuse of Southey. Croft, however, supplied the material for the exhaustive account of Chatterton in A. Kippis's *Biographia Britannica* (vol. iv., 1789). In 1788 he addressed a letter to William Pitt on the subject of a new dictionary. He criticized Samuel Johnson's efforts, and in 1790 he claimed to have collected 11,000 words used by excellent authorities but omitted by Johnson. Two years later he issued proposals for a revised edition of Johnson's *Dictionary*, but subscribers were lacking and his 200 vols. of MS. remained unused. Croft was a good scholar and linguist, and the author of some curious books in French.

The Love Letters of Mr H. and Miss R. 1775-1779 were edited from Croft's book by Mr Gilbert Burgess (1895). See also John Nichols's *Illustrations* ... (1828), v. 202-218.

CROFT, SIR JAMES

(d. 1590), lord deputy of Ireland, belonged to an old family of Herefordshire, which county he represented in parliament in 1541. He was made governor of Haddington in 1549, and became lord deputy of Ireland in 1551. There he effected little beyond gaining for himself the reputation of a conciliatory disposition. Croft was all his life a double-dealer. He was imprisoned in the Tower for treason in the reign of Mary, but was released and treated with consideration by Elizabeth after her accession. He was made governor of Berwick, where he was visited by John Knox in 1559, and where he busied himself actively on behalf of the Scottish Protestants, though in 1560 he was suspected, probably with good reason, of treasonable correspondence with Mary of Guise, the Catholic regent of Scotland; and for ten years he was out of public employment. But in 1570 Elizabeth, who showed the greatest forbearance and favour to Sir James Croft, made him a privy councillor and controller of her household. He was one of the commissioners for the trial of Mary queen of Scots, and in 1588 was sent on a diplomatic mission to arrange peace with the duke of Parma. Croft established private relations with Parma, for which on his return he was sent to the Tower. He was released before the end of 1589, and died on the 4th of September 1590.

Croft's eldest son, Edward, was put on his trial in 1589 on the curious charge of having contrived the death of the earl of Leicester by witchcraft, in revenge for the earl's supposed hostility to Sir James Croft. Edward Croft was father of Sir Herbert Croft (d. 1622), who became a Roman Catholic and wrote several controversial pieces in defence of that faith. His son Herbert Croft (1603-1691), bishop of Hereford, after being for some time, like his father, a member of the Roman church, returned to the church of England about 1630, and about ten years later was chaplain to Charles I., and obtained within a few years a prebend's stall at Worcester, a canonry of Windsor, and the deanery of Hereford, all of which preferments he lost during the Civil War and Commonwealth. By Charles II. he was made bishop of Hereford in 1661. Bishop Croft was the author of many books and pamphlets, several of them against the Roman Catholics; and one of his works, entitled *The Naked Truth, or the True State of the Primitive Church* (London, 1675), was very celebrated in its day, and gave rise to prolonged controversy. The bishop died in 1691. His son Herbert was created a baronet in 1671, and was the ancestor of Sir Herbert Croft (*q.v.*), the 18th century writer.

BIBLIOGRAPHY.—See Richard Bagwell, *Ireland under the Tudors*, vol. i. (3 vols., London, 1885); David Lloyd, *State Worthies from the Reformation to the Revolution* (2 vols., London, 1766); John Strype, *Annals of the Reformation* (Oxford, 1824), which contains an account of the trial of Edward Croft; S. L. Lee's art. "Croft, Sir James," in *Dict. of National Biography*, vol. xiii.; and for Bishop Croft see Anthony à Wood, *Athenae Oxonienses* (ed. Bliss, 1813-1820); John Le Neve, *Fasti Ecclesiae Anglicanae* (ed. by T. D. Hardy, Oxford, 1854).

CROFT WILLIAM

(1678-1727), English composer, was born in 1678, at Nether Ettington in Warwickshire. He received his musical education in the Chapel Royal under Dr Blow. He early obtained the place of organist of St Anne's, Soho, and in 1700 was admitted a gentleman extraordinary of the Chapel Royal. In 1707 he was appointed joint-organist with Blow; and upon the death of the latter in 1708 he became solo organist, and also master of the children and composer of the Chapel Royal, besides being made organist of Westminster Abbey. In 1712 he wrote a brief introduction on the history of English church music to a collection of the words of anthems which he had edited under the title of *Divine Harmony*. In 1713 he obtained his degree of doctor of music in the university of Oxford. In 1724 he published an edition of his choral music in 2 vols. folio, under the name of *Musica Sacra, or Select Anthems in score, for two, three, four, five, six, seven and eight voices, to which is added the Burial Service, as it is occasionally performed in Westminster Abbey*. This handsome work included a portrait of the composer and was the first of the kind executed on pewter plates and in score. John Page, in his *Harmonia Sacra*, published in 1800 in 3 vols. folio, gives seven of Croft's anthems. Of instrumental music, Croft published six sets of airs for two violins and a bass, six sonatas for two flutes, six solos for a flute and bass. He died at Bath on the 14th of August 1727, and was buried in the north aisle of Westminster Abbey, where a monument was erected to his memory by his friend and admirer Humphrey Wyrley Birch. Burney in his *History of Music* devotes several pages of his third volume (pp. 603-612) to Dr Croft's life, and criticisms of some of his anthems. During the earlier period of his life Croft wrote much for the theatre, including overtures and incidental music for *Courtship à la mode* (1700), *The Funeral* (1702) and *The Lying Lover* (1703).

CROFTER

A term used, more particularly in the Highlands and islands of Scotland, to designate a tenant who rents and cultivates a small holding of land or "croft." This Old English word, meaning originally an enclosed field, seems to correspond to the Dutch *kroft*, a field on high ground or downs. The ultimate origin is unknown. By the Crofters' Holdings (Scotland) Act 1886, a crofter is defined as the tenant of a holding who resides on his holding, the annual rent of which does not exceed £30 in money, and which is situated in a crofting parish. The wholesale clearances of tenants from their crofts during the 19th century, in violation of, as the tenants claimed, an implied security of tenure, has led in the past to much agitation on the part of the crofters to secure consideration of their grievances. They have been the subject of royal commissions and of considerable legislation, but the effect of the Crofters Act of 1886, with subsequent amending acts, has been to improve their condition markedly, and much of the agitation has now died out. A history of the legislation dealing with the crofters is given in the article SCOTLAND.

CROKER, JOHN WILSON

(1780-1857), British statesman and author, was born at Galway on the 20th of December 1780, being the only son of John Croker, the surveyor-general of customs and excise in Ireland. He was educated at Trinity College, Dublin, where he graduated in 1800. Immediately afterwards he was entered at Lincoln's Inn, and in 1802 he was called to the Irish bar. His interest in the French Revolution led him to collect a large number of valuable documents on the subject, which are now in the British Museum. In 1804 he published anonymously *Familiar Epistles to J. F. Jones, Esquire, on the State of the Irish Stage*, a series of caustic criticisms in verse on the management of the Dublin theatres. The book ran through five editions in one year. Equally successful was the *Intercepted Letter from Canton* (1805), also anonymous, a satire on Dublin society. In 1807 he published a pamphlet on *The State of Ireland, Past and Present*, in which he advocated Catholic emancipation.

In the following year he entered parliament as member for Downpatrick, obtaining the seat on petition, though he had been unsuccessful at the poll. The acumen displayed in his Irish pamphlet led Spencer Perceval to recommend him in 1808 to Sir Arthur Wellesley, who had just been appointed to the command of the British forces in the Peninsula, as his deputy in the office of chief secretary for Ireland. This connexion led to a friendship which remained unbroken till Wellington's death. The notorious case of the duke of York in connexion with his abuse of military patronage furnished him with an opportunity for distinguishing himself. The speech which he delivered on the 14th of March 1809, in answer to the charges of Colonel Wardle, was regarded as the most able and ingenious defence of the duke that was made in the debate; and Croker was appointed to the office of secretary to the Admiralty, which he held without interruption under various administrations for more than twenty years. He proved an excellent public servant, and made many improvements which have been of permanent value in the organization of his office. Among the first acts of his official career was the exposure of a fellow-official who had misappropriated the public funds to the extent of £200,000.

In 1827 he became the representative of the university of Dublin, having previously sat successively for the boroughs of Athlone, Yarmouth (Isle of Wight), Bodmin and Aldeburgh. He was a determined opponent of the Reform Bill, and vowed that he would never sit in a reformed parliament; his parliamentary career accordingly terminated in 1832. Two years earlier he had retired from his post at the admiralty on a pension of £1500 a year. Many of his political speeches were published in pamphlet form, and they show him to have been a vigorous and effective, though somewhat unscrupulous and often virulently personal, party debater. Croker had been an ardent supporter of Peel, but finally broke with him when he began to advocate the repeal of the Corn Laws. He is said to have been the first to use (Jan. 1830) the term "conservatives." He was for many years one of the leading contributors on literary and historical subjects to the *Quarterly Review*, with which he had been associated from its foundation. The rancorous spirit in which many of his articles were written did much to embitter party feeling. It also reacted unfavourably on Croker's reputation as a worker in the department of pure literature by bringing political animosities into literary criticism. He had no sympathy with the younger school of poets who were in revolt against the artificial methods of the 18th century, and he was responsible for the famous *Quarterly* article on Keats. It is, nevertheless, unjust to judge Croker by the criticisms which Macaulay brought against his *magnum opus*, his edition of Boswell's *Life of Johnson* (1831). With all its defects the work had merits which Macaulay was of course not concerned to point out, and Croker's researches have been of the greatest value to subsequent editors. There is little doubt that Macaulay had personal reasons for his attack on Croker, who had more than once exposed in the House the fallacies that lay hidden under the orator's brilliant rhetoric. Croker made no immediate reply to Macaulay's attack, but when the first two volumes of the *History* appeared he took the opportunity of pointing out the inaccuracies that abounded in the work. Croker was occupied for several years on an annotated edition of Pope's works. It was left unfinished at the time of his death, but it was afterwards completed by the Rev. Whitwell Elwin and Mr W. J. Courthope. He died at St Albans Bank, Hampton, on the 10th of August 1857.

Croker was generally supposed to be the original from which Disraeli drew the character of "Rigby" in *Coningsby*, because he had for many years had the sole management of the estates of the marquess of Hertford, the "Lord Monmouth" of the story; but the comparison is a great injustice to the sterling worth of Croker's character.

> The chief works of Croker not already mentioned were his *Stories for Children from the History of England* (1817), which provided the model for Scott's *Tales of a Grandfather*; *Letters on the Naval War with America*; *A Reply to the Letters of Malachi Malagrowther* (1826); *Military Events of the French Revolution of 1830* (1831); a translation of Bassompierre's *Embassy to England* (1819); and several lyrical pieces of some merit, such as the *Songs of Trafalgar* (1806) and *The Battles of Talavera* (1809). He also edited the *Suffolk Papers* (1823), *Hervey's Memoirs of the Court of George II.* (1817), the *Letters of Mary Lepel, Lady Hervey* (1821-1822), and *Walpole's Letters to Lord Hertford* (1824). His memoirs, diaries and correspondence were edited by Louis J. Jennings in 1884 under the title of *The Croker Papers* (3 vols.).

CROKER, RICHARD

(1843-), American politician, was born at Blackrock, Ireland, on the 24th of November 1843. He was taken to the United States by his parents when two years old, and was educated in the public schools of New York City, where he eventually became a member of Tammany Hall and active in its politics. He was an alderman from 1868 to 1870, a coroner from 1873 to 1876, a fire commissioner in 1883 and 1887, and city chamberlain from 1889 to 1890. After the fall of John Kelly he became the leader of Tammany Hall (*q.v.*), and for some time almost completely controlled the organization. His greatest political success was his bringing about the election of Robert A. van Wyck as first mayor of greater New York in 1897, and during van Wyck's administration Croker is popularly supposed to have dominated completely the government of the city. After Croker's failure to "carry" the city in the presidential election of 1900 and the defeat of his mayoralty candidate, Edward M. Shepard, in 1901, he resigned from his position of leadership in Tammany, and retired to a country life in England and Ireland. In 1907 he won the Derby with his race-horse Orby.

CROKER, THOMAS CROFTON

(1798-1854), Irish antiquary and humorist, was born in Cork on the 15th of January 1798. He was apprenticed to a merchant, but in 1819, through the interest of John Wilson Croker, who was, however, no relation of his, he became a clerk in the Admiralty. Moore was indebted to him in the production of his *Irish Melodies* for "many curious fragments of ancient poetry." In 1825 he produced his most popular book, the *Fairy Legends and Traditions of the South of Ireland*, which he followed up by the publication of his *Legends of the Lakes* (1829), his *Adventures of Barney Mahoney* (1852), and an edition of the *Popular Songs of Ireland* (1839). In 1827 he was made a member of the Irish Academy; in 1839 and 1840 he helped to found the Camden and Percy Societies, and in 1843 the British Archaeological Association. He wrote *Narratives Illustrative of the Contests in Ireland in 1641 and 1688* (1841), for the Camden Society, *Historical Songs of Ireland*, &c. (1841), for the Percy Society, and several other works. He was also a member of the Hakluyt and the Antiquarian Society. He died in London on the 8th of August 1854.

CROLL, JAMES

(1821-1890), Scottish man of science, was born of a peasant family at Little Whitefield, in the parish of Cargill, in Perthshire, on the 2nd of January 1821. He was regarded as an unpromising boy, but a trifling circumstance aroused a passion for reading, and he made great progress in self-education. He was apprenticed to a wheelwright at Collace in Perthshire, but being debarred by ill-health from manual labour, he became successively a shop-keeper and an insurance agent. In 1859 he was made keeper of the Andersonian Museum in Glasgow, a humble appointment, which, however, gave him congenial occupation. In 1857, being deeply impressed by the metaphysics of Jonathan Edwards, he had published an anonymous volume entitled *The Philosophy of Theism*; but his connexion with the Museum induced him to take up physical science, and from 1861 onwards he studied with such perseverance that he was enabled to contribute papers to the *Philosophical Magazine* and other journals. For that magazine in 1864 he wrote his celebrated essay "On the Physical Cause of the Changes of Climate during Geological Epochs." This led to his receiving an appointment on the Scottish Geological Survey in 1867, and for thirteen years he took charge of the Edinburgh Office. In 1875 he summed up his researches upon the ancient condition of the earth in his *Climate and Time, in their Geological Relations*, in which he contends that terrestrial revolutions are due in a measure to cosmical causes. This theory excited warm controversy. Croll's replies to his opponents are collected in his *Climate and Cosmology* (1885). He had been compelled by ill-health to withdraw from the public service in 1880; yet, working under the greatest difficulties, and harassed by the inadequacy of his retiring pension, he managed to produce *Stellar Evolution*, discussing, among other things, the age of the sun, in 1889; and *The Philosophical Basis of Evolution*, partly a critique of Herbert Spencer's philosophy, in 1890. He died on the 15th of December 1890. The soundness of Croll's astronomical theory regarding the glacial period has since been criticized by E. P. Culverwell in the *Geological Magazine* for 1895, and by others; and it is now generally abandoned. Nevertheless it must be admitted that his character as a scientific worker under great discouragements was nothing less than heroic. The hon. degree of LL.D. was conferred on him in 1876 by the university of St Andrews; and he was elected F.R.S. in the same year.

An *Autobiographical Sketch of James Croll, with Memoir of his Life and Work*, was prepared by J. C. Irons, and published in 1896.

CROLY, GEORGE

(1780-1860), British divine and author, son of a Dublin physician, was born on the 17th of August 1780. He was educated at Trinity College, Dublin, and after ordination was appointed to a small curacy in the north of Ireland. About 1810 he came to London, and occupied himself with literary work. A man of restless energy, he claims attention by his extraordinary versatility. He wrote dramatic criticisms for a short-lived periodical called the *New Times*; he was one of the earliest contributors to *Blackwood's Magazine*; and to the *Literary Gazette* he contributed poems, reviews and essays on all kinds of subjects. In 1819 he married Margaret Helen Begbie. Efforts to secure an English living for Croly were frustrated, according to the *Gentleman's Magazine* (Jan. 1861), because Lord Eldon confounded him with a Roman Catholic of the same name. Excluding his contributions to the daily and weekly press his chief works were:—*Paris in 1815* (1817), a poem in imitation of *Childe Harold; Catiline* (1822), a tragedy lacking in dramatic force; *Salathiel: A Story of the Past, the Present and the Future* (1829), a successful romance of the "Wandering Jew" type; *The Life and Times of his late Majesty George the Fourth* (1830); *Marston; or, The Soldier and Statesman* (1846), a novel of modern life; *The Modern Orlando* (1846), a satire which owes something to *Don Juan*; and some biographies, sermons and theological works.

Croly was an effective preacher, and continued to hope for preferment from the Tory leaders, to whom he had rendered considerable services by his pen; but he eventually received, in 1835, the living of St Stephen's, Walbrook, London, from a Whig patron, Lord Brougham, with whose family he was connected. In 1847 he was made afternoon lecturer at the Foundling hospital, but this appointment proved unfortunate. He died suddenly on the 24th of November 1860, in London.

His *Poetical Works* (2 vols.) were collected in 1830. For a list of his works see Allibone's *Critical Dictionary of English Literature* (1859).

CROMAGNON RACE

The name given by Paul Broca to a type of mankind supposed to be represented by remains found by Lartet, Christy and others, in France in the Cromagnon cave at Les Eyzies, Tayac district, Dordogne. At the foot of a steep rock near the village this small cave, nearly filled with debris, was found by workmen in 1868. Towards the top of the loose strata three human skeletons were unearthed. They were those of an old man, a young man and a woman, the latter's skull bearing the mark of a severe wound. The skulls presented such special characteristics that Broca took them as types of a race. Palaeolithic man is exclusively long-headed, and the dolichocephalic appearance of the crania (they had a mean cephalic index of 73.34) supported the view that the "find" at Les Eyzies was palaeolithic. It is, however, inaccurate to state that brachycephaly appears at once with the neolithic age, dolichocephaly even of a pronounced type persisting far into neolithic times. The Cromagnon race may thus be, as many anthropologists believe it, early neolithic, a type of man who spread over and inhabited a large portion of Europe at the close of the Pleistocene period. Some have sought to find in it the substratum of the present populations of western Europe. Quatrefages identifies Cromagnon man with the tall, long-headed, fair Kabyles (Berbers) who still survive in various parts of Mauritania. He suggests the introduction of the Cromagnon from Siberia, "arriving in Europe simultaneously with the great mammals (which were driven by the cold from Siberia), and no doubt following their route."

See A. H. Keane's *Ethnology* (1896); Mortillet, *Le Préhistorique* (1900); Sergi, *The Mediterranean Race* (1901); Lord Avebury, *Prehistoric Times*, p. 317 of 1900 edition.

CROMARTY, GEORGE MACKENZIE

1st Earl of (1630-1714), Scottish statesman, was the eldest son of Sir John Mackenzie, Bart., of Tarbat (d. 1654), and belonged to the same family as the earls of Seaforth. In 1654 he joined the rising in Scotland on behalf of Charles II. and after an exile of six years he returned to his own country and took some part in public affairs after the Restoration. In 1661 he became a lord of session as Lord Tarbat, but having been concerned in a vain attempt to overthrow Charles II.'s secretary, the earl of Lauderdale, he was dismissed from office in 1664. A period of retirement followed until 1678 when Mackenzie was appointed lord justice general of Scotland; in 1681 he became lord clerk register and a lord of session for the second time, and from 1682 to 1688 he was the chief minister of Charles II. and James II. in Scotland, being created viscount of Tarbat in 1685. In 1688, however, he deserted James and soon afterwards made his peace with William III., his experience being very serviceable to the new government in settling the affairs of Scotland. From 1692 to 1695 Tarbat was again lord clerk register, and having served for a short time as a secretary of state under Queen Anne he was created earl of Cromarty in 1703. He was again lord justice general from 1704 to 1710. He warmly supported the union between England and Scotland, writing some pamphlets in favour of this step, and he died on the 17th of August 1714. Cromarty was a man of much learning, and among his numerous writings may be mentioned his *Account of the conspiracies by the earls of Gowry and R. Logan* (Edinburgh, 1713).

The earl's grandson George, 3rd earl of Cromarty (*c.* 1703-1766), succeeded his father John, the 2nd earl, in February 1731. In 1745 he joined Charles Edward, the young pretender, and he served with the Jacobites until April 1746 when he was taken prisoner in Sutherlandshire. He was tried and sentenced to death, but he obtained a conditional pardon although his peerage was forfeited. He died on the 28th of September 1766.

This earl's eldest son was John Mackenzie, Lord Macleod (1727-1789), who shared his father's fortunes in 1745 and his fate in 1746. Having pleaded guilty at his trial Macleod was pardoned on condition that he gave up all his rights in the estates of the earldom, and he left England and entered the Swedish army. In this service he rose to high rank and was made Count Cromarty. The count returned to England in 1777 and was successful in raising, mainly among the Mackenzies, two splendid battalions of Highlanders, the first of which, now the Highland Light Infantry, served under him in India. In 1784 he regained the family estates and he died on the 2nd of April 1789. Macleod wrote an account of the Jacobite rising of 1745, and also one of a campaign in Bohemia in which he took part in 1757; both are printed in Sir W. Fraser's *Earls of Cromartie* (Edinburgh, 1876).

Macleod left no children, and his heir was his cousin, Kenneth Mackenzie (d. 1796), a grandson of the 2nd earl, who also died childless. The estates then passed to Macleod's sister, Isabel (1725-1801), wife of George Murray, 6th Lord Elibank. In 1861 Isabel's descendant, Anne (1829-1888), wife of George, 3rd duke of Sutherland, was created countess of Cromartie with remainder to her second son Francis (1852-1893), who became earl of Cromartie in 1888. In 1895, two years after the death of Francis, his daughter Sibell Lilian (b. 1878) was granted by letters patent the title of countess of Cromartie.

CROMARTY

A police burgh and seaport of the county of Ross and Cromarty, Scotland. Pop. (1901) 1242. It is situated on the southern shore of the mouth of Cromarty Firth, 5 m. E. by S. of Invergordon on the opposite coast, with which there is daily communication by steamer, and 9 m. N.E. of Fortrose, the most convenient railway station. Before the union of the shires of Ross and Cromarty, it was the county town of Cromartyshire, and is one of the Wick district group of parliamentary burghs. Its name is variously derived from the Gaelic *crom*, crooked, and *bath*, bay, or *ard*, height, meaning either the "crooked bay," or the "bend between the heights" (the high rocks, or Sutors, which guard the entrance to the Firth), and gave the title to the earldom of Cromarty. The principal buildings are the town hall and the Hugh Miller Institute. The harbour, enclosed by two piers, accommodates the herring fleet, but the fisheries, the staple industry, have declined. The town, however, is in growing repute as a midsummer resort. The thatched house with crow-stepped gables in Church Street, in which Hugh Miller the geologist was born, still stands, and a statue has been erected to his memory. To the east of the burgh is Cromarty House, occupying the site of the old castle of the earls of Ross. It was the birthplace of Sir Thomas Urquhart, the translator of Rabelais.

Cromarty, formerly a county in the north of Scotland, was incorporated with Ross-shire in 1889 under the designation of the county of Ross and Cromarty. The nucleus of the county consisted of the lands of Cromarty in the north of the peninsula of the Black Isle. To this were added from time to time the various estates scattered throughout Ross-shire—the most considerable of which were the districts around Ullapool and Little Loch Broom on the Atlantic coast, the area in which Ben Wyvis is situated, and a tract to the north of Loch Fannich—which had been acquired by the ancestors of Sir George Mackenzie (1630-1714), afterwards Viscount Tarbat (1685) and 1st earl of Cromarty (1703). Desirous of combining these sporadic properties into one shire, Viscount Tarbat was enabled to procure their annexation to his sheriffdom of Cromarty in 1685 and 1698, the area of the enlarged county amounting to nearly 370 sq. m. (See ROSS AND CROMARTY.)

CROMARTY FIRTH

An arm of the North Sea, belonging to the county of Ross and Cromarty, Scotland. From the Moray Firth it extends inland in a westerly and then south-westerly direction for a distance of 19 m. Excepting at the Bay of Nigg, on the northern shore, and Cromarty Bay, on the southern, where it is about 5 m. wide (due N. and S.), and at Alness Bay, where it is 2 m. wide, it has an average width of 1 m. and a depth varying from 5 to 10 fathoms, forming one of the safest and most commodious anchorages in the north of Scotland. Besides other streams it receives the Conon, Peffery, Skiack and Alness, and the principal places on its shores are Dingwall near the head, Cromarty near the mouth, Kiltearn, Invergordon and Kilmuir on the north. The entrance is guarded by two precipitous rocks—the one on the north 400 ft., that on the south 463 ft. high—called the Sutors from a fancied resemblance to a couple of shoemakers (*Scotice*, souter), bending over their lasts. There are ferries at Cromarty, Invergordon and Dingwall.

CROME, JOHN

(1769-1821), English landscape painter, founder and chief representative of the "Norwich School," often called Old Crome, to distinguish him from his son, was born at Norwich, on the 21st of December 1769. His father was a weaver, and could give him only the scantiest education. His early years were spent in work of the humblest kind; and at a fit age he became apprentice to a house-painter. To this step he appears to have been led by an inborn love of art and the desire to acquaint himself by any means with its materials and processes. During his apprenticeship he sometimes painted signboards, and devoted what leisure time he had to sketching from nature. Through the influence of a rich art-loving friend he was enabled to exchange his occupation of house-painter for that of drawing-master; and in this he was engaged throughout his life. He took great delight in a collection of Dutch pictures to which he had access, and these he carefully studied. About 1790 he was introduced to Sir William Beechey, whose house in London he frequently visited, and from whom he gathered additional knowledge and help in his art. In 1805 the Norwich Society of Artists took definite shape, its origin being traceable a year or two further back. Crome was its president and the largest contributor to its annual exhibitions. Among his pupils were James Stark, Vincent, Thirtle and John Bernay (Barney) Crome (1794-1842), his son. J. S. Cotman, too, a greater artist than any of these, was associated with him. Crome continued to reside at Norwich, and with the exception of his short visits to London had little or no communication with the great artists of his own time. He first exhibited at the Royal Academy in 1806; but in this and the following twelve years he exhibited there only fourteen of his works. With very few exceptions Crome's subjects are taken from the familiar scenery of his native county. Fidelity to nature was his dominant aim. "The bit of heath, the boat, and the slow water of the flattish land, trees most of all—the single tree in elaborate study, the group of trees, and how the growth of one affects that of another, and the characteristics of each,"—these, says Frederick Wedmore (*Studies in English Art*), are the things to which he is most constant. He still remains, says the same critic, of many trees the greatest draughtsman, and is especially the master of the oak. His most important works are—"Mousehold Heath, near Norwich," now in the National Gallery; "Clump of Trees, Hautbois Common"; "Oak at Poringland"; the "Willow"; "Coast Scene near Yarmouth"; "Bruges, on the Ostend River"; "Slate Quarries"; the "Italian Boulevards"; and the "Fishmarket at Boulogne." He executed a good many etchings, and the great charm of these is in the beautiful and faithful representation of trees. Crome enjoyed a very limited reputation during his life, and his pictures were sold at low prices; but since his death they have been more and more appreciated, and have given him a high place among English painters of landscape. He died at Norwich on the 22nd of April 1821. His son, J. B. Crome, was his assistant in teaching, and his best pictures were in the same style, his moonlight effects being much admired.

A collection of "Old" Crome's etchings, entitled *Norfolk Picturesque Scenery*, was published in 1834, and was re-issued with a memoir by Dawson Turner in 1838, but in this issue the prints were retouched by other hands.

CROMER, EVELYN BARING

1st Earl (1841-), British statesman and diplomatist, was born on the 26th of February 1841, the ninth son of Henry Baring, M.P., by Cecilia Anne, eldest daughter of Admiral Windham of Felbrigge Hall, Norfolk. Having joined the Royal Artillery in 1858, he was appointed in 1861 A.D.C. to Sir Henry Storks, high commissioner of the Ionian Islands, and acted as secretary to the same chief during the inquiry into the Jamaica outbreak in 1865. Gazetted captain in 1870, he went in 1872 as private secretary to his cousin Lord Northbrook, Viceroy of India, where he remained until 1876, when he became major, received the C.S.I., and was appointed British commissioner of the Egyptian public debt office. Up to this period Major Baring had given no unusual signs of promise, and the appointment of a comparatively untried major of artillery as the British representative on a Financial Board composed of representatives of all the great powers was considered a bold one. Within a very short time it was recognized that the Englishman, though keeping himself carefully in the background, was unmistakably the predominant factor on the board. He was mainly responsible for the searching report, issued in 1878, of the commission of inquiry that had been instituted into the financial methods of the Khedive Ismail; and when that able and unscrupulous Oriental had to submit to an enforced abdication in 1879, it was Major Baring who became the British controller-general and practical director of the Dual Control. Had he remained in Egypt, the whole course of Egyptian history might have been altered, but his services were deemed more necessary in India, and under Lord Ripon he became financial member of council in June 1880. He remained there till 1883, leaving an unmistakable mark on the Indian financial system, and then, having been rewarded by the K.C.S.I., he was appointed British agent and consul-general in Egypt and a minister plenipotentiary in the diplomatic service.

Sir Evelyn Baring was at that time only a man of forty-two, who had gained a reputation for considerable financial ability, combined with an abruptness of manner and a certain autocracy of demeanour which, it was feared, would impede his success in a position which required considerable tact and diplomacy. It was a friendly colleague who wrote—

> "The virtues of
> Patience are known,
> But I think that, when
> put to the touch,
> The people of Egypt will
> own, with a groan,
> There's an Evil in
> Baring too much."

When he arrived in Cairo in 1883 he found the administration of the country almost non-existent. Ismail had ruled with all the vices, but also with all the advantages, of autocracy. Disorder in the finances, brutality towards the people, had been combined with public tranquillity and the outer semblance of civilization. Order, at least, reigned from the Sudan to the Mediterranean, and such trivial military disturbances as had occurred had been of Ismail's own devising and for his own purposes. Tewfik, who had succeeded him, had neither the inclination nor character to be a despot. Within three years his government had been all but overthrown, and he was only khedive by the grace of British bayonets. Government by bayonets was not in accord with the views of the House of Commons, yet Ismail's government by the kourbash could not be restored. The British government, under Mr Gladstone, desired to establish in Egypt a sort of constitutional government; and as there existed no single element of a constitution, they had sent out Lord Dufferin (the first marquess of Dufferin) to frame one. That gifted nobleman, in the delightful lucidity of his picturesque report, left nothing to be desired except the material necessary to convert the flowing periods into political entities.[1] In the absence of that, the constitution was still-born, and Sir Evelyn Baring arrived to find, not indeed a clean slate, but a worn-out papyrus, disfigured by the efforts of centuries to describe in hieroglyph a method of rule for a docile people.

From that date the history of Sir Evelyn Baring, who became Baron Cromer in 1892, G.C.B. in 1895, viscount in 1897, and earl in 1901, is the history of Egypt, and requires the barest mention of its salient points here. From the outset he realized that the task he had to perform could only be effected piecemeal and in detail, and his very first measure was one which, though severely criticized at the time, has been justified by events, and which in any case showed that he shirked no responsibility, and was capable of adopting heroic methods. He counselled the abandonment, at least temporarily, by Egypt of its authority in the Sudan provinces, already challenged by the mahdi. His views were shared by the British ministry of the day and the policy of abandonment enforced upon the Egyptian government. At the same time it was decided that efforts should be made to relieve the Egyptian garrisons in the Sudan and this resolve led to the mission of General C. G. Gordon (*q.v.*) to Khartum. Lord Cromer subsequently told the story of Gordon's mission at length, making clear the measure of responsibility resting upon him as British agent. The proposal to employ Gordon came from the British government and twice Sir Evelyn rejected the suggestion. Finally, mistrusting his own judgment, for he did not consider Gordon the proper person for the mission, Baring yielded to pressure from Lord Granville. Thereafter he gave Gordon all the support possible, and in the critical matter of the proposed despatch of Zobeir to Khartum, Baring—after a few days' hesitation—cordially endorsed Gordon's request. The request was refused by the British government—and the catastrophe which followed at Khartum rendered inevitable.

The Sudan crisis being over, for the time, Sir Evelyn Baring set to work to reorganize Egypt itself. This work he attacked in detail. The very first essential was to regulate the financial situation; and in Egypt, where the entire revenue is based on the production of the soil, irrigation was of the first importance. With the assistance of Sir Colin Scott Moncrieff, in the public works department, and Sir Edgar Vincent, as financial adviser, these two great departments were practically put in order before he gave more than superficial attention to the rest. The ministry of justice was the next department seriously taken in hand, with the assistance of Sir John Scott, while the army had been reformed under Sir Evelyn Wood, who was succeeded by Sir Francis (afterwards Lord) Grenfell. Education, the ministry of the interior, and gradually every other department, came to be reorganized, or, more correctly speaking, formed, under Lord Cromer's carefully persistent direction, until it may be said to-day that the Egyptian administration can safely challenge comparison with that of any other state. In the meantime the rule of the mahdi and his successor, the khalifa, in the temporarily abandoned provinces of the Sudan, had been weakened by internal dissensions; the Italians from Massawa, the Belgians from the Congo State, and the French from their West African possessions, had gradually approached nearer to the valley of the Nile; and the moment had arrived at which Egypt must decide either to recover her position in the Sudan or allow the Upper Nile to fall into hands hostile to Great Britain and her position in Egypt. Lord Cromer was as quick to recognize the moment for action and to act as he had fifteen years earlier been prompt to recognize the necessity of abstention. In March-September 1896 the first advance was made to Dongola under the Sirdar, Sir Herbert (afterwards Lord) Kitchener; between July 1897 and April 1898 the advance was pushed forward to the Atbara; and on the 2nd of September 1898, the battle of Omdurman finally crushed the power of the khalifa and restored the Sudan to the rule of Egypt and Great Britain. In the negotiations which resulted in the Anglo-French Declaration of the 8th of April 1904, whereby France bound herself not to obstruct in any manner the action of Great Britain in Egypt and the Egyptian government acquired financial freedom, Lord Cromer took an active part. He also successfully guarded the interests of Egypt and Great Britain in 1906 when Turkey attempted by encroachments in the Sinai Peninsula to obtain a strategic position on the Suez Canal. To have effected all this in the face of the greatest difficulties—political, national and international—and at the same time to have raised the credit of the country from a condition of bankruptcy to an equality with that of the first European powers, entitles Lord Cromer to a very high place among the greatest administrators and statesmen that the British empire has produced. In April 1907, in consequence of the state of his health, he resigned office, having held the post of British agent in Egypt for twenty-four years. In July of the same year parliament granted £50,000 out of the public funds to Lord Cromer in recognition of his "eminent services" in Egypt. In 1908 he published, in two volumes, *Modern Egypt*, in which he gave an impartial narrative of events in Egypt and the Sudan since 1876, and dealt with the results to Egypt of the British occupation of the country. Lord Cromer also took part in the political controversies at home, joining himself to the free-trade wing of the Unionist party.

Lord Cromer married in 1876 Ethel Stanley, daughter of Sir Rowland Stanley Errington, eleventh baronet, but was left a widower with two sons in 1898; and in 1901 he married Lady Katherine Thynne, daughter of the 4th marquess of Bath.

---

1 In 1892 Lord Dufferin wrote to Lord Cromer: "These institutions were a good deal ridiculed at the time, but as it was then uncertain how long we were going to remain, or rather how soon the Turks might not be reinvested with their ancient supremacy, I desired to erect some sort of barrier, however feeble, against their intolerable tyranny." In 1906 Lord Cromer bore public testimony to the good results of the measures adopted on Lord Dufferin's "statesmanlike initiative." Such results were, however, only possible in consequence of the continuance of the British occupation.

CROMER

A watering-place in the northern parliamentary division of Norfolk, England, 139 m. N.E. by N. from London by the Great Eastern railway; served also by the Midland and Great Northern joint line. Pop. of urban district (1901) 3781. Standing on cliffs of considerable elevation, the town has repeatedly suffered from ravages of the sea. A wall and esplanade extend along the bottom of the cliffs, and there is a fine stretch of sandy beach. There is also a short pier. The church of St Peter and St Paul is Perpendicular (largely restored) with a lofty tower. On a site of three acres stands the convalescent home of the Norfolk and Norwich hospital. There is an excellent golf course. The herring, cod, lobster and crab fisheries are prosecuted. The village of Sheringham (pop. of urban district, 2359), lying to the west, is also frequented by visitors. A so-called Roman camp, on an elevation overlooking the sea, is actually a modern beacon.

CROMORNE

Also **CRUMHORNE**[1] (Ger. *Krummhorn*; Fr. *tournebout*), a wind instrument of wood in which a cylindrical column of air is set in vibration by a reed. The lower extremity is turned up in a half-circle, and from this peculiarity it has gained the French name *tournebout*. The reed of the cromorne, like that of the bassoon, is formed by a double tongue of cane adapted to the small end of a conical brass tube or crook, the large end fitting into the main bore of the instrument. It presents, however, this difference, that it is not, like that of the bassoon, in contact with the player's lips, but is covered by a cap pierced in the upper part with a raised slit against which the performer's lips rest, the air being forced through the opening into the cap and setting the reed in vibration. The reed itself is therefore not subject to the pressure of the lips. The compass of the instrument is in consequence limited to the simple fundamental sounds produced by the successive opening of the lateral holes. The length of the cromornes is inconsiderable in proportion to the deep sounds produced by them, which arises from the fact that these instruments, like all tubes of cylindrical bore provided with reeds, have the acoustic properties of the stopped pipes of an organ. That is to say, theoretically they require only half the length necessary for the open pipes of an organ or for conical tubes provided with reeds, to produce notes of the same pitch. Moreover, when, to obtain an harmonic, the column of air is divided, the cromorne will not give the octave, like the oboe and bassoon, but the twelfth, corresponding in this peculiarity with the clarinet and all stopped pipes or bourdons. In order, however, to obtain an harmonic on the cromorne, the cap would have to be discarded, for a reed only overblows to give the harmonic overtones when pressed by the lips. With the ordinary boring of eight lateral holes the cromorne possesses a limited compass of a ninth. Sometimes, however, deeper sounds are obtained by the addition of one or more keys. By its construction the cromorne is one of the oldest wind instruments; it is evidently derived from the Gr. aulos[2] and the Roman tibia, which likewise consisted of a simple cylindrical pipe of which the air column was set in vibration, at first by a double reed, and, we have reason to believe, later by a single reed (see AULOS and CLARINET). The Phrygian aulos was sometimes curved (see Tib. ii. i. 85 *Phrygio tibia curva sono*; Virgil, *Aen.* xi. 737 *curva choros indixit tibia Bacchi*).[3]

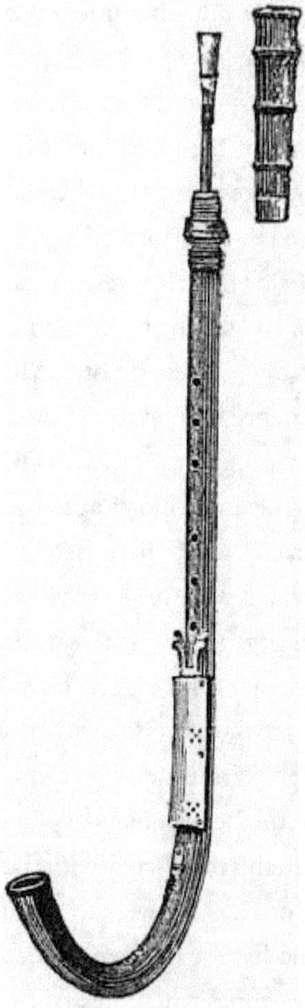

Bass Tournebout.

Notwithstanding the successive improvements that were introduced in the manufacture of wind instruments, the cromorne scarcely ever varied in the details of its construction. Such as we see it represented in the treatise by Virdung[4] we find it again about the epoch of its disappearance.[5] The cromornes existed as a complete family from the 15th century, consisting, according to Virdung, of four instruments; Praetorius[6] cites five—the deep bass, the bass, the tenor or alto, the cantus or soprano and the high soprano, with compass as shown. A band, or, to use the expression of Praetorius, an "accort" of cromornes comprised 1 deep bass, 2 bass, 3 tenor, 2 cantus, 1 high soprano = 9.

Mersenne[7] explains the construction of the cromorne, giving careful illustrations of the instrument with and without the cap. From him we learn that these instruments were made in England, where they were played in concert in sets of four, five and six. Their scheme of construction and especially the reed and cap is very similar to that of the chalumeau of the musette (see BAG-PIPE), but its timbre is by no means so pleasant. Mersenne's cromornes have ten fingerholes, Nos. 7 and 8 being duplicates for right and left-handed players. They were probably sometimes used, as was the case with the hautbois de Poitou (see BAG-PIPE), without the cap, when an extended compass was required.

The cromornes were in very general use in Europe from the 14th to the 17th century, and are to be found in illustrations of pageants, as for instance in the magnificent collection of woodcuts designed by Hans Burgmair, a pupil of Albrecht Dürer, representing the triumph of the emperor Maximilian,[8] where a bass and a tenor Krumbhorn player figure in the procession among countless other musicians. In the inventory of the wardrobe, &c., belonging to Henry VIII. at

> Westminster, made during the reign of Edward VI., we find eighteen crumhornes (see British Museum, Harleian MS. 1419, ff. 202b and 205). The cromornes did not always form an orchestra by themselves, but were also used in concert with other instruments and notably with flutes and oboes, as in municipal bands and in the private bands of princes. In 1685 the orchestra of the Neue Kirche at Strassburg comprised two tournebouts or cromornes, and until the middle of the 18th century these instruments formed part of the court band known as "Musique de la Grande Écurie" in the service of the French kings. They are first mentioned in the accounts for the year 1662, together with the tromba-marina, although the instrument was already highly esteemed in the 16th century. In that year five players of the cromorne were enrolled among the musicians of the Grande Écurie du Roi;[9] they received a yearly salary of 120 livres, which various supplementary allowances brought up to about 330 livres. In 1729 one of the cromorne players sold his appointment for 4000 francs. This was a sign of the failing popularity of the instrument. The duties of the cromorne and tromba-marina players consisted in playing in the great *divertissements* and at court functions and festivals in honour of royal marriages, births and thanksgivings.

Cromornes have become of extreme rarity and are not to be found in all collections. The Paris Conservatoire possesses one large bass cromorne of the 16th century, the Kgl. Hochschule für Musik,[10] Berlin, a set of seven, and the Ambroser Sammlung, Vienna, a cromorne in E□.[11] The museum of the Conservatoire Royal de Musique at Brussels has the good fortune to possess a complete family which is said to have belonged to the duke of Ferrara, Alphonso II. d'Este, a prince who reigned from 1559 to 1597. The soprano (cantus or discant) has the same compass as above, while those of the alto, the tenor (furnished with a key) and the bass are as shown.

The bass (see figure), besides having two keys, is distinguished from the others by two contrivances like small bolts, which slide in grooves and close the two holes that give the lowest notes of the instrument. The use of these bolts, placed at the extremity of the tournebout and out of reach of the fingers of the instrumentalist, renders necessary the assistance of a person whose sole mission is to attend to them during the performance. E. van der Straeten[12] mentions a key belonging to a large cromorne bearing the date 1537, of which he gives a large drawing. A cromorne appears in a musical scene with a trumpet in Hermann Finck's *Practica Musica*.[13]

The "Platerspil," of which Virdung gives a drawing, is only a kind of cromorne. It is characterized by having, instead of a cap to cover the reed, a spherical receiver surrounding the reed, to which the tube for insufflation is adapted. The Platerspiel is also frequently classified among bagpipes. In the *Cantigas di Sante Maria*,[14] a MS. of the 13th century preserved in the Escorial, Madrid, two instruments of this type are represented. One of these has two straight, parallel pipes, slightly conical; the other is frankly conical with wide bore turned up at the end.

Other instruments belonging by their most important characteristics of cylindrical bore and double reed to the same family as the cromorne, although the bore was somewhat differently disposed, are the racket bassoon and the sourdine or sordelline. The latter was introduced into the orchestra by Cavaliere in his opera *Rappresentazione di anima e di corpo*, and is described by Giudotto[15] in his edition of the score as "Flauti overo due tibie all' antica che noi chiamiamo sordelline," a description which tallies with what has been said above concerning the aulos and tibia.

(V. M. and K. S.)

1 Crumhorne need not be regarded as a corruption of the German, since the two words of which it is composed were both in use in medieval England. *Crumb* = curved; *crumbe* = hook, bend; *crome* = a staff with a hook at the end of it. See Stratmann's *Middle English Dictionary* (1891), and Halliwell, *Dictionary of Archaic and Provincial Words* (London, 1881).

2 See A. Howard, "Aulos or Tibia," *Harvard Studies*, iv. (Boston, 1893).

3 See also A. A. Howard, op. cit., "Phrygian Aulos," pp. 35-38.

4 *Musica getutscht und auszgezogen* (Basel, 1511).

5 See Diderot and d'Alembert's *Encyclopédie* (Paris, 1751-1780), t. 5, "Lutherie," pl. ix.

6 *Organographia* (Wolfenbüttel, 1618).

7 *L'Harmonie universelle* (Paris, 1636-1637), book v. pp. 289 and 290. Cf. "Musette," pp. 282-287 and 305.

8 See "Triumphzug des Kaisers Maximilian I." Beilage zum II. Band des *Jahrb. der Sammlungen des Allerhöchsten Kaiserhauses* (Vienna, 1884-1885), pl. 20. Explanatory text and part i. in Band i. of the same publication, 1883-1884. A French edition with 135 plates was also published in Vienna by A. Schmidt, and in London by J. Edwards (1796). See also Dr August Reissmann, *Illustrierte Geschichte der deutschen Musik* (Leipzig, 1881), where a few of the plates are reproduced.

9 See J. Écorcheville, "Quelques documents sur la musique de la grande écurie du roi," *Sammelband d. Intern. Musik. Ges.* Jahrg. ii., Heft 4 (1901, Leipzig, London, &c.), pp. 630-632.

10 Oskar Fleischer, *Führer* (Berlin, 1892), p. 29, Nos. 400 to 406.

11 For an illustration see Captain C. R. Day, *Descriptive Catalogue* (London, 1891), pl. iv. E. and p. 99.

12 *Histoire de la musique aux Pays-Bas avant le XIXe siècle* (Brussels, 1867-1888), vol. vii. p. 336, and description, p. 333 et seq.

13 Wittenberg, 1556; reproduced by A. Reissmann, op. cit., pp. 233 and 226.

14 Reproduced in Riaño's *Notes on Early Spanish Music* (London, 1887), pp. 119-127.

15 See Hugo Goldschmidt, "Das Orchester der italienischen Oper im 17. Jahrh." *Sammelband der Intern. Musikgesellschaft*, Jahrg. ii., Heft 1 (Leipzig, 1900), p. 24.

CROMPTON, SAMUEL

(1753-1827), English inventor, was born on the 3rd of December 1753 at Firwood near Bolton-le-Moors, Lancashire. While yet a boy he lost his father, and had to contribute to the family resources by spinning yarn. The defects of the spinning jenny imbued him with the idea of devising something better, and for five or six years the effort absorbed all his spare time and money, including what he earned by playing the violin at the Bolton theatre. About 1779 he succeeded in producing a machine which span yarn suitable for use in the manufacture of muslin, and which was known as the muslin wheel or the Hall-in-the-Wood wheel (from the name of the house in which he and his family resided), and later as the spinning mule. After his marriage in 1780 a good demand arose for the yarn which he himself made at Hall-in-the-Wood, but the prying to which his methods were subjected drove him, in the absence of means to take out a patent, to the choice of destroying his machine or making it public. He adopted the latter alternative on the promise of a number of manufacturers to pay him for the use of the mule, but all he received was about £60. He then resumed spinning on his own account, but with indifferent success. In 1800 a sum of £500 was raised for his benefit by subscription, and when in 1809 Edmund Cartwright, the inventor of the power-loom obtained £10,000 from parliament, he determined also to apply for a grant. In 1811 he made a tour in the manufacturing districts of Lancashire and Scotland to collect evidence showing how extensively his mule was used, and in 1812 parliament allowed him £5000. With the aid of this money he embarked in business, first as a bleacher and then as a cotton merchant and spinner, but again without success. In 1824 some friends, without his knowledge, bought him an annuity of £63. He died at Bolton on the 26th of June 1827.

CROMPTON

An urban district of Lancashire, England, 2½ m. N. of Oldham, within the parliamentary borough of Oldham. Pop. (1901) 13,427. At Shaw, a populous village included within it, is a station on the Lancashire & Yorkshire railway. Cotton mills and the collieries of the neighbourhood employ the large industrial population.

CROMWELL, HENRY

(1628-1674), fourth son of Oliver Cromwell, was born at Huntingdon on the 20th of January 1628, and served under his father during the latter part of the Civil War. His active life, however, was mainly spent in Ireland, whither he took some troops to assist Oliver early in 1650, and he was one of the Irish representatives in the Little, or Nominated, Parliament of 1653. In 1654 he was again in Ireland, and after making certain recommendations to his father, now lord protector, with regard to the government of that country, he became major-general of the forces in Ireland and a member of the Irish council of state, taking up his new duties in July 1655. Nominally Henry was subordinate to the lord-deputy, Charles Fleetwood, but Fleetwood's departure for England in September 1655 left him for all practical purposes the ruler of Ireland. He moderated the lord-deputy's policy of deporting the Irish, and unlike him he paid some attention to the interests of the English settlers; moreover, again unlike Fleetwood, he appears to have held the scales evenly between the different Protestant sects, and his undoubted popularity in Ireland is attested by Clarendon. In November 1657 Henry himself was made lord-deputy; but before this time he had refused a gift of property worth £1500 a year, basing his refusal on the grounds of the poverty of the country, a poverty which was not the least of his troubles. In 1657 he advised his father not to accept the office of king, although in 1634 he had supported a motion to this effect; and after the dissolution of Cromwell's second parliament in February 1658 he showed his anxiety that the protector should act in a moderate and constitutional manner. After Oliver's death Henry hailed with delight the succession of his brother Richard to the office of protector, but although he was now appointed lieutenant and governor general of Ireland, it was only with great reluctance that he remained in that country. Having rejected proposals to assist in the restoration of Charles

II., Henry was recalled to England in June 1659 just after his brother's fall; quietly obeying this order he resigned his office at once. Although he lost some property at the Restoration, he was allowed after some solicitation to keep the estate he had bought in Ireland. His concluding years were passed at Spinney Abbey in Cambridgeshire; he was unmolested by the government, and he died on the 23rd of March 1674. In 1653 Henry married Elizabeth (d. 1687), daughter of Sir Francis Russell, and he left five sons and two daughters.

CROMWELL, OLIVER

(1599-1658), lord protector of England, was the 5th and only surviving son of Robert Cromwell of Huntingdon and of Elizabeth Steward, widow of William Lynn. His paternal grandfather was Sir Henry Cromwell of Hinchinbrook, a leading personage in Huntingdonshire, and grandson of Richard Williams, knighted by Henry VIII., nephew of Thomas Cromwell, earl of Essex, Henry VIII.'s minister, whose name he adopted. His mother was descended from a family named Styward in Norfolk, which was not, however, connected in any way, as has been often asserted, with the royal house of Stuart. Oliver was born on the 25th of April 1599, was educated under Dr Thomas Beard, a fervent puritan, at the free school at Huntingdon, and on the 23rd of April 1616 matriculated as a fellow-commoner at Sidney Sussex College, Cambridge, then a hotbed of puritanism, subsequently studying law in London. The royalist anecdotes relating to his youth, including charges of ill-conduct, do not deserve credit, the entries in the register of St John's, Huntingdon, noting Oliver's submission on two occasions to church censure being forgeries; but it is not improbable that his youth was wild and possibly dissolute.[1] According to Edmund Waller he was "very well read in the Greek and Roman story." Burnet declares he had little Latin, but he was able to converse with the Dutch ambassador in that language. According to James Heath in his *Flagellum*, "he was more famous for his exercises in the fields than in the schools, being one of the chief match-makers and players at football, cudgels, or any other boisterous game or sport." On the 22nd of August 1620 he married Elizabeth, daughter of Sir James Bourchier, a city merchant of Tower Hill, and of Felstead in Essex; and his father having died in 1617 he settled at Huntingdon and occupied himself in the management of his small estate. In 1628 he was returned to parliament as member for the borough, and on the 11th of February

1629 he spoke in support of puritan doctrine, complaining of the attempt by the king to silence Dr Beard, who had raised his voice against the "flat popery" inculcated by Dr Alabaster at Paul's Cross. He was also one of the members who refused to adjourn at the king's command till Sir John Eliot's resolutions had been passed.

During the eleven years of government without parliament very little is recorded of Cromwell. His name is not connected with the resistance to the levy of ship-money or to the action of the ecclesiastical courts, but in 1630 he was one of those fined for refusing to take up knighthood. The same year he was named one of the justices of the peace for his borough; and on the grant of a new charter showed great zeal in defending the rights of the commoners, and succeeded in procuring an alteration in the charter in their favour, exhibiting much warmth of temper during the dispute and being committed to custody by the privy council for angry words spoken against the mayor, for which he afterwards apologized. He also defended the rights of the commoners of Ely threatened by the "adventurers" who had drained the Great Level, and he was nicknamed afterwards by a royalist newspaper "Lord of the Fens." He was again later the champion of the commoners of St Ives in the Long Parliament against enclosures by the earl of Manchester, obtaining a commission of the House of Commons to inquire into the case, and drawing upon himself the severe censure of the chairman, the future Lord Clarendon, by his "impetuous carriage" and "insolent behaviour," and by the passionate vehemence he imparted into the business. Bishop Williams, a kinsman of Cromwell's, relates at this time that he was "a common spokesman for sectaries, and maintained their part with great stubbornness"; and his earliest extant letter (in 1635) is an appeal for subscriptions for a puritan lecturer. There appears to be no foundation for the statement that he was stopped by an order of council when on the point of abandoning England for America, though there can be little doubt that the thoughts of emigration suggested themselves to his mind at this period. He viewed the "innovations in religion" with abhorrence. According to Clarendon he told the latter in 1641 that if the Grand Remonstrance had not passed "he would have sold all he had the next morning and never have seen England more." In 1631 he converted his landed property into money, and John Hampden, his cousin, a patentee of Connecticut in 1632, was on the point of emigrating. Cromwell was perhaps arrested in his project by his succession in 1636 to the estate of his uncle Sir Thomas Steward, and to his office of farmer of the cathedral tithes at Ely, whither he now removed. Meanwhile, like Bunyan and many other puritans, Cromwell had been passing through a trying period of mental and religious change and struggle, beginning with deep melancholy and religious doubt and depression, and ending with "seeing light" and with enthusiastic and convinced faith, which remained henceforth the chief characteristic and impulse in his career.

He represented Cambridge in the Short and Long Parliaments of 1640, and at once showed extraordinary zeal and audacity in his opposition to the government, taking a large share in business and serving on numerous and important ***Cromwell's first parliamentary efforts.*** committees. As the cousin of Hampden and St. John he was intimately associated with the leaders of the parliamentary party. His sphere of action, however, was not in parliament. He was not an orator, and though he could express himself forcibly on occasion, his speech was incoherent and devoid of any of the arts of rhetoric. Clarendon notes on his first appearance in parliament that "he seemed to have a person in no degree gracious, no ornament of discourse, none of those talents which use to reconcile the affections of the standers by; yet as he grew into place and authority his parts seemed to be renewed." He supported stoutly the extreme party of opposition to the king, but did not take the lead except on a few less important occasions, and was apparently silent in the debates on the Petition of Right, the Grand Remonstrance and the Militia. His first recorded intervention in debate in the Long Parliament was on the 9th of November 1640, a few days after the meeting of the House, when he delivered a petition from the imprisoned John Lilburne. He was described by Sir Philip Warwick on this occasion:—"I came into the House one morning well clad and perceived a gentleman speaking whom I knew not, very ordinarily apparelled; for it was a plain cloth suit which seemed to have been made by an ill country tailor; his linen was plain and not very clean; ... his stature was of a good size; his sword stuck close to his side; his countenance swollen and reddish; his voice sharp and untunable and his eloquence full of fervour ... I sincerely profess it much lessened my reverence as to that great council for he was very much hearkened unto." On the 30th of December he moved to the second reading of Strode's bill for annual parliaments. His chief interest from the first, however, lay in the religious question. He belonged to the Root and Branch party, and spoke in favour of the petition of the London citizens for the abolition of episcopacy on the 9th of February 1641, and pressed upon the House the Root and Branch Bill in May. On the 6th of November he carried a motion entrusting the trainbands south of the Trent to the command of the earl of Essex. On the 14th of January 1642, after the king's attempt to seize the five members, he moved for a committee to put the kingdom in a posture of defence. He contributed £600 to the proposed Irish campaign and £500 for raising forces in England—large sums from his small estate—and on his own initiative in July 1642 sent arms of the value of £100 down to Cambridge, seized the magazine there in August, and prevented the king's commission of array from being executed in the county, taking these important steps on his own authority and receiving subsequently indemnity by vote of the House of Commons. Shortly afterwards he joined Essex with sixty horse, and was present at Edgehill, where his troop was one of the few not routed by Rupert's charge, Cromwell himself being mentioned among those officers who "never stirred from their troops but fought till the last minute."

During the earlier part of the year 1643 the military position of Charles was greatly superior to that of the parliament. Essex was inactive near Oxford; in the west Sir Ralph Hopton had won a series of victories, and in the north ***Beginning of Civil War.*** Newcastle defeated the Fairfaxes at Adwalton Moor, and all Yorkshire except Hull was in his hands. It seemed likely that the whole of the north would be laid open and the royalists be able to march upon London and join Charles and Hopton there. This stroke, which would most probably have given the victory to the king, was prevented by the "Eastern Association," a union of Norfolk, Suffolk, Essex, Cambridgeshire and Hertfordshire, constituted in December 1642 and augmented in 1643 by Huntingdonshire and Lincolnshire, of which Cromwell was the leading spirit. His zeal and energy met everywhere with conspicuous success. In January 1643 he seized the royalist high sheriff of Hertfordshire in the act of proclaiming the king's commission of array at St Albans; in February he was at Cambridge taking measures for the defence of the town; in March suppressing royalist risings at Lowestoft and Lynn; in April those of Huntingdon, when he also recaptured Crowland from the king's party. In May he defeated a greatly superior royalist force at Grantham, proceeding afterwards to Nottingham in accordance with Essex's plan of penetrating into Yorkshire to relieve the Fairfaxes; where, however, difficulties, arising from jealousies between the officers, and the treachery of John Hotham, whose arrest Cromwell was instrumental in effecting, obliged him to retire again to the association, leaving the Fairfaxes to be defeated at Adwalton Moor. He showed extraordinary energy, resource and military talent in stemming the advance of the royalists, who now followed up their victories by advancing into the association; he defeated them at Gainsborough on the 28th of July, and managed a masterly retreat before overwhelming numbers to Lincoln, while the victory on the 11th of October at Winceby finally secured the association, and maintained the wedge which prevented the junction of the royalists in the north with the king in the south.

One great source of Cromwell's strength was the military reforms he had initiated. At Edgehill he had observed the inferiority of the parliamentary to the royalist horse, composed as it was of soldiers of fortune and the dregs ***Cromwell's soldiers.*** of the populace. "Do you think," he had said, "that the spirits of such base, mean fellows will ever be able to encounter gentlemen that have honour and courage and resolution in them? You must get men of a spirit that is likely to go as far as gentlemen will go or you will be beaten still." The royalists were fighting for a great cause. To succeed the parliamentary soldiers must also be inspired by some great principle, and this was now found in religion. Cromwell chose his own troops, both officers and privates, from the "religious men," who fought not for pay or for adventure, but for their faith. He declared, when answering a complaint that a certain captain in his regiment was a better preacher than fighter, that he who prayed best would fight best, and that he knew nothing could "give the like courage and confidence as the knowledge of God in Christ will." The superiority of these men—more intelligent than the common soldiers, better disciplined, better trained, better armed, excellent horsemen and fighting for a great cause—not only over the other parliamentary troops but over the royalists, was soon observed in battle. According to Clarendon the latter, though frequently victorious in a charge, could not rally afterwards, "whereas Cromwell's troops if they prevailed, or though they were beaten and routed, presently rallied again and stood in good order till they received new orders"; and the king's military successes dwindled in proportion to the gradual preponderance of Cromwell's troops in the parliamentary army. At first these picked men only existed in Cromwell's own troop, which, however, by frequent additions became the nucleus of a regiment, and by the time of the New Model included about 11,000 men.

In July 1643 Cromwell had been appointed governor of the Isle of Ely; on the 22nd of January 1644 he became second in command under the earl of Manchester as lieutenant-general of the Eastern Association, and on the 16th of February 1644 a member of the Committee of Both Kingdoms with greatly increased influence. In March he took Hillesden House in Buckinghamshire; in May was at the siege of Lincoln, when he repulsed Goring's attempt to relieve the town, and subsequently took part in Manchester's campaign in the north. At Marston Moor (*q.v.*) on the 2nd of July he commanded all the horse of the Eastern Association, with some Scottish troops; and though for a time disabled by a wound in the neck, he charged and routed Rupert's troops opposed to him, and subsequently went to the support of the Scots, who were hard pressed by the enemy, and converted what appeared at one time a defeat into a decisive victory. It was on this occasion that he earned the nickname of "Ironsides," applied to him now by Prince Rupert, and afterwards to his soldiers, "from the impenetrable strength of his troops which could by no means be broken or divided."

The movements of Manchester after Marston Moor were marked by great apathy. He was one of the moderate party who desired an accommodation with the king, and was opposed to Cromwell's sectaries. He remained at Lincoln, did nothing to prevent the defeat of Essex's army in the west, and when he at last advanced south to join Essex's and Waller's troops his management of the army led to the failure of the attack upon the king at Newbury on the 27th of October 1644. He delayed supporting the infantry till too late, and was repulsed; he allowed the royal army to march past his outposts; and a fortnight afterwards, without any attempt to prevent it, and greatly to Cromwell's vexation, permitted the moving of the king's artillery and the relief of Donnington Castle by Prince Rupert. "If you beat the king ninety-nine times," Manchester urged at Newbury, "yet he is king still and so will his posterity be after him; but if the king beat us once we shall all be hanged and our posterity be made slaves." "My lord," answered Cromwell, "if this be so, why did we take up arms at first? This is against fighting ever hereafter. If so let us make peace, be it ever so base." The contention brought to a crisis the struggle between the moderate Presbyterians and the Scots on the one side, who decided to maintain the monarchy and fought for an accommodation and to establish Presbyterianism in England, and on the other the republicans who would be satisfied with nothing less than the complete overthrow of the king, and the Independents who regarded the establishment of Presbyterianism as an evil almost as great as that of the Church of England. On the 25th of November Cromwell charged Manchester with "unwillingness to have the war prosecuted to a full victory"; which Manchester answered by accusing Cromwell of having used expressions against the nobility, the Scots and Presbyterianism; of desiring to fill the army of the Eastern Association with Independents to prevent any accommodation; and of having vowed if he met the king in battle he would as lief fire his pistol at him as at anybody else. The lords and the Scots vehemently took Manchester's part; but the Commons eventually sided with Cromwell, appointed Sir Thomas Fairfax general of the New Model Army, and passed two self-denying ordinances, the second of which, ordering all members of both houses to lay down their commissions within forty days, was accepted by the lords on the 3rd of April 1645.

Meanwhile Cromwell had been ordered on the 3rd of March by the House to take his regiment to the assistance of Waller, under whom he served as an admirable subordinate. "Although he was blunt," says Waller, "he did not bear himself with pride or disdain. As an officer he was obedient and did never dispute my orders or argue upon them." He returned on the 19th of April, and on the 23rd was sent to Oxfordshire to prevent a junction between Charles and Prince Rupert, in which he succeeded after some small engagements and the storming of Blechingdon House. His services were felt to be too valuable to be lost, and on the 10th of May his command was prolonged for forty days. On the 28th he was sent to Ely for the defence of the eastern counties against the king's advance; and on the 10th of June, upon Fairfax's petition, he was named by the Commons lieutenant-general, joining Fairfax on the 13th with six hundred horse. At the decisive battle of Naseby (the 14th ***The battle of Naseby.*** of June 1645) he commanded the parliamentary right wing and routed the cavalry of Sir Marmaduke Langdale, subsequently falling upon and defeating the royalist centre, and pursuing the fugitives as far as the outskirts of Leicester. At Langport again, on the 10th of July 1645, his management of the troops was largely instrumental in gaining the victory. As the king had no longer a field army, the war after Naseby resolved itself into a series of sieges which Charles had no means of raising. Cromwell was present at the sieges of Bridgwater, Bath, Sherborne and Bristol; and later, in command of four regiments of foot and three of horse, he was employed in clearing Wiltshire and Hampshire of the royalist garrisons. He took Devizes and Laycock House, Winchester and Basing House, and rejoined Fairfax in October at Exeter, and accompanied him to Cornwall, where he assisted in the defeat of Hopton's forces and in the suppression of the royalists in the west. On the 9th of January 1646 he surprised Lord Wentworth's brigade at Bovey Tracey, and was present with Fairfax at the fall of Exeter on the 9th of April. He then went to London to give an account of proceedings to the parliament, was thanked for his services and rewarded with the estate of the marquess of Worcester. He was present again with Fairfax at the capitulation of Oxford on the 24th of June, which practically terminated the Civil War, when he used his influence in favour of granting lenient terms. He then removed with his family from Ely to Drury Lane, London, and about a year later to King Street, Westminster.

The war being now over, the great question of the establishment of Presbyterianism or Independency had to be decided. Cromwell, without naming himself an adherent of any denomination, fought vigorously for Independency as a policy. In 1644 he had remonstrated at the removal by Crawford of an anabaptist lieutenant-colonel. "The state," he said, "in choosing men to serve it, takes no notice of their opinions. If they be willing faithfully to serve it, that satisfies. Take heed of being sharp ... against those to whom you can object little but that they square not with you in every opinion concerning matters of religion." He had patronized Lilburne and welcomed all into his regiment, and the Independents had spread from his troops throughout the whole army. But while the sectarians were in a vast majority in the army, the parliament was equally strong in Presbyterianism and opposed to toleration. The proposed disbandment of the army in February 1647 would have placed the soldiers entirely in the power of the parliament; while the negotiations of the king, first with the Scots and then with the parliament, appeared to hazard all the fruits of victory. The petition from the army to the parliament for arrears of pay was suppressed and the petitioners declared enemies of the state. In consequence the army organized a systematic opposition, and elected representatives styled Agitators or Agents to urge their claims.

Cromwell, though greatly disliking the policy of the Presbyterians, yet gave little support at first to the army in resisting parliament. In May 1647 in company with Skippon, Ireton and Fleetwood, he visited the army, inquired ***Parliament and the army.*** into and reported on the grievances, and endeavoured to persuade them to submit to the parliament. "If that authority falls to nothing," he said, "nothing can follow but confusion." The Presbyterians, however, now engaged in a plan for restoring the king under their own control, and by the means of a Scottish army, forced on their policy, and on the 27th of May ordered the immediate disbandment of the army, without any guarantee for the payment of arrears. A mutiny was the consequence. The soldiers refused to disband, and on the 3rd of June Cromwell, whom, it was believed, the parliament intended to arrest, joined the army. "If he would not forthwith come and lead them," they had told him, "they would go their own way without him." The supremacy of the army without a guiding hand meant anarchy, that of the Presbyterians the outbreak of another civil war.

Possession of the king's person now became an important consideration. On the 31st of May 1647 Cromwell had ordered Cornet Joyce to prevent the king's removal by the parliament or the Scots from Holmby, and Joyce by his own authority and with the king's consent brought him to Newmarket to the headquarters of the army. Cromwell soon restored order, and the representative council, including privates as well as officers chosen to negotiate with the parliament, was subordinated to the council of war. The army with Cromwell then advanced towards London. In a letter to the city, possibly written by Cromwell himself, the officers repudiated any wish to alter the civil government or upset the establishment of Presbyterianism, but demanded religious toleration. Subsequently, in the declaration of the 14th of June, arbitrary power either in the parliament or in the king was denounced, and demand was made for a representative parliament, the speedy termination of the actual assembly, and the recognition of the right to petition. Cromwell used his influence in restraining the more eager who wished to march on London immediately, and in avoiding the use of force by which nothing permanent could be effected, urging that "whatsoever we get by treaty will be firm and durable. It will be conveyed over to posterity." The army faction gradually gathered strength in the parliament. Eleven Presbyterian leaders impeached by the army withdrew of their own accord on the 26th of June, and the parliament finally yielded. Fairfax was appointed sole commander-in-chief on the 19th of July, the soldiers levied to oppose the army were dismissed, and the command of the city militia was again restored to the committee approved by the army. These votes, however, were cancelled later, on the 26th of July, under the pressure of the royalist city mob which invaded the two Houses; but the two speakers, with eight peers and fifty-seven members of the Commons, themselves joined the army, which now advanced to London, overawing all resistance, escorting the fugitive members in triumph to Westminster on the 6th of August, and obliging the parliament on the 20th to cancel the last votes, with the threat of a regiment of cavalry drawn up by Cromwell in Hyde Park.

Cromwell and the army now turned with hopes of a settlement to Charles. On the 4th of July Cromwell had had an interview with the king at Caversham. He was not insensible to Charles's good qualities, was touched by the paternal affection he showed for his children, and is said to have declared that Charles "was the uprightest and most conscientious man of his three kingdoms." The *Heads of the Proposals*, which, on Charles raising objections, had been modified by the influence of Cromwell and Ireton, demanded the control of the militia and the choice of ministers by parliament for ten years, a religious toleration, and a council of state to which much of the royal control over the army and foreign policy would be delegated. These proposals without doubt largely diminished the royal power, and were rejected by Charles with the hope of maintaining his sovereign rights by "playing a game," to use his own words, *i.e.* by negotiating simultaneously with army and parliament, by inflaming their jealousies and differences, and finally by these means securing his restoration with his full prerogatives unimpaired. On the 9th of September Charles refused once mere the *Newcastle Propositions* offered him by the parliament, and Cromwell, together with Ireton and Vane, obtained the passing of a motion for a new application; but the terms asked by the parliament were higher than before and included a harsh condition—the exclusion from pardon of all the king's leading adherents, besides the indefinite establishment of Presbyterianism and the refusal of toleration to the Roman Catholics and members of the Church of England.

Meanwhile the failure to come to terms with Charles and provide a settlement appeared to threaten a general anarchy. Cromwell's moderate counsels created distrust in his good faith amongst the soldiers, who accused him of "prostituting the liberties and persons of all the people at the foot of the king's interest." The agitators demanded immediate settlement by force by the army. The extreme republicans, anticipating Rousseau, put forward the *Agreement of the People*. This was strongly opposed by Cromwell, who declared the very consideration of it had dangers, that it would bring upon the country "utter confusion" and "make England like Switzerland." Universal suffrage he rejected as tending "very much to anarchy," spoke against the hasty abolition of either the monarchy or the Lords, and refused entirely to consider the abstract principles brought into the debate. Political problems were not to be so resolved, but practically. With Cromwell as with Burke the question was "whether the spirit of the people of this nation is prepared to go along with it." The special form of government was not the important point, but its possibility and its acceptability. The great problem was to found a stable government, an authority to keep order. If every man should fight for the best form of government the state would come to desolation. He reproached the soldiers for their insubordination against their officers, and the army for its rebellion against the parliament. He would lay hold of anything "if it had but the force of authority," rather than have none. Cromwell's influence prevailed and these extreme proposals were laid aside.

Meanwhile all hopes of an accommodation with Charles were dispelled by his flight on the 11th of November from Hampton Court to Carisbroke Castle in the Isle of Wight, his object being to negotiate independently with the ***Flight of the king.*** Scots, the parliament and the army. His action, however, in the event, diminished rather than increased his chances of success, owing to the distrust of his intentions which it inspired. Both the army and the parliament gave cold replies to his offers to negotiate; and Charles, on the 27th of December 1647, entered into the *Engagement* with the Scots by which he promised the establishment of Presbyterianism for three years, the suppression of the Independents and their sects, together with privileges for the Scottish nobles, while the Scots undertook to invade England and restore him to his throne. This alliance, though the exact terms were not known to Cromwell—"the attempt to vassalize us to a foreign nation," to use his own words—convinced him of the uselessness of any plan for maintaining Charles on the throne; though he still appears to have clung to monarchy, proposing in January 1648 the transference of the crown to the prince of Wales. A week after the signing of the treaty he supported a proposal for the king's deposition, and the vote of *No Addresses* was carried. Meanwhile the position of Charles's opponents had been considerably strengthened by the suppression of a dangerous rebellion in November 1647 by Cromwell's intervention, and by the return of troops to obedience. Cromwell's difficulties, however, were immense. His moderate and trimming attitude was understood neither by the extreme Independents nor by the Presbyterians. He made one attempt to reconcile the disputes between the army and the politicians by a conference, but ended the barren discussion on the relative merits of aristocracies, monarchies and democracies, interspersed with Bible texts, by throwing a cushion at the speaker's head and running downstairs. On the 19th of January 1648 Cromwell was accused of high treason by Lilburne. Plots were formed for his assassination. He was overtaken by a dangerous illness, and on the 2nd of March civil war in support of the king broke out.

Cromwell left London in May to suppress the royalists in Wales, and took Pembroke Castle on the 11th of July. Meanwhile behind his back the royalists had risen all over England, the fleet in the Downs had declared for Charles, and the Scottish army under Hamilton had invaded the north. Immediately on the fall of Pembroke Cromwell set out to relieve Lambert, who was slowly retreating before Hamilton's superior forces; he joined him near Knaresborough on the 12th of August, and started next day in pursuit of Hamilton in Lancashire, placing himself at Stonyhurst near Preston, cutting off Hamilton from the north and his allies, and defeating him in detail on the 17th, 18th and 19th at Preston and at Warrington. He then marched north into Scotland, following the forces of Monro, and established a new government of the Argyle faction at Edinburgh; replying to the Independents who disapproved of his mild treatment of the Presbyterians, that he desired "union and right understanding between the godly people, Scots, English, Jews, Gentiles, Presbyterians, Anabaptists and all; ... a more glorious work in our eyes than if we had gotten the sacking and plunder of Edinburgh ... and made a conquest from the Tweed to the Orcades."

The incident of the Second Civil War and the treaty with the Scots exasperated Cromwell against the king. On his return to London he found the parliament again negotiating with Charles, and on the eve of making a treaty which ***Cromwell supports the Remonstrance.*** Charles himself had no intention of keeping and regarded merely as a means of regaining his power, and which would have thrown away in one moment all the advantages gained during years of bloodshed and struggle. Cromwell therefore did not hesitate to join the army in its opposition to the parliament, and supported the Remonstrance of the troops (20th of November 1648), which included the demand for the king's punishment as "the grand author of all our troubles," and justified the use of force by the army if other means failed. The parliament, however, continued to negotiate, and accordingly Charles was removed by the army to Hurst Castle on the 1st of December, the troops occupied London on the 2nd; while on the 6th and 7th Colonel Pride "purged" the House of Commons of the Presbyterians. Cromwell was not the originator of this act, but showed his approval of it by taking his seat among the fifty or sixty Independent members who remained.

The disposal of the king was now the great question to be decided. During the next few weeks Cromwell appears to have made once more attempts to come to terms with Charles; but the king was inflexible in his refusal to part with the essential powers of the monarchy, or with the Church; and at the end of December it was resolved to bring him to trial. The exact share which Cromwell had in this decision and its sequel is obscure, and the later accounts of the regicides when on their trial at the Restoration, ascribing the whole transaction to his initiation and agency, cannot be altogether accepted. But it is plain that, once convinced of the necessity for the king's execution, he was the chief instrument in overcoming all scruples among his judges, and in resisting the protests and appeals of the Scots. To Algernon Sidney, who refused to take part in proceedings on the plea that neither the king nor any man could be tried by such a court, Cromwell replied, "I tell you, we will cut off his head with the crown upon it."

The execution of the king took place on the 30th of January 1649. This event, the turning-point in Cromwell's career, casts a shadow, from one point of view, over the whole of his future statesmanship. He himself never repented ***The execution of Charles I.*** of the act, regarding it, on the contrary, as "one which Christians in after times will mention with honour and all tyrants in the world look at with fear," and as one directly ordained by God. Opinions, no doubt, will always differ as to the wisdom or authority of the policy which brought Charles to the scaffold. On the one hand, there was no law except that of force by which an offence could be attributed to the sovereign, the anointed king, the source of justice. The ordinance establishing the special tribunal for the trial was passed by a remnant of the House of Commons alone, from which all dissentients were excluded by the army. The tribunal was composed, not of judges—for all unanimously refused to sit on it—but of fifty-two men drawn from among the king's enemies. The execution was a military and not a national act, and at the last scene on the scaffold the triumphant shouts of the soldiery could not overwhelm the groans and sobs raised by the populace. Whatever crimes might be charged against Charles, his past conduct might appear to be condoned by the act of negotiating with him. On the other hand, the execution seemed to Cromwell the only alternative to anarchy, or to a return to despotism and the abandonment of all they had fought for. Cromwell had exhausted every expedient for arriving at an arrangement with the king by which the royal authority might be preserved, and the repeated perfidy and inexhaustible shiftiness of Charles had proved the hopelessness of such attempts. The results produced by the king's execution were far-reaching and permanent. It is true that Puritan austerity and the lack of any strong central authority after Oliver's death produced a reaction which temporarily restored Charles's dynasty to the throne; but it is not less true that the execution of the king, at a later time when all over Europe absolute monarchies "by divine right" were being established on the ruins of the ancient popular constitutions, was an object lesson to all the world; and it produced a profound effect, not only in establishing constitutional monarchy in Great Britain after James II., with the dread of his father's fate before him, had abdicated by flight, but in giving the impulse to that revolt against the idea of "the divinity that doth hedge a king" which culminated in the Revolution of 1789, and of which the mighty effects are still evident in Europe and beyond.

The king and the monarchy being now destroyed in England, Cromwell had next to turn his attention to the suppression of royalism in Ireland and in Scotland. In Ireland Ormonde had succeeded in uniting the English and the ***Cromwell in Ireland.*** Irish in a league against the supporters of the parliament, and only a few scattered forts held out for the Commonwealth, while the young king was every day expected to land and complete the conquest of the island. Accordingly in March 1649 Cromwell was appointed lord-lieutenant and commander-in-chief for its reduction.

But before starting he was called upon to suppress disorder at home. He treated the Levellers with some severity and showed his instinctive dislike to revolutionary proposals. "Did not that levelling principle," he said, "tend to the reducing of all to an equality? What was the purport of it but to make the tenant as liberal a fortune as the landlord, which I think if obtained would not have lasted long." Equally characteristic was his treatment of the mutinous army, in which he suppressed a rebellion in May. He landed at Dublin on the 13th of August. Before his arrival the Dublin garrison had defeated Ormonde with a loss of 5000 men, and Cromwell's work was limited to the capture of detached fortresses. On the 10th of September he stormed Drogheda, and by his order the whole of its 2800 defenders were put to the sword without quarter. Cromwell, who was as a rule especially scrupulous in protecting non-combatants from violence, justified his severity in this case by the cruelties perpetrated by the Irish in the rebellion of 1641, and as being necessary on military and political grounds in that it "would tend to prevent the effusion of blood for the future, which were the satisfactory grounds of such actions which otherwise cannot but work remorse and regret." After the fall of Drogheda Cromwell sent a few troops to relieve Londonderry, and marched himself to Wexford, which he took on the 11th of October, and where similar scenes of cruelty were repeated; every captured priest, to use Cromwell's own words, being immediately "knocked on the head," though the story of the three hundred women slaughtered in the market-place has no foundation.

The surrender of Trim, Dundalk and Ross followed, but at Waterford Cromwell met with a stubborn resistance and the advent of winter obliged him to raise the siege. Next year Cromwell penetrated into Munster. Cashel, Cahir and several castles fell in February, and Kilkenny in March; Clonmel repulsing the assault with great loss, but surrendering on the 10th of May 1650. Cromwell himself sailed a fortnight later, leaving the reduction of the island, which was completed in 1652, to his generals. The re-settlement of the conquered and devastated country was now organized on the Tudor and Straffordian basis of colonization from England, conversion to Protestantism, and establishment of law and order. Cromwell thoroughly approved of the enormous scheme of confiscation and colonization, causing great privations and sufferings, which was carried out. The Roman Catholic landowners lost their estates, all or part according to their degree of guilt, and these were distributed among Cromwell's soldiers and the creditors of the government; Cromwell also invited new settlers from home and from New England, two-thirds of the whole land of Ireland being thus transferred to new proprietors. The suppression of Roman Catholicism was zealously pursued by Cromwell; the priests were hunted down and imprisoned or exiled to Spain or Barbados, the mass was everywhere forbidden, and the only liberty allowed was that of conscience, the Romanist not being obliged to attend Protestant services.

These methods, together with education, "assiduous preaching ... humanity, good life, equal and honest dealing with men of different opinion," Cromwell thought, would convert the whole island to Protestantism. The law was ably and justly administered, and Irish trade was admitted to the same privileges as English, enjoying the same rights in foreign and colonial trade; and no attempt was made to subordinate the interests of the former to the latter, which was the policy adopted both before and after Cromwell's time, while the union of Irish and English interests was further recognized by the Irish representation at Westminster in the parliaments of 1654, 1656 and 1659. These advantages, however, scarcely benefited at all the Irish Roman Catholics, who were excluded from political life and from the corporate towns; and Cromwell's union meant little more than the union of the English colony in Ireland with England. A just administration, too, did not compensate for unjust laws or produce contentment; the policy of conversion and colonization was unsuccessful, the descendants of many of Cromwell's soldiers becoming merged in the Roman Catholic Irish, and the union with England, political and commercial, being extinguished at the Restoration. Cromwell's land settlement—modified by the restoration under Charles II. of about one-third of the estates to the royalists—survived, and added to the difficulties with which the English government was afterwards confronted in Ireland.

Meanwhile Cromwell had hurried home to deal with the royalists in Scotland. He urged Fairfax to attack the Scots at once in their own country and to forestall their invasion; but Fairfax refused and resigned, and ***The battles of Dunbar and Worcester.*** Cromwell was appointed by parliament, on the 26th of June 1650, commander-in-chief of all the forces of the Commonwealth. He entered Scotland in July, and after a campaign in the neighbourhood of Edinburgh which proved unsuccessful in drawing out the Scots from their fortresses, he retreated to Dunbar to await reinforcements from Berwick. The Scots under Leslie followed him, occupied Doon Hill commanding the town, and seized the passes between Dunbar and Berwick which Cromwell had omitted to secure. Cromwell was outmanœuvred and in a perilous situation, completely cut off from England and from his supplies except from the sea. But Leslie descended the hill to complete his triumph, and Cromwell immediately observed the disadvantages of his antagonist's new position, cramped by the hill behind and separated from his left wing. A stubborn struggle on the next day, the 3rd of September, gave Cromwell a decisive victory. Advancing, he occupied Edinburgh and Leith. At first it seemed likely that his victories and subsequent remonstrances would effect a peace with the Scots; but by 1651 Charles II. had succeeded in forming a new union of royalists and presbyterians, and another campaign became inevitable. Some delay was caused in beginning operations by Cromwell's dangerous illness, during which his life was despaired of; but in June he was confronting Leslie entrenched in the hills near Stirling, impregnable to attack and refusing an engagement. Cromwell determined to turn his antagonist's position. He sent 14,000 men into Fifeshire and marched to Perth, which he captured on the 2nd of August, thus cutting off Leslie from the north and his supplies. This movement, however, left open the way to England, and Charles immediately marched south, in reality thus giving Cromwell the wished-for opportunity of crushing the royalists finally and decisively. Cromwell followed through Yorkshire, and uniting with Lambert and Harrison at Evesham proceeded to attack the royalists at Worcester; where on the 3rd of September after a fierce struggle the great victory, "the crowning mercy" which terminated the Civil War, was obtained over Charles.

Monk completed the subjugation of Scotland by 1654. The settlement here was made on more moderate lines than in Ireland. The estates of only twenty-four leaders of the defeated cause were forfeited by Cromwell, and the national church was left untouched though deprived of all powers of interference with the civil government, the general assembly being dissolved in 1653. Large steps were made towards the union of the two kingdoms by the representation of Scotland in the parliament at Westminster; free trade between the two countries was established, the administration of justice greatly improved, vassalage and heritable jurisdictions abolished, and security and good order maintained by the council of nine appointed by the Protector. In 1658 the improved condition of Scotland was the subject of Cromwell's special congratulation in addressing parliament. But as in Ireland so Cromwell's policy in Scotland was unpopular and was only upheld by the maintenance of a large army, necessitating heavy taxation and implying the loss of the national independence. It also vanished at the Restoration.

On the 12th of September 1651 Cromwell made his triumphal entry into London at the conclusion of his victorious campaigns; and parliament granted him Hampton Court as a residence with £4000 a year. These triumphs, however, had all been obtained by force of arms; the more difficult task now awaited Cromwell of governing England by parliament and by law. As Milton wrote:—

> "Cromwell! our chief of men,
> who through a cloud
> Not of war only, but detractions rude,
> Guided by faith and
> matchless fortitude,
> To peace and truth thy glorious
> way hast ploughed,
> ... Peace hath her victories
> No less renowned than war."

Cromwell's moderation and freedom from imperiousness were acknowledged even by those least friendly to his principles. Although the idol of his victorious army, and in a position enabling him to exercise autocratic power, he laboured unostentatiously for more than a year and a half as a member of the parliament, whose authority he supported to the best of his ability. While occupied with work on committees and in administration he pressed forward several schemes of reform, including a large measure of law reform prepared by a commission presided over by Matthew Hale, and the settlement of the church; but very little was accomplished by the parliament, which seemed to be almost exclusively taken up with the maintenance and increase of its own powers; and Cromwell's dissatisfaction, and that of the army which increased every day, was intensified by the knowledge that the parliament, instead of dissolving for a new election, was seeking to perpetuate its tenure of power. At length, in April 1653, a "bill for a new representation" was discussed, which provided for the retention of their seats by the existing members without re-election, so that they would also be the sole judges of the eligibility of the rest. This measure, which placed the whole powers of the state—executive, legislative, military and judicial—in the hands of one irresponsible and permanent chamber, "the horridest arbitrariness that ever was exercised in the world," Cromwell and the army determined to resist at all costs. On the 15th of April they proposed that the parliament should appoint a provisional government and dissolve itself. This compromise was refused by the parliament, which proceeded on the 20th to press through its last stages the "bill for a new representation." Cromwell hastened to the House, and at the last moment, on the bill being put to the vote, whispering to Harrison, "This is the time; I must do it," he rose, and after alluding to the ***Cromwell expels the Long Parliament.*** former good services of the parliament, proceeded to overwhelm the members with reproaches. Striding up and down the House in a passion, he made no attempt to control himself, and turning towards individuals as he hurled significant epithets at each, he called some "whoremasters," others "drunkards, corrupt, unjust, scandalous to the profession of the Gospel." "Perhaps you think," he exclaimed, "that this is not parliamentary language; I confess it is not, neither are you to expect any such from me." In reply to a complaint of his violence he cried, "Come, come, I will put an end to your prating. You are no parliament, I say you are no parliament. I will put an end to your sitting." By his directions Harrison then fetched in a small band of Cromwell's musketeers and compelled the speaker Lenthall to vacate the chair. Looking at the mace he said, "What shall we do with this bauble?" and ordered a soldier to take it away. The members then trooped out, Cromwell crying after them, "It is you that have forced me to this; for I have sought the Lord night and day that He would rather slay me than put me upon the doing this work." He then snatched the obnoxious bill from the clerk, put it under his cloak, and commanding the doors to be locked went back to Whitehall. In the afternoon he dissolved the council in spite of John Bradshaw's remonstrances, who said, "Sir, we have heard what you did at the House this morning...; but you are mistaken to think that the parliament is dissolved, for no power under heaven can dissolve them but themselves; therefore take you notice of that." Cromwell had no patience with formal pedantry of this sort; and in point of strict legality "The Rump" of the Long Parliament had little better title to authority than the officers who expelled it from the House. After this Cromwell had nothing left but the army with which to govern, and "henceforth his life was a vain attempt to clothe that force in constitutional forms, and make it seem something else so that it might become something else."[2]

By the dissolution of the Long Parliament Cromwell as commander-in-chief was left the sole authority in the state. He determined immediately to summon another parliament. This was the "Little" or "Barebones Parliament," consisting of one hundred and forty persons selected by the council of officers from among those nominated by the congregations in each county, which met on the 4th of July 1653. This assembly, however, soon showed itself impracticable and incapable, and on the 12th of December the speaker, followed by the more moderate members, marched to Whitehall and returned their powers to Cromwell, while the rest were expelled by the army.

Cromwell, who had no desire to exercise arbitrary power and whose main object therefore was to devise some constitutional limit to the authority which circumstances had placed in his hands, now accepted the written constitution drawn up by some of the officers, called the *Instrument of Government*, the earliest example of a "fixed government" based on "fundamentals," or constitutional guarantees, and the only example of it in English history. Its authors had wished Oliver to assume the title of king, but this he repeatedly refused; and in the instrument he was named Protector, a parliament was established, limited in powers but whose measures were not restricted by the Protector's veto unless they contravened the constitution, the Protector's executive power being also limited by the council. The Protector and the council together were given a life tenure of office, with a large army and a settled revenue sufficient for public needs in time of peace; while the clauses relating to religion "are remarkable as laying down for the first time with authority a principle of toleration,"[3] though this toleration did not apply to Roman Catholics and Anglicans. On the 16th of December 1653 Cromwell was installed in his new office, dressed as a civilian in a plain black coat instead of in scarlet as a general, in order to demonstrate that military government had given place to civil; for he approached his task in the same spirit that had prompted his declaration to the Little Parliament of his wish "to divest the sword of all power in the Civil administration."

In the interval between his nomination as Protector and the summoning of his first parliament in September 1654, Cromwell was empowered together with his council to legislate by ordinances; and eighty-two were issued in all, dealing ***The government of the Protector.*** with numerous and various reforms and including the reorganization of the treasury, the settlement of Ireland and Scotland and the union of the three kingdoms, the relief of poor prisoners, and the maintenance of the highways. These ordinances in many instances showed the hand of the true statesman. Cromwell was essentially a conservative reformer; in his attempts to purge the court of chancery of its most flagrant abuses, and to settle the ecclesiastical affairs of the nation, he showed himself anxious to retain as much of the existing system as could be left untouched without doing positive evil. He was out-voted by his council on the question of commutation of tithes, and his enlightened zeal for reforming the "wicked and abominable" sentences of the criminal law met with complete failure. Most of these ordinances were subsequently confirmed by parliament, and, "on the whole, this body of dictatorial legislation, abnormal in form as it is, in substance was a real, wise and moderate set of reforms."[4] His ordinances for the "Reformation of Manners," the product of the puritan spirit, had but a transitory effect. The Long Parliament had ordered a strict observance of Sunday, punished swearing severely, and made adultery a capital crime; Cromwell issued further ordinances against duelling, swearing, race-meetings and cock-fights—the last as tending to the disturbance of the public peace and the encouragement of "dissolute practices to the dishonour of God." Cromwell himself was no ascetic and saw no harm in honest sport. He was exceedingly fond of horses and hunting, leaping ditches prudently avoided by the foreign ambassadors. Baxter describes him as full of animal spirits, "naturally of such a vivacity, hilarity and alacrity as another man is when he hath drunken a cup of wine too much," and notes his "familiar rustic carriage with his soldiers in sporting." He was fond of music and of art, and kept statues in Hampton Court Gardens which scandalized good puritans. He preferred that Englishmen should be free rather than sober by compulsion. Writing to the Scottish clergy, and rejecting their claim to suppress dissent in order to extirpate error, he said, "Your pretended fear lest error should step in is like the man who would keep all wine out of the country lest men should be drunk. It will be found an unjust and unwise jealousy to deprive a man of his natural liberty upon a supposition he may abuse it. When he doth abuse it, judge." It is probable that very little of this moral legislation was enforced in practice, though special efforts were made under the government of the major-generals. Cromwell expected more results from the effects of education and culture. A part of the revenue of confiscated church lands was allotted to the maintenance of schools, and the question of national education was seriously taken in hand by the Commonwealth. Cromwell was especially interested in the universities. In 1649 he had been elected D.C.L. at Oxford, and in 1651 chancellor of the University, an office which he held till 1657, when he was succeeded by his son Richard. He founded a new readership in Divinity, and presented Greek MSS. to the Bodleian. He appointed visitors for the universities and great public schools, and defended the universities from the attacks of the extreme sectaries who clamoured for their abolition, even Clarendon allowing that Oxford "yielded a harvest of extraordinary good and sound knowledge in all parts of learning." In 1657 he founded a new university at Durham, which was suppressed at the Restoration. He patronized learning. Milton and Marvell were his secretaries. He allowed the royalists Hobbes and Cowley to return to England, and lived in friendship with the poet Waller.

Cromwell's religious policy included the maintenance of a national church, a policy acceptable to the army but much disliked by the Scots, who wanted the church to control the state, not the state the church. He ***Cromwell's church policy.*** improved the incomes of poor livings by revenues derived from episcopal estates and the fines of delinquents. An important feature of his church government was the appointment on the 20th of March 1654 of the "Triers," thirty-eight clerical and lay commissioners, who decided upon the qualifications of candidates for livings, and without whose recommendation none could be appointed;

while an ordinance of August 1654 provided for the removal of the unfit, the latter class including besides immoral persons those holding "popish" or blasphemous opinions, those publicly using the English Prayer Book, and the disaffected to the government. Religious toleration was granted, but with the important exception that some harsh measures were enacted against Anglicans and Roman Catholics, to neither of whom was liberty of worship accorded. The acts imposing fines for recusancy, repealed in 1650, were later executed with great severity. In 1655 a proclamation was issued for administering the laws against the priests and Jesuits, and some executions were carried out. Complete toleration in fact was only extended to Protestant nonconformists, who composed the Cromwellian established church, and who now meted out to their antagonists the same treatment which they themselves were later to receive under the *Clarendon Code* of Charles II.

Cromwell himself, however, remained throughout a staunch and constant upholder of religious toleration. "I had rather that Mahommedanism were permitted amongst us," he avowed, "than that one of God's children should ***His religious toleration.*** be persecuted." Far in advance of his contemporaries on this question, whenever his personal action is disclosed it is invariably on the side of forbearance and of moderation. It is probable, from the absence of evidence to the contrary, that much of this severe legislation was never executed, and it was without doubt Cromwell's restraining hand which moderated the narrow persecuting spirit of the executive. In practice Anglican private worship appears to have been little interfered with; and although the recusant fines were rigorously exacted, the same seems to have been the case with the private celebration of the mass. Bordeaux, the French envoy in England, wrote that, in spite of the severe laws, the Romanists received better treatment under the Protectorate than under any other government. Cromwell's strong personal inclination towards toleration is clearly seen in his treatment of the Jews and Quakers. He was unable, owing to the opposition of the divines and of the merchants, to secure the full recognition of the right to reside in England of the former who had for some time lived in small numbers and traded unnoticed and untroubled in the country; but he obtained an opinion from two judges that there was no law which forbade their return, and he gave them a private assurance of his protection, with leave to celebrate their private worship and to possess a cemetery.

Cromwell's policy in this instance was not overturned at the Restoration, and the great Jewish immigration into England with all its important consequences may be held to date practically from these first concessions made by Cromwell. His personal intervention also alleviated the condition of the Quakers, much persecuted at this time. In an interview in 1654 the sincerity and enthusiasm of George Fox had greatly moved Cromwell and had convinced him of their freedom from dangerous political schemes. He ordered Fox's liberation, and in November 1657 issued a general order directing that Quakers should be treated with leniency, and be discharged from confinement. Doctrines directly attacking Christianity Cromwell regarded, indeed, as outside toleration and to be punished by the civil power, but at the same time he mitigated the severity of the penalty ordained by the law. In general the toleration enjoyed under Cromwell was probably far larger than at any period since religion became the contending ground of political parties, and certainly greater than under his immediate successors. Lilburne and the anabaptists, and John Rogers and the Fifth Monarchy men, were prosecuted only on account of their direct attacks upon the government, and Cromwell in his broad-minded and tolerant statesmanship was himself in advance of his age and his administration. He believed in the spiritual and unseen rather than in the outward and visible unity of Christendom.

In foreign policy Cromwell's chief aims appear to have been to support and extend the Protestant faith, to promote English trade, and to prevent a Stuart restoration by foreign aid—the religious mission of England in the world, ***Foreign policy.*** her commercial interests, and her political independence being indissolubly connected in his mind. The beginning of his rule inherited a war with France and Holland; the former consequent on

Cromwell's failure to obtain terms for the Huguenots or the cession of Dunkirk, and the latter—for which he was not responsible—the result of commercial rivalry, of disputes concerning the rights of neutrals, of bitter memories of Dutch misdeeds in the East Indies, and of dynastic causes arising from the stadtholder, William II. of Orange, having married Mary, daughter of Charles I. In 1651 the Dutch completed a treaty with Denmark to injure English trade in the Baltic; to which England replied the same year by the Navigation Act, which suppressed the Dutch trade with the English colonies and the Dutch fish trade with England, and struck at the Dutch carrying trade. War was declared in May 1652 after a fight between Blake and Tromp off Dover, and was continued with signal victories and defeats on both sides till 1654. The religious element, however, which predominated in Cromwell's foreign policy inclined him to peace, and in April of that year terms were arranged by which England on the whole was decidedly the gainer. The Dutch acknowledged the supremacy of the English flag in the British seas, which Tromp had before refused; they accepted the Navigation Act, and undertook privately to exclude the princes of Orange from the command of their forces. The Protestant policy was further followed up by treaties with Sweden and Denmark which secured the passage of the Sound for English ships on the same conditions as the Dutch, and a treaty with Portugal which liberated English subjects from the Inquisition and allowed commerce with the Portuguese colonies. The two great Roman Catholic powers now both bid for Cromwell's alliance. Cromwell wisely inclined towards France, for Spain was then a greater menace than France alike to the Protestant cause and to the growth of British trade in the western hemisphere; but as no concessions could be gained from either France or Spain, the year 1654 closed without a treaty being made with either. In December 1654 Penn and Venables sailed for the West Indies with orders to attack the Spanish colonies and the French shipping; and for the first time since the Plantagenets an English fleet appeared in the Mediterranean, where Blake upheld the supremacy of the English flag, made a treaty with the dey of Algiers, destroyed the castles and ships of the dey of Tunis at Porto Farina on the 4th of April 1655, and liberated the English prisoners captured by the pirates.

The incident of the massacre of the Protestant Vaudois at this time decided Cromwell's policy in favour of France. In response to Cromwell's splendid championship of the persecuted people—which has been well described as "one of the noblest memories of England"—France undertook to put pressure upon Savoy, in consequence of which the persecution ceased for a time; but Cromwell's intervention had less practical effect than has generally been supposed, though "never was the great conception of a powerful state having duties along with interests more magnanimously realized."[5] The treaty of Pinerolo withdrew the edict ordering the persecutions, but they were soon afterwards renewed, and in 1658 formed the subject of another remonstrance by Cromwell to Louis XIV. in his last extant public letter before his death. The treaty of Westminster (24th of October 1655) dealt chiefly with commercial subjects, and contained a clause promising the expulsion from France of political exiles. Meanwhile the West Indian expedition had been defeated at Hispaniola, and war was declared by Spain, who now promised help to Charles II. for regaining his throne. Cromwell sent powerful English fleets to watch the coast of Spain and to prevent communications with the West Indies and America; on the 8th of September 1656 a fleet of treasure ships was destroyed off Cadiz by Stayner, and on the 20th of April 1657 Blake performed his last exploit in the destruction of the whole Spanish fleet of sixteen treasure ships in the harbour of Santa Cruz in Teneriffe. These naval victories were followed by a further military alliance with France against Spain, termed the treaty of Paris (the 23rd of March 1657). Cromwell furnished 6000 men with a fleet to join in the attack upon Spain in Flanders, and obtained as reward Mardyke and Dunkirk, the former being captured and handed over on the 3rd of October 1657, and the latter after the battle of the Dunes on the 4th of June 1658, when Cromwell's Ironsides were once more pitted against English royalists fighting for the Spaniards.

Such was the character of Cromwell's policy abroad. The inspiring principle had been the defence and support of Protestantism, the question with Cromwell being "whether the Christian world should be all popery." He desired England to be everywhere the protector of the oppressed and the upholder of "true religion." His policy was in principle the policy of Elizabeth, of Gustavus Adolphus, and—in the following generation—of William of Orange. He appreciated, without over-estimating, the value of England's insular position. "You have accounted yourselves happy," he said in January 1658, "in being environed by a great ditch from all the world beside. Truly you will not be able to keep your ditch nor your shipping unless you turn your ships and shipping into troops of horse and companies of foot, and fight to defend yourselves on *terra firma*." He did not regard himself merely as the trustee of the national resources. These were not to be employed for the advancement of English interests alone. "God's interest in the world," he declared, "is more extensive than all the people of these three nations. God has brought us hither to consider the work we may do in the world as well as at home." In 1653 he had made the astonishing proposal to the Dutch that England and Holland should divide the habitable globe outside Europe between them, that all states maintaining the Inquisition should be treated as enemies by both the proposed allies, and that the latter "should send missionaries to all peoples willing to receive them, to inculcate the truth of Jesus Christ and the Holy Gospel." Great writers like Milton and Harrington supported Cromwell's view of the duty of a statesman; the poet Waller acclaimed Cromwell as "the world's protector"; but the London tradesmen complained of the loss of their Spanish trade and regarded Holland and not Spain as the national enemy. But Cromwell's dream of putting himself at the head of European Protestantism never even approached realization. War broke out between the Protestant states of Sweden, Denmark, Holland and Brandenburg, with whom religion was entirely subordinated to individual aims and interests, and who were far from rising to Cromwell's great conceptions; while the Vaudois were soon subjected to fresh persecutions. On the other hand, Cromwell could justly boast "there is not a nation in Europe but is very willing to ask a good understanding with you." He raised England to a predominant position among the Powers of Europe, and anticipated the triumphs of the elder Pitt. "It was hard to discover," wrote Clarendon, "which feared him most, France, Spain or the Low Countries." The vigour and success with which he organized the national resources and upheld the national honour, asserted the British sovereignty of the seas, defended the oppressed, and caused his name to be feared and respected in foreign courts where that of Stuart was despised and neglected, command praise and admiration equally from contemporaries and from modern critics, from his friends and from his opponents. "He once more joined us to the continent," wrote Marvell, while Dryden describes him as teaching the British lion to roar. "Cromwell's greatness at home," said Clarendon, "was a mere shadow of his greatness abroad." "It is strange," wrote Pepys in 1667 under a different régime, "how everybody nowadays reflect upon Oliver and commend him, what brave things he did, and made all the neighbour princes fear him." To Cromwell more than to any other British ruler belongs the credit of having laid the foundation of England's maritime supremacy and of her over-sea empire.

Cromwell's colonial policy aimed definitely at the recognition and extension of the British empire. By March 1652 the whole of the territory governed by the Stuarts had submitted to the authority of the Commonwealth, and the Navigation ***Cromwell and the empire.*** Act of the 9th of October 1651, by which colonial goods could only be imported to England in British ships and all foreign trade to the colonies was restricted to products of the exporting country, sought to bind the colonies to England and to support the interests of the shipowners and merchants, and therefore of the English maritime supremacy, the act being, moreover, memorable as the first public measure which treated the colonies as a whole and as an integral part of Great Britain. The hindrance, however, to the general development of trade which the act involved aroused at once loud complaints, to which Cromwell turned a deaf ear, continuing to seize Dutch ships trading in forbidden goods. In the internal administration of the colonies Cromwell interfered very little, maintaining specially friendly relations with the New Englanders, and showing no jealousy of their desire for self-government. The war with France, Holland and Spain offered opportunities of gaining additional territory. A small expedition sent by Cromwell in February 1654 to capture New Amsterdam (New York) from the Dutch was abandoned on the conclusion of peace, and the fleet turned to attack the French colonies; Major Robert Sedgwick taking with a handful of men the fort of St John's, Port Royal or Annapolis, and the French fort on the river Penobscot, the whole territory from this river to the mouth of the St Lawrence remaining British territory till its cession in 1667. In December 1654 Cromwell despatched Penn and Venables with a fleet of thirty-eight ships and 2500 soldiers to the West Indies, their numbers being raised by recruits at the islands to 7000 men. The attack on Hispaniola, however, was a disastrous failure, and though a landing at Jamaica and the capture of the capital, Santiago de la Vega, was effected, the expedition was almost annihilated by disease; and Penn and Venables returned to England, when Cromwell threw them into the Tower. Cromwell, however, persevered, reminding Fortescue, who was left in command, that the war was one against the "Roman Babylon," that they were "fighting the Lord's battles"; and he sent out reinforcements under Sedgwick, offering inducements to the New Englanders to migrate to Jamaica. In spite of almost insuperable difficulties the colony took root, trade began, the fleet lay in wait for the Spanish treasure ships, the settlements of the Spaniards were raided, and their repeated attempts to retake the island were successfully resisted. In 1658 Colonel Edward Doyley, the governor, gained a decisive victory over thirty companies of Spanish foot, and sent ten of their flags to Cromwell. The Protector, however, did not live to witness the final triumph of his undertaking, which gave to England, as he had wished, "the mastery of those seas," ensuring the English colonies against Spanish attacks, and being maintained and followed up at the Restoration.

Meanwhile, the first parliament of the Protectorate had met in September 1654. A scheme of electoral reform had been carried by which members were taken from the small and corrupt boroughs and given to the large hitherto ***Parliamentary difficulties.*** unrepresented towns, and which provided for thirty representatives from Scotland and from Ireland. Instead, however, of proceeding with the work of practical legislation, accepting the Instrument of Government without challenge as the basis of its authority, the parliament immediately began to discuss and find fault with the constitution and to debate about "Fundamentals." About a hundred members who refused to engage not to attempt to change the form of government were excluded on the 12th of September. The rest sat on, discussing the constitution, drawing up lists of damnable heresies and of incontrovertible articles of faith, producing plans for the reduction of the army and demanding for themselves its control. Incensed by the dilatory and factious proceedings of the House, Cromwell dismissed the parliament on the 22nd of January 1655. Various dangerous plots against his government and person were at this time rife. Vane, Ludlow, Robert Overton, Harrison and Major Wildman, the head of the Levellers, were all arrested, while the royalist rising under Penruddock was crushed in Devonshire. Other attacks upon his authority were met with the same resort to force. The judges and lawyers began to question the legality of his ordinances, and to doubt their competency to convict royalist prisoners of treason. A merchant named Cony refused to pay customs not imposed by parliament, his counsel declaring their levy by ordinance to be contrary to Magna Carta, and Chief Justice Rolle resigning in order to avoid giving judgment. Cromwell was thus inevitably drawn farther along the path of arbitrary government. He arrested the persons who refused to pay taxes, and sent Cony's lawyers to the Tower. Hitherto he had been scrupulously impartial in raising the best men to the judicial bench, including the illustrious Matthew Hale, but he now appointed compliant judges, and, alluding to Magna Carta in terms impossible to transcribe for modern readers, declared that "it should not control his actions which he knew were for the safety of the Commonwealth." The country was now divided ***The major-generals.*** into twelve districts each governed by a major-general, to whom was entrusted the duty of maintaining order, stamping out disaffection and plots, and executing the laws relating to public morals. They had power to transport royalists and those who could not produce good characters, and supported themselves by a special tax of 10% on the incomes of the royalist gentry. Enormous numbers of ale-houses were closed—a proceeding which excited intense resentment and was probably no slight cause of the royalist reaction. Still more serious an encroachment upon the constitution perhaps even than the institution of the major-generals was Cromwell's tampering with the municipal franchise by confiscating the charters, depriving the burgesses, now hostile to his government, of their parliamentary votes, and limiting the franchise to the corporation; thereby corrupting the national liberties at their very source, and introducing an evil precedent only too readily followed by Charles II. and James II.

It was in these embarrassed and perilous circumstances that Cromwell summoned a new parliament in the summer of 1656. In spite of the influence and interference of the major-generals a large number of members hostile to the ***Refusal of the crown.*** government were returned, of whom Cromwell's council immediately excluded nearly a hundred. The major-generals were the object of general attack, while the special tax on the royalists was declared unjust, and the bill for its continuation rejected by a large majority. An attempt at the assassination of Cromwell by Miles Sindercombe added to the general feeling of anxiety and unrest. The military rule excited universal hostility; there was an earnest desire for a settled and constitutional government, and the revival of the monarchy in the person of Cromwell appeared the only way of obtaining it. On the 23rd of February 1657 the *Remonstrance* offering Cromwell the crown was moved by Sir Christopher Packe in the parliament and violently resisted by the officers and the army party, one hundred officers waiting upon Cromwell on the 27th to petition against his acceptance of it. On the 25th of March the *Remonstrance*, now termed the *Petition and Advice*, and including a new scheme of government, was passed by a majority of 123 to 62 in spite of the opposition of the officers; and on the 31st it was presented to Cromwell in the Banqueting House at Whitehall whence Charles I. had stepped out on to the scaffold. Cromwell replied by requesting a brief delay to ask counsel of God and his own heart. On the 8th of May about thirty officers presented a petition to parliament against the revival of the monarchy, and Fleetwood, Desborough and Lambert threatened to lay down their commissions. Accordingly Cromwell the same day refused the crown definitely, greatly to the astonishment both of his followers and his enemies, who considered his decision a fatal neglect of an opportunity of consolidating his rule and power. In particular, his acceptance of the crown would have guaranteed his followers, under the act of Henry VII., from liability in the future to the charge of high treason for having given allegiance to himself as a *de facto* king. Cromwell himself, however, seems to have regarded the question of title as of secondary importance, as merely (to use his own words) "a feather in the hat," "a shining bauble for crowds to gaze at or kneel to." "Your father," wrote Sir Francis Russell to Henry Cromwell, "hath of late made more wise men fools than ever; he laughs and is merry, but they hang down their heads and are pitifully out of countenance."

On the 25th of May the petition was presented to Cromwell again, with the title of Protector substituted for that of King, and he now accepted it. On the 26th of June 1657 he was once more installed as Protector, this time, however, with regal ceremony in contrast with the simple formalities observed on the first occasion, the heralds proclaiming his accession in the same manner as that of the kings. Cromwell's government seemed now established on the firmer footing of law and national approval, he himself obtaining the powers though not the title of a constitutional monarch, with a permanent revenue of £1,300,000 for the ordinary expenses of the administration, the command of the forces, the right to nominate his successor and, subject to the approval of parliament, the members of the council and of the new second chamber now established, while at the same time the freedom of parliament was guaranteed in its elections. Difficulties, however, appeared immediately the parliament got to work. The republicans hostile to the Protectorate, excluded before, now returned, took the places vacated by strong supporters of Cromwell who had been removed to the Lords, and attacked the authority of the new chamber, opened communications with the disaffected in the city and army, protested against unparliamentary taxation and arbitrary imprisonment, and demanded again the supremacy of parliament. In consequence Cromwell summoned both Houses to his presence on the 4th of February 1658, and having pointed out the perils to which they were once more exposing the state, dissolved parliament, dismissing the members with the words, "let God be judge between me and you."

During the period following the dissolution Cromwell's power appeared outwardly at least to be at its height. The revolts of royalists and sectaries against his government had been easily suppressed, and the various attempts to assassinate him, contemptuously referred to by Cromwell as "little fiddling things," were anticipated and prevented by an excellent system of police and spies, and by his bodyguard of 160 men. The victory at Dunkirk increased his reputation, while Louis XIV. showed his respect for the ruler of England by the splendid reception given to the Protector's envoy, Lord Fauconberg, and by a complimentary mission despatched to England.

The great career, the incidents of which we have been following, was now, however, drawing to a close. Cromwell's health had long been impaired by the hardships of campaigning. Now at the age of 58 he was already old, and his firm, strong signature had become feeble and trembling. The responsibilities and anxieties of government unassisted by parliament, and the continued struggle against the force of anarchy, weighed upon him and exhausted his physical powers. "It has been hitherto," Cromwell said, "a matter of, I think, but philosophical discourse, that a great place, a great authority, is a great burthen. I know it is." "I can say in the presence of God, in comparison of whom we are but like poor creeping ants upon the earth, I would have lived under my woodside to have kept a flock of sheep rather than undertook such a government as this." "I doubt not to say," declared his steward Maidston, "it drank up his spirits, of which his natural constitution afforded a vast stock, and brought him to his grave."

Domestic bereavements added further causes of grief and of weakened vitality. On the 6th of February 1658 he lost his favourite daughter, Elizabeth Claypole, and he was much cast down by the shock of his bereavement and of her long sufferings. Shortly afterwards he fell ill of an intermittent fever, but seemed to recover. On the 20th of August George Fox met him riding at the head of his guards in the park at Hampton Court, but declared "he looked like a dead man." The next day he again fell ill and was removed from Hampton Court to Whitehall, where his condition became worse. The anecdotes believed and circulated by the royalists that Cromwell died in all the agonies of remorse and fear are entirely false. On the 31st of August ***Death.*** he seemed to rally, and one who slept in his bedchamber and who heard him praying, declared, "a public spirit to God's cause did breathe in him to the very last." During the next few days he grew weaker and resigned himself to death. "I would," he said, "be willing to be further serviceable to God and his people, but my work is done." For the first time doubts as to his spiritual state seemed to have troubled him. "Tell me is it possible to fall from grace?" he asked the attendant minister. "No, it is not possible," the latter replied. "Then," said Cromwell, "I am safe, for I know that I was once in grace." He refused medicine to induce sleep, declaring "it is not my design to drink or to sleep, but my design is to make what haste I can to be gone." Towards the morning of the 3rd of September he again spoke, "using divers holy expressions, implying much inward consolation and peace," together with "some exceeding self-debasing words, annihilating and judging himself." He died on the afternoon of the same day, his day of triumph, the anniversary both of Dunbar and of Worcester. His body was privately buried in the chapel of Henry VII. in Westminster Abbey, the public funeral taking place on the 23rd of November, with great ceremony and on the same scale as that of Philip II.

of Spain, and costing the enormous sum of £60,000. At the Restoration his body was exhumed, and on the 30th of January 1661, the anniversary of the execution of Charles I., it was drawn on a sledge from Holborn to Tyburn, together with the bodies of Ireton and Bradshaw, accompanied by "the universal outcry and curses of the people." There it was hanged on a gallows, and in the evening taken down, when the head was cut off and set up upon Westminster Hall, where it remained till as late as 1684, the trunk being thrown into a pit underneath the gallows. According to various legends Cromwell's last burial place is stated to be Westminster Abbey, Naseby Field or Newburgh Abbey; but there appears to be no evidence to support them, or to create any reasonable doubt that the great Protector's dust lies now where it was buried, in the neighbourhood of the present Connaught Square.

As a military commander Cromwell was as prompt as Gustavus, as ardent as Condé, as exact as Turenne. These, moreover, were soldiers from their earliest years. Condé's fame was established in his twenty-second year, Gustavus ***Cromwell's military genius.*** was twenty-seven and Turenne thirty-three at the beginning of their careers as commanders-in-chief. Cromwell, on the other hand, was forty-three when he fought in his first battle. In less than two years he had taken his rank as one of the great cavalry leaders of history. His campaigns of 1648 and 1651 placed him still higher as a great commander. Worcester, his crowning victory, has been indicated by a German critic as the prototype of Sédan. Yet his early military education could have consisted at most of the perusal of the *Swedish Intelligencer* and the practice of riding. It is not, therefore, strange that Cromwell's first essays in war were characterised more by energy than technical skill. It was some time before he realized the spirit of cavalry tactics, of which he was later so complete a master. At first he speaks with complacence of a *mêlée*, and reports that he and his men "agreed to charge" the enemy. But before long he came to understand, as no other commander of the age save Gustavus understood it, the value of true "shock-action." Of Marston Moor he writes, "we never charged but we routed them"; and thereafter his battles were decided by the shock of closed squadrons, the fresh impulse of a second and even a third line, and above all by the unquestioning discipline and complete control over their horses to which he trained his men. This gave them not merely greater steadiness, but, what was far more important, the power of rallying and reforming for a second effort. The Royalist cavalry was disorganized by victory as often as by defeat, and illustrated on numerous fields the now discredited maxim that cavalry cannot charge twice in one day. Cromwell shares with Frederick the Great the credit of founding the modern cavalry spirit. As a horsemaster he was far superior to Murat. His marches in the eastern campaign of 1643 show a daily average at one time of 28 m. as against the 21 of Murat's cavalry in the celebrated pursuit after Jena. And this result he achieved with men of less than two years' service, men, too, more heavily equipped and worse mounted than the veterans of the *Grande Armée*. It has been said that his battles were decided by shock action; the real emphasis should be laid upon the word "decided." The swift, unhesitating charge was more than unusual in the wars of the time, and was possible only because of the peculiar earnestness of the men who fought the English war. The professional soldiers of the Continent could rarely be brought to force a decision; but the English, contending for a cause, were imbued with the spirit of the modern "nation in arms"; and having taken up arms wished to decide the quarrel by arms. This feeling was not less conspicuous in the far-ranging rides, or raids, of the Cromwellian cavalry. At one time, as in the case of Blechingdon, they would perform strange exploits worthy of the most daring hussars; at another their speed and tenacity paralyses armies. Not even Sheridan's horsemen in 1864-65 did their work more effectively than did the English squadrons in the Preston campaign. Cromwell appreciated this feeling at its exact worth, and his pre-eminence in the Civil War was due to this highest gift of a general, the power of feeling the pulse of his army. Resolution, vigour and clear sight marked his conduct as a commander-in-chief. He aimed at nothing less than the annihilation of the enemy's forces, which Clausewitz was the first to define, a hundred and fifty years later, as the true objective of military operations. Not merely as exemplifying the tactical envelopment, but also as embodying the central idea of grand strategy, was Worcester the prototype of Sédan. The contrast between a campaign of Cromwell's and one of Turenne's is far more than remarkable, and the observation of a military critic who maintains that Cromwell's art of war was two centuries in advance of its time, finds universal acceptance.

At a time when throughout the rest of Europe armies were manœuvring against one another with no more than a formal result, the English and Scots were fighting decisive battles; and Cromwell's battles were more decisive than those of any other leader. Until his fiery energy made itself felt, hardly any army on either side actually suffered rout; but at Marston Moor and Naseby the troops of the defeated party were completely dissolved, while at Worcester the royalist army was annihilated. Dunbar attested his constancy and gave proof that Cromwell was a master of the tactics of all arms. Preston was an example like Austerlitz of the two stages of a battle as defined by Napoleon, the first *flottante*, the second *foudroyante*.

Cromwell's strategic manœuvres, if less adroit than those of Turenne or Montecucculi, were, in accordance with his own genius and the temper of his army, directed always to forcing a decisive battle. That he was also capable of strategy of the other type was clear from his conduct of the Irish War. But his chief work was of a different kind and done on a different scale. The greatest feat of Turenne was the rescue of one province in 1674-1675; Cromwell, in 1648 and again in 1651, had two-thirds of England and half of Scotland for his theatre of war. Turenne levelled down his methods to suit the ends which he had in view. The task of Cromwell was far greater. Any comparison between the generalship of these two great commanders would therefore be misleading, for want of a common basis. It is when he is contrasted with other commanders, not of the age of Louis XIV., but of the Civil War, that Cromwell's greatness is most conspicuous. Whilst others busied themselves with the application of the accepted rules of the

Dutch, the German, and other formal schools of tactical thought, Cromwell almost alone saw clearly into the heart of the questions at issue, and evolved the strategy, the tactics, and the training suited to the work to which he had set his hand.

Cromwell's career as a statesman has been already traced in its different spheres, and an endeavour has been made to show the breadth and wisdom of his conceptions and at the same time the cause of the immediate failure of his ***Cromwell's statesmanship.*** constructive policy. Whether if Cromwell had survived he would have succeeded in gradually establishing legal government is a question which can never be answered. His administration as it stands in history is undoubtedly open to the charge that after abolishing the absolutism of the ancient monarchy he substituted for it, not law and liberty, but a military tyranny far more despotic than the most arbitrary administration of Charles I. The statement of Vane and Ludlow, when they refused to acknowledge Cromwell's government, that it was "in substance a re-establishment of that which we all engaged against," was true. The levy of ship money and customs by Charles sinks into insignificance beside Cromwell's wholesale taxation by ordinances; the inquisitional methods of the major-generals and the unjust and exceptional taxation of royalists outdid the scandals of the extra-legal courts of the Stuarts; the shipment of British subjects by Cromwell as slaves to Barbados has no parallel in the Stuart administration; while the prying into morals, the encouragement of informers, the attempt to make the people religious by force, were the counterpart of the Laudian system, and Cromwell's drastic treatment of the Irish exceeded anything dreamed of by Strafford. He discovered that parliamentary government after all was not the easy and plain task that Pym and Vane had imagined, and Cromwell had in the end no better justification of his rule than that which Strafford had suggested to Charles I.,—"parliament refusing (to give support and co-operation in carrying on the government) you are acquitted before God and man." The fault was no doubt partly Cromwell's own. He had neither the patience nor the tact for managing loquacious parliamentary pedants. But the chief responsibility was not his but theirs. John Morley (*Oliver Cromwell*, p. 297) has truly observed of the execution of Charles I., that it was "an act of war, and was just as defensible or just as assailable, and on the same grounds, as the war itself." The parliamentary party took leave of legality when they took up arms against the sovereign, and it was therefore idle to dream of a formally legal sanction for any of their subsequent revolutionary proceedings. An entirely fresh start had to be made. A new foundation had to be laid on which a new system of legality might be reared. It was for this that Cromwell strove. If the Rump or the Little Parliament had in a business-like spirit assumed and discharged the functions of a constituent assembly, such a foundation might have been provided. It was only when five years had passed since the death of the king without any "settlement of the nation" being arrived at, that Cromwell at last accepted a constitution drafted by his military officers, and attempted to impose it on the parliament. And it was not until the parliament refused to acknowledge the Instrument as the required starting point for the new legality, that Cromwell in the last resort took arbitrary power into his hands as the only method remaining for carrying on the government. For much as he hated arbitrariness, he hated anarchy still more. While therefore Cromwell's administration became in practice little different from that of Strafford, the aims and ideals of the two statesmen had nothing in common. It is therefore profoundly true, as observed by S. R. Gardiner (*Cromwell*, p. 315), that "what makes Cromwell's biography so interesting in his perpetual effort to walk in the paths of legality—an effort always frustrated by the necessities of the situation. The man—it is ever so with the noblest—was greater than his work." The nature of Cromwell's statesmanship is to be seen rather in his struggles against the retrograde influences and opinions of his time, in the many political reforms anticipated though not originated or established by himself, and in his religious, perhaps fanatical, enthusiasm, than in the outward character of his administration, which, however, in spite of its despotism shows itself in its inner spirit of justice, patriotism and self-sacrifice, so immeasurably superior to that of the Stuarts.

Cromwell's personal character has been inevitably the subject of unceasing controversy. According to Clarendon he was "a brave bad man," with "all the wickedness against which damnation is pronounced and for which hell fire ***Personal character.*** is prepared." Yet he cannot deny that "he had some virtues which have caused the memory of some men in all ages to be celebrated"; and admits that "he was not a man of blood," and that he possessed "a wonderful understanding in the natures and humour of men," and "a great spirit, an admirable circumspection and sagacity and a most magnanimous resolution." According to contemporary republicans he was a mere selfish adventurer, sacrificing the national cause "to the idol of his own ambition." Richard Baxter thought him a good man who fell before a great temptation. The writers of the next century generally condemned him as a mixture of knave, fanatic and hypocrite, and in 1839 John Forster endorsed Landor's verdict that Cromwell lived a hypocrite and died a traitor. These crude ideas of Cromwell's character were extinguished by Macaulay's irresistible logic, by the publication of Cromwell's letters by Carlyle in 1845, which showed Cromwell clearly to be "not a man of falsehoods, but a man of truth"; and by Gardiner, whom, however, it is somewhat difficult to follow when he represents Cromwell as "a typical Englishman." In particular that conception which regarded "ambition" as the guiding motive in his career has been dispelled by a more intimate and accurate knowledge of his life; this shows him to have been very little the creator of his own career, which was largely the result of circumstances outside his control, the influence of past events and of the actions of others, the pressure of the national will, the natural superiority of his own genius. "A man never mounts so high," Cromwell said to the French ambassador in 1647, "as when he does not know where he is going." "These issues and events," he said in 1656, "have not been forecast, but were providences in things." His "hypocrisy" consists principally in the Biblical language he employed, which with Cromwell, as with many of his contemporaries, was the most natural way of expressing his feelings, and in the ascription of every incident to the direct intervention of God's providence, which was really Cromwell's sincere belief and conviction. In later times Cromwell's character and administration have been the subject of almost too indiscriminate eulogy, which has found tangible shape in the statue erected to his memory at Westminster in 1899. Here Cromwell's effigy stands in the midst of the sanctuaries of the law, the church, and the parliament, the three foundations of the state which he subverted, and in sight of Whitehall where he destroyed the monarchy in blood. Yet Cromwell's monument is not altogether misplaced in such surroundings, for in him are found the true principles of piety, of justice, of liberty and of governance.

John Maidston, Cromwell's steward, gives the "character of his person." "His body was compact and strong, his stature under six foot (I believe about two inches), his head so shaped as you might see it a storehouse and a shop both of a vast treasury of natural parts." "His temper exceeding fiery, as I have known, but the flame of it, ... kept down for the most part, was soon allayed with those moral endowments he had. He was naturally compassionate towards objects in distress even to an effeminate measure; though God had made him a heart wherein was left little room for fear, ... yet did he exceed in tenderness towards sufferers. A larger soul I think hath seldom dwelt in a house of clay than his was. I believe if his story were impartially transmitted and the unprejudiced world well possessed with it, she would add him to her nine worthies." By his wife Elizabeth Bourchier, Cromwell had four sons, Robert (who died in 1639), Oliver (who died in 1644 while serving in his father's regiment), Richard, who succeeded him as Protector, and Henry. He also had four daughters. Of these Bridget was the wife successively of Ireton and Fleetwood, Elizabeth married John Claypole, Mary was wife of Thomas Belasyse, Lord Fauconberg; and Frances was the wife of Sir Robert Rich, and secondly of Sir John Russell. The last male descendant of the Protector was his great-great-grandson, Oliver Cromwell of Cheshunt, who died in 1821.

By the female line, through his children Henry, Bridget and Frances, the Protector has had numerous descendants, and is the ancestor of many well-known families.[6]

BIBLIOGRAPHY.—A detailed bibliography, with the chief authorities for particular periods, will be found in the article in the *Dict. of Nat. Biography*, by C. H. Firth (1888). The following works may be mentioned: S. R. Gardiner's *Hist. of England* (1883-1884) and of the *Great Civil War* (1886), *Cromwell's Place in History* (1897), *Oliver Cromwell* (1901), and *History of the Commonwealth and Protectorate* (1894-1903); *Cromwell*, by C. H. Firth (1900); *Oliver Cromwell*, by J. Morley (1904); *The Last Years of the Protectorate, 1656-1658*, 2 vols., by C. H. Firth (1909); *Oliver Cromwell*, by Fred. Harrison (1903); *Letters and Speeches of Oliver Cromwell*, by T. Carlyle, ed. by S. C. Lomas, with an introd. by C. H. Firth (the best edition, rejecting the spurious Squire papers, 1904); *Oliver Cromwell*, by F. Hoenig (1887); *Oliver Cromwell, the Protector*, by R. F. D. Palgrave (1890); *Oliver Cromwell ... and the Royalist Insurrection ... of March 1655*, by the same author (1903); *Oliver Cromwell*, by Theodore Roosevelt (1900); *Oliver Cromwell*, by R. Pauli (tr. 1888); *Cromwell, a Speech delivered at the Cromwell Tercentenary Celebration 1899*, by Lord Rosebery (1900); *The Two Protectors*, by Sir Richard Tangye (valuable for its illustrations, 1899); *Life of Sir Henry Vane*, by W. W. Ireland (1905); *Die Politik des Protectors Oliver Cromwell in der Auffassung und Tätigkeit ... des Staatssekretärs John Thurloe*, by Freiherr v. Bischofshausen (1899); *Cromwell as a Soldier*, by T. S. Baldock (1899); *Cromwell's Army*, by C. H. Firth (1902); *The Diplomatic Relations between Cromwell and Charles X. of Sweden*, by G. Jones (1897); *The Interregnum*, by F. A. Inderwick (dealing with the legal aspect of Cromwell's rule, 1891); *Administration of the Royal Navy*, by M. Oppenheim (1896); *History of the English Church during the Civil Wars*, by W. Shaw (1900); *The Protestant Interest in Cromwell's Foreign Relations*, by J. N. Bowman (1900); *Cromwell's Jewish Intelligencies* (1891), *Crypto-Jews under the Commonwealth* (1894), *Menasseh Ben Israel's Mission to Oliver Cromwell* (1901), by L. Wolf.

(P. C. Y.; C. F. A.; R. J. M.)

---

1 *Life of Sir H. Vane*, by W. W. Ireland, 222.
2 C. H. Firth, Cromwell, p. 324.
3 John Morley, Oliver Cromwell, p. 393.
4 Frederic Harrison, *Oliver Cromwell*, p. 214.
5 John Morley, *Oliver Cromwell*, p. 483.
6 Frederic Harrison, *Cromwell*, p. 34.

CROMWELL, RICHARD

(1626-1712), lord protector of England, eldest surviving son of Oliver Cromwell and of Elizabeth Bourchier, was born on the 4th of October 1626. He served in the parliamentary army, and in 1647 was admitted a member of Lincoln's Inn. In 1649 he married Dorothy, daughter of Richard Mayor, or Major, of Hursley in Hampshire. He represented Hampshire in the parliament of 1654, and Cambridge University in that of 1656, and in November 1655 was appointed one of the council of trade. But he was not brought forward by his father or prepared in any way for his future greatness, and lived in the country occupied with field sports, till after the institution of the second protectorate in 1657 and the recognition of Oliver's right to name his successor. On the 18th of July he succeeded his father as chancellor of the university of Oxford, on the 31st of December he was made a member of the council of state, and about the same time obtained a regiment and a seat in Cromwell's House of Lords. He was received generally as his father's successor, and was nominated by him as such on his death-bed. He was proclaimed on the 3rd of September 1658, and at first his accession was acclaimed with general favour both at home and abroad. Dissensions, however, soon broke out between the military faction and the civilians. Richard's elevation, not being "general of the army as his father was," was distasteful to the officers, who desired the appointment of a commander-in-chief from among themselves, a request refused by Richard. The officers in the council, moreover, showed jealousy of the civil members, and to settle these difficulties and to provide money a parliament was summoned on the 27th of January 1659, which declared Richard protector, and incurred the hostility of the army by criticizing severely the arbitrary military government of Oliver's last two years, and by impeaching one of the major-generals. A council of the army accordingly established itself in opposition to the parliament, and demanded on the 6th of April a justification and confirmation of former proceedings, to which the parliament replied by forbidding meetings of the army council without the permission of the protector, and insisting that all officers should take an oath not to disturb the proceedings in parliament. The army now broke into open rebellion and assembled at St James's. Richard was completely in their power; he identified himself with their cause, and the same night dissolved the parliament. The Long Parliament (which re-assembled on the 7th of May) and the heads of the army came to an agreement to effect his dismissal; and in the subsequent events Richard appears to have played a purely passive part, refusing to make any attempt to keep his power or to forward a restoration of the monarchy. On the 25th of May his submission was communicated to the House. He retired into private life, heavily burdened with debts incurred during his tenure of office and narrowly escaping arrest even before he quitted Whitehall. In the summer of 1660 he left England for France, where he lived in seclusion under the name of John Clarke, subsequently removing elsewhere, either (for the accounts differ) to Spain, to Italy, or to Geneva. He was long regarded by the government as a dangerous person, and in 1671 a strict search was made for him but without avail. He returned to England about 1680 and lived at Cheshunt, in the house of Sergeant Pengelly, where he died on the 12th of July 1712, being buried in Hursley church in Hampshire. Richard Cromwell was treated with general contempt by his contemporaries, and invidiously compared with his great father. According to Mrs Hutchinson he was "gentle and virtuous but a peasant in his nature and became not greatness." He was nevertheless a man of respectable abilities, of an irreproachable private character, and a good speaker.

Bibliography.—See the article in the *Dict. of Nat. Biography*, and authorities there cited; Noble's *Memoirs of the Protectoral House of Cromwell* (1787); *Memoirs of the Protector ... and of his Sons*, by O. Cromwell (1820); *The Two Protectors*, by Sir R. Tangye (1899); *Kebleland and a Short Life of Richard Cromwell*, by W. T. Warren (1900); *Letters and*

Speeches of O. Cromwell, by T. Carlyle (1904); *Eng. Hist. Review*, xiii. 93 (letters) and xviii. 79; *Cal. of State Papers, Domestic, Lansdowne MSS.* in British Museum.

(P. C. Y.)

CROMWELL, THOMAS

Earl of Essex (1485?-1540), born probably not later than 1485 and possibly a year or two earlier, was the only son of Walter Cromwell, *alias* Smyth, a brewer, smith and fuller of Putney. His grandfather, John Cromwell, seems to have belonged to the Nottinghamshire family, of whom the most distinguished member was Ralph, Lord Cromwell (1394?-1456), lord treasurer; and he migrated from Norwell, Co. Notts, to Wimbledon some time before 1461. John's son, Walter, seems to have acquired the *alias* Smyth from being apprenticed to his uncle, William Smyth, "armourer," of Wimbledon. He was of a turbulent, vicious disposition, perpetually being fined in the manor-court for drunkenness, for evading the assize of beer, and for turning more than his proper number of beasts on to Putney Common. Once he was punished for a sanguinary assault, and his connexion with Wimbledon ceased in 1514 when he "falsely and fraudulently erased the evidences and terrures of the lord." Till that time he had flourished like the bay-tree.

Under these circumstances the absence of Thomas Cromwell's name from the Wimbledon manor rolls is almost a presumption of respectability. Perhaps it would be safer to attribute it to Cromwell's absence from the manor. He is said to have quarrelled with his father—no great crime considering the father's character—and fled to Italy, where he served as a soldier in the French army at the battle of the Garigliano (Dec. 1503). He escaped from the battle-field to Florence, where he was befriended by the banker Frescobaldi, a debt which he appears to have repaid with superabundant interest later on. He is next heard of at Antwerp as a trader, and about 1510 he was induced to accompany a Bostonian to Rome in quest of some papal indulgences for a Boston gild; Cromwell secured the boon by the timely present of some choice sweetmeats to Julius II. In 1512 there is some slight evidence that he was at Middelburg, and also in London, engaged in business as a merchant and solicitor. His marriage must have taken place about the same time, judging from the age of his son Gregory. His wife was Elizabeth Wykes, daughter of a well-to-do shearman of Putney, whose business Cromwell carried on in combination with his own.

For about eight years after 1512 we hear nothing of Cromwell. A letter to him from Cicely, marchioness of Dorset, in which he is seen in confidential business relations with her ladyship, is probably earlier than 1520, and it is possible that Cromwell owed his introduction to Wolsey to the Dorset family. On the other hand, it is stated that his cousin, Robert Cromwell, vicar of Battersea under the cardinal, gave Thomas the stewardship of the archiepiscopal estate of York House. At any rate he was advising Wolsey on legal points in 1520, and from that date he occurs frequently not only as mentor to the cardinal, but to noblemen and others when in difficulties, especially of a financial character; he made large sums as a money-lender.

In 1523 Cromwell emerges into public life as a member of parliament. The official returns for this election are lost and it is not known for what constituency he sat, but we have a humorous letter from Cromwell describing its proceedings, and a remarkable speech which he wrote and perhaps delivered, opposing the reckless war with France and indicating a sounder policy which was pursued after Wolsey's fall. If, he said, war was to be waged, it would be better to secure Boulogne than advance on Paris; if the king went in person and were killed without leaving a male heir, he hinted there would be civil war; it would be wiser to attempt a union with Scotland, and in any case the proposed subsidy would be a fatal drain on the resources of the realm. Neither Henry nor Wolsey was so foolish as to resent this criticism, and Cromwell lost nothing by it. He was made a collector of the subsidy he had opposed—a doubtful favour perhaps—and in 1524 was admitted at Gray's Inn; but he now became the most confidential servant of the cardinal. In 1525 he was Wolsey's agent in the dissolution of the smaller monasteries which were designed to provide the endowments for Wolsey's foundations at Oxford and

Ipswich, a task which gave Cromwell a taste and a facility for similar enterprises on a greater scale later on. For these foundations Cromwell drew up the necessary deeds, and he was receiver-general of cardinal's college, constantly supervising the workmen there and at Ipswich. His ruthless vigour and his accessibility to bribes earned him such unpopularity that there were rumours of his projected assassination or imprisonment. All this constituted a further bond of sympathy between him and his master, and Cromwell grew in Wolsey's favour until his fall. His wife had died in 1527 or 1528, and in July 1529 he made his will, in which one of the chief beneficiaries was his nephew, Richard Williams, alias Cromwell, the great-grandfather of the protector.

Wolsey's disgrace reduced Cromwell to such despair that Cavendish once found him in tears and at his prayers "which had been a strange sight in him afore." Many of the cardinal's servants had been taken over by the king, but Cromwell had made himself particularly obnoxious. However, he rode to court from Esher to "make or mar," as he himself expressed it, and offered his services to Norfolk. Possibly he had already paved the way by the pensions and grants which he induced Wolsey to make through him, out of the lands and revenues of his bishoprics and abbeys, to nobles and courtiers who were hard pressed to keep up the lavish style of Henry's court. Cromwell could be most useful to the government in parliament, and the government, represented by Norfolk, undertook to use its influence in procuring him a seat, on the natural understanding that Cromwell should do his best to further government business in the House of Commons. This was on the 2nd of November 1529; the elections had been made, and parliament was to meet on the morrow. A seat was, however, found or made for Cromwell at Taunton. He signalized himself by a powerful speech in opposition to the bill of attainder against Wolsey which had already passed the Lords. The bill was thrown out, possibly with Henry's connivance, though no theory has yet explained its curious history so completely as the statement of Cavendish and other contemporaries, that its rejection was due to the arguments of Cromwell.

Doubtless he championed his fallen chief not so much for virtue's sake as for the impression it would make on others. He did not feel called upon to accompany Wolsey on his exile from the court.

Cromwell had now, according to Cardinal Pole, whose story has been too readily accepted, been converted into an "emissary of Satan" by the study of Machiavelli's *Prince*. In the one interview which Pole had with Cromwell, the latter, so Pole wrote ten years later in 1539, recommended him to read a new Italian book on politics, which Pole says he afterwards discovered was Machiavelli's *Prince*. But this discovery was not made for some years: the *Prince* was not published until 1532, three years after the conversation; there is evidence that Cromwell was not acquainted with it until 1537 or 1539, and there is nothing in the *Prince* bearing on the precise point under discussion by Pole and Cromwell. On the other hand, the point is discussed in Castiglione's *Il Cortegiano* which had just been published in 1528, and of which Cromwell promised to lend Bonner a copy in 1530. The *Cortegiano* is the antithesis of the *Prince*; and there is little doubt that Pole's account is the offspring of an imagination heated by his own perusal of the Prince in 1538, and by Cromwell's ruin of the Pole family at the same time; until then he had failed to see in Cromwell the Machiavellian "emissary of Satan."

Equally fanciful is Pole's ascription of the whole responsibility for the Reformation to Cromwell's suggestion. It was impossible for Pole to realize the substantial causes of that perfectly natural development, and it was his cue to represent Henry as having acted at the diabolic suggestion of Satan's emissary. In reality the whole programme, the destruction of the liberties and confiscation of the wealth of the church by parliamentary agency, had been indicated before Cromwell had spoken to Henry. The use of Praemunire had been applied to Wolsey; laymen had supplanted ecclesiastics in the chief offices of state; the plan of getting a divorce without papal intervention had been the original idea, which Wolsey had induced the king to abandon, and it had been revived by Cranmer's suggestion about the universities. The root idea of the supreme authority of the king had been asserted in Tyndale's *Obedience of a Christian Man* published in 1528, which Anne Boleyn herself had brought to Henry's notice: "this," he said, "is a book for me and all kings to read," and Campeggio had felt compelled to warn him against these notions, of which Pole imagines that he had never heard until they were put into his head by Cromwell late in 1530. In the same way Cromwell's influence over the government from 1529-1533 has been grossly exaggerated. It was not till 1531 that he was admitted to the privy council nor till 1534 that he was made secretary, though he had been made master of the Jewel-House, clerk of the Hanaper and master of the Wards in 1532, and chancellor of the exchequer (then a minor office) in 1533. It is not till 1533 that his name is as much as mentioned in the correspondence of any foreign ambassador resident in London. This obscurity has been attributed to deliberate suppression: but no secrecy was made about Cranmer's suggestion, and it was not Henry's habit to assume a responsibility which he could devolve upon others. It is said that Cromwell's life would not have been safe, had he been known as the author of this policy; but that is not a consideration which would have appealed to Henry, and he was just as able to protect his minister in 1530 as he was in 1536. Cromwell, in fact, was not the author of that policy, but he was the most efficient instrument in its execution.

He was Henry's parliamentary agent, but even in this capacity his power has been overrated, and he is supposed to have invented those parliamentary complaints against the clergy, which were transmuted into the legislation of 1532. But the complaints were old enough; many of them had been heard in parliament nearly twenty years before, and there is ample evidence to show that the petition against the clergy represents the "infinite clamours" of the Commons against the Church, which the House itself resolved should be "put in writing and delivered to the king." The actual drafting of the statute, as of all the Reformation Acts between 1532 and 1539, was largely Cromwell's work; and the success with which parliament was managed during this period was also due to him. It was not an easy task, for the House of Commons more than once rejected government measures, and members were heard to threaten Henry VIII. with the fate of Richard III.; they even complained of Cromwell's reporting their proceedings to the king. That was his business rather than conveying imaginary royal orders to the House. "They be contented," he wrote in one of these reports, "that deed and writing shall be treason," but words were only to be misprision: they refused to include an heir's rebellion or disobedience in the bill "as rebellion is already treason, and disobedience is no cause of forfeiture of inheritance." There was, of course, room for manipulation, which Cromwell extended to parliamentary elections; but parliamentary opinion was a force of which he had to take account, and not a negligible quantity.

From the date of his appointment as secretary in 1534, Cromwell's biography belongs to the history of England, but it is necessary to define his personal attitude to the revolution in which he was the king's most conspicuous agent. He was included by Foxe in his *Book of Martyrs* to the Protestant faith: more recent historians regard him as a sacrilegious ruffian. Now, there were two cardinal principles in the Protestantism of the 16th century—the supremacy of the temporal sovereign over the church in matters of government, and the supremacy of the Scriptures over the Church in matters of faith. There is no room for doubt as to the sincerity of Cromwell's belief in the first of these two articles: he paid at his own expense for an English translation of Marsiglio of Padua's *Defensor Pacis*, the classic medieval advocate of that doctrine; he had a scheme for governing England by means of administrative councils nominated by the king to the detriment of parliament; and he urged upon Henry the adoption of the maxim of the Roman civil law—*quod principi placuit legis habet vigorem*. He wanted, in his own words, "one body politic" and no rival to the king's authority; and he set the divine right of kings against the divine right of the papacy. There is more doubt about the sincerity of Cromwell's attachment to the second article; it is true that he set up a Bible in every parish church, and regarded them as invaluable; and the correspondents who unbosom themselves to him are all of a Protestant way of thinking. But Protestantism was the greatest support of absolute monarchy. Hence its value in Cromwell's eyes. Of religious conviction there is in him little trace, and still less of the religious temperament. He was a polished representative of the callous, secular middle class of that most irreligious age. Sentiment found no place, and feeling little, in his composition; he used the axe with as little passion as the surgeon does the knife, and he operated on some of the best and noblest in the land. He saw that it was wiser to proscribe a few great opponents than to fall on humbler prey; but he set law above justice, and law to him was simply the will of the state.

In 1534 Cromwell was appointed master of the Rolls, and in 1535 chancellor of Cambridge University and visitor-general of the monasteries. The policy of the Dissolution has been theoretically denounced, but practically approved in every civilized state, Catholic as well as Protestant. Every one has found it necessary, sooner or later, to curtail or to destroy its monastic foundations; only those which delayed the task longest have generally lagged farthest behind in national progress. The need for reform was admitted by a committee of cardinals appointed by Paul III. in 1535, and it had been begun by Wolsey. Cromwell was not affected by the iniquities of the monks except as arguments for the confiscation of their property. He had boasted that he would make Henry VIII. the richest prince in Christendom; and the monasteries, with their direct dependence on the pope and their cosmopolitan organization, were obstacles to that absolute authority of the national state which was Cromwell's ideal. He had learnt how to visit monasteries under Wolsey, and the visitation of 1535 was carried out with ruthless efficiency. During the storm which followed, Henry took the management of affairs into his own hands, but Cromwell was rewarded in July 1536 by being knighted, created lord privy seal, Baron Cromwell, and vicar-general and viceregent of the king in "Spirituals."

In this last offensive capacity he sent a lay deputy to preside in Convocation, taking precedence of the bishops and archbishops, and issued his famous Injunctions of 1536 and 1538; a Bible was to be provided in every church; the *Paternoster*, Creed and Ten Commandments were to be recited by the incumbent in English; he was to preach at least once a quarter, and to start a register of births, marriages and deaths. During these years the outlook abroad grew threatening because of the alliance, under papal guarantee, between Charles V. and Francis I.; and Cromwell sought to counterbalance it by a political and theological union between England and the Lutheran princes of Germany. The theological part of the scheme broke down in 1538 when Henry categorically refused to concede the three reforms demanded by the Lutheran envoys. This was ominous, and the parliament of 1539, into which Cromwell tried to introduce a number of personal adherents, proved thoroughly reactionary. The temporal peers were unanimous in favour of the Six Articles, the bishops were divided, and the Commons for the most part agreed with the Lords. Cromwell, however, succeeded in suspending the execution of the act, and was allowed to proceed with his one independent essay in foreign policy. The friendship between Francis and Charles was apparently getting closer; Pole was exhorting them to a crusade against a king who was worse than the Turk; and anxious eyes searched the Channel in 1539 for signs of the coming Armada. Under these circumstances Henry acquiesced in Cromwell's negotiations for a marriage with Anne of Cleves. Anne, of course, was not a Lutheran, and the state religion in Cleves was at least as Catholic as Henry's own. But her sister was married to the elector of Saxony, and her brother had claims on Guelders, which Charles V. refused to recognize. Guelders was to the emperor's dominions in the Netherlands what Scotland was to England, and had often been used by France in the same way, and an alliance between England, Guelders, Cleves and the Schmalkaldic League would, Cromwell thought, make Charles's position in the Netherlands almost untenable. Anne herself was the weak point in the argument; Henry conceived an invincible repugnance to her from the first; he was restrained from an immediate breach with his new allies only by fear of Francis and Charles. In the spring of 1540 he was reassured on that score; no attack on him from that quarter was impending; there was a rift between the two Catholic sovereigns, and there was no real need for Anne and her German friends.

From that moment Cromwell's fate was sealed; the Lords loathed him as an upstart even more than they had loathed Wolsey; he had no church to support him; Norfolk and Gardiner detested him from pique as well as on principle, and he had no friend in the council save Cranmer. As lay vicegerent he had given umbrage to nearly every churchman, and he had put all his eggs in the one basket of royal favour, which had now failed him. Cromwell did not succumb without an effort, and a desperate struggle ensued in the council. In April the French ambassador wrote that he was tottering to his fall; a few days later he was created earl of Essex and lord great chamberlain, and two of his satellites were made secretaries to the king; he then despatched one bishop to the Tower, and threatened to send five others to join him. At last Henry struck as suddenly and remorselessly as a beast of prey; on the 10th of June Norfolk accused him of treason; the whole council joined in the attack, and Cromwell was sent to the Tower. A vast number of crimes was laid to his charge, but not submitted for trial. An act of attainder was passed against him without a dissentient voice, and after contributing his mite towards the divorce of Anne, he was beheaded on Tower Hill on the 28th of July, repudiating all heresy and declaring that he died in the Catholic faith.

In estimating Cromwell's character it must be remembered that his father was a blackguard, and that he himself spent the formative years of his life in a vile school of morals. A ruffian he doubtless was, as he says, in his youth, and he was the last man to need the tuition of Machiavelli. Nevertheless he civilized himself to a certain extent; he was not a drunkard nor a forger like his father; from personal immorality he seems to have been singularly free; he was a kind master, and a stanch friend; and he possessed all the outward graces of the Renaissance period. He was not vindictive, and his atrocious acts were done in no private quarrel, but in what he conceived to be the interests of his master and the state. Where those interests were concerned he had no heart and no conscience and no religious faith; no man was more completely blighted by the 16th century worship of the state.

The authorities for the early life of Cromwell are the Wimbledon manor rolls, used by Mr John Phillips of Putney in *The Antiquary* (1880), vol. ii., and the *Antiquarian Mag.* (1882), vol. ii.; Pole's *Apologia*, i. 126; Bandello's *Novella*, xxxiv.; Chapuys' letter to Granvelle, 21 Nov. 1535; and Foxe's *Acts and Mon.* From 1522 see *Letters and Papers of Henry VIII.*, vols. iii.-xvi.; Cavendish's *Life of Wolsey*; Hall's *Chron.*; Wriothesley's *Chron.* These and practically all other available sources have been utilized in R. B. Merriman's *Life and Letters of Thomas Cromwell* (2 vols., 1902). For Cromwell and Machiavelli see Paul van Dyke's *Renascence Portraits* (1906), App.

(A. F. P.)

CRONJE, PIET ARNOLDUS

(*c.* 1840-), Boer general, was born about 1840 in the Transvaal and in 1881 took part in the first Boer War in the rank of commandant. He commanded in the siege of the British garrison at Potchefstroom, though he was unable to force their surrender until after the conclusion of the general armistice. The Boer leader was at this time accused of withholding knowledge of this armistice from the garrison (see POTCHEFSTROOM). He held various official positions in the years 1881-1899, and commanded the Boer force which compelled the surrender of the Jameson raiders at Doornkop (Jan. 2, 1896). In the war of 1899 Cronje was general commanding in the western theatre of war, and began the siege of Kimberley. He opposed the advance of the British division under Lord Methuen, and fought, though without success, three general actions at Belmont, Graspan and Modder River. At Magersfontein, early in December 1899, he completely repulsed a general attack made upon his position, and thereby checked for two months the northward advance of the British column. In the campaign of February 1900, Cronje opposed Lord Roberts's army on the Magersfontein battleground, but he was unable to prevent the relief of Kimberley; retreating westward, he was surrounded near Paardeberg, and, after a most obstinate resistance, was forced to surrender with the remnant of his army (Feb. 27, 1900). As a prisoner of war Cronje was sent to St Helena, where he remained until released after the conclusion of peace (see TRANSVAAL: *History*).

CROOKES, SIR WILLIAM

(1832-), English chemist and physicist, was born in London on the 17th of June 1832, and studied chemistry at the Royal College of Chemistry under A. W. von Hofmann, whose assistant he became in 1851. Three years later he was appointed an assistant in the meteorological department of the Radcliffe observatory, Oxford, and in 1855 he obtained a chemical post at Chester. In 1861, while conducting a spectroscopic examination of the residue left in the manufacture of sulphuric acid, he observed a bright green line which had not been noticed previously, and by following up the indication thus given he succeeded in isolating a new element, thallium, a specimen of which was shown in public for the first time at the exhibition of 1862. During the next eight years he carried out a minute investigation of this metal and its properties. While determining its atomic weight, he thought it desirable, for the sake of accuracy, to weigh it in a vacuum, and even in these circumstances he found that the balance behaved in an anomalous manner, the metal appearing to be heavier when cold than when hot. This phenomenon he explained as a "repulsion from radiation," and he expressed his discovery in the statement that in a vessel exhausted of air a body tends to move away from another body hotter than itself. Utilizing this principle he constructed the radiometer (*q.v.*), which he was at first disposed to regard as a machine that directly transformed light into motion, but which was afterwards perceived to depend on thermal action. Thence he was led to his famous researches on the phenomena produced by the discharge of electricity through highly exhausted tubes (sometimes known as "Crookes' tubes" in consequence), and to the development of his theory of "radiant matter" or matter in a "fourth state," which led up to the modern electronic theory. In 1883 he began an inquiry into the nature and constitution of the rare earths. By repeated fractionations he was able to divide yttrium into distinct portions which gave different spectra when exposed in a high vacuum to the spark from an induction coil. This result he considered to be due, not to any removal of impurities, but to an actual splitting-up of the yttrium molecule into its constituents, and he ventured to draw the provisional conclusion that the so-called simple bodies are in reality compound molecules, at the same time suggesting that all the elements have been produced by a process of evolution from one primordial stuff or "protyle." A later result of this method of investigation was the discovery of a new member of the rare earths, monium or victorium, the spectrum of which is characterized by an isolated group of lines, only to be detected photographically, high up in the ultra-violet; the existence of this body was announced in his presidential address to the British Association at Bristol in 1898. In the same address he called attention to the conditions of the world's food supply, urging that with the low yield at present realized per acre the supply of wheat would within a comparatively short time cease to be equal to the demand caused by increasing population, and that since nitrogenous manures are essential for an increase in the yield, the hope of averting starvation, as regards those races for whom wheat is a staple food, depended on the ability of the chemist to find an artificial method for fixing the nitrogen of the air. An authority on precious stones, and especially the diamond, he succeeded in artificially making some minute specimens of the latter gem; and on the discovery of radium he was one of the first to take up the study of its properties, in particular inventing the spinthariscope, an instrument in which the effects of a trace of radium salt are manifested by the phosphorescence produced on a zinc sulphide screen. In addition to many other researches besides those here mentioned, he wrote or edited various books on chemistry and chemical technology, including *Select Methods of Chemical Analysis*, which went through a number of editions; and he also gave a certain amount of time to the investigation of psychic phenomena, endeavouring to effect some measure of correlation between them and ordinary physical laws. He was knighted in 1897, and received the Royal (1875), Davy (1888), and Copley (1904) medals of the Royal Society, besides filling the offices of president of the Chemical Society and of the Institution of Electrical Engineers. He married Ellen, daughter of W. Humphrey, of Darlington, and their golden wedding was celebrated in 1906.

CROOKSTON

A city and the county-seat of Polk county, Minnesota, U.S.A., on the Red Lake river in the Red River valley, about 300 m. N.W. of Minneapolis, and about 25 m. E. of Grand Forks, North Dakota. Pop. (1890) 3457; (1900) 5359; (1905, state census) 6794, 2049 being foreign-born, including 656 from Norway (2 Norwegian weeklies are published), 613 from Canada, 292 from Sweden; (1910 U.S. census) 7559. Crookston is served by the Great Northern and the Northern Pacific railways. It has a Carnegie library, and the St Vincent and Bethesda hospitals, and is the seat of a Federal Land Office and of a state agricultural high school (with an experimental farm). Dams on the Red Lake river provide a fine water-power, and among the city's manufactures are lumber, leather, flour, farm implements, wagons and bricks. The city is situated in a fertile farming region, and is a market for grain, potatoes and other agricultural products, and lumber. Crookston was settled about 1872, was incorporated in 1879, received its first city charter in 1883, and adopted a new one in 1906. It was named in honour of William Crooks, an early settler.

CROP

(a word common in various forms, such as Germ. *Kropf*, to many Teutonic languages for a swelling, excrescence, round head or top of anything; it appears also in Romanic languages derived from Teutonic, in Fr. as *croupe*, whence the English "crupper"; and in Ital. *groppo*, whence English "group"), the *ingluvies*, or pouched expansion of a bird's oesophagus, in which the food remains to undergo a preparatory process of digestion before being passed into the true stomach. From the meaning of "top" or "head," as applied to a plant, herb or flower, comes the common use of the word for the produce of cereals or other cultivated plants, the wheat-crop, the cotton-crop and the like, and generally, "the crops"; more particular expressions are the "white-crop," for such grain crops as barley or wheat, which whiten as they grow ripe and "green-crop" for such as roots or potatoes which do not, and also for those which are cut in a green state, like clover (see AGRICULTURE). Other uses, more or less technical, of the word are, in leather-dressing, for the whole untrimmed hide; in mining and geology, for the "outcrop" or appearance at the surface of a vein or stratum and, particularly in tin mining, of the best part of the ore produced after dressing. A "hunting-crop" is a short thick stock for a whip, with a small leather loop at one end, to which a thong may be attached. From the verb "to crop," *i.e.* to take off the top of anything, comes "crop" meaning a closely cut head of hair, found in the name "croppy" given to the Roundheads at the time of the Great Rebellion, to the Catholics in Ireland in 1688 by the Orangemen, probably with reference to the priests' tonsures, and to the Irish rebels of 1798, who cut their hair short in imitation of the French revolutionaries.

CROPSEY, JASPER FRANCIS

(1823-1900), American landscape painter, was born at Rossville, Staten Island, New York, on the 18th of February 1823. After practising architecture for several years, he turned his attention to painting, studying in Italy from 1847 to 1850. In 1851 he was elected a member of the National Academy of Design. From 1857 to 1863 he had a studio in London, and after his return to America enjoyed a considerable vogue, particularly as a painter of vivid autumnal effects, along the lines of the Hudson River school. He was one of the original members of the American Water Color Society. He continued actively in this profession until within a few days of his death, at Hastings-on-Hudson, New York, on the 22nd of June 1900. He made the architectural designs for the stations of the elevated railways in New York City.

CROQUET

(from Fr. *croc*, a crook, or crooked stick), a lawn game played with balls, mallets, hoops and two pegs. The game has been evolved, according to some writers, from the *paille-maille* which was played in Languedoc at least as early as the 13th century. Under the name of *le jeu de la crosse*, or *la crosserie*, a similar game was at the same period immensely popular in Normandy, and especially at Avranches, but the object appears to have been to send the ball as far as possible by driving it with the mallet (see *Sports et jeux d'adresse*, 1904, p. 203). Pall Mall, a fashionable game in England in the time of the Stuarts, was played with a ball and a mallet, and with two hoops or a hoop and a peg, the game being won by the player who ran the hoop or hoops and touched the peg under certain conditions in the fewest strokes. Croquet certainly has some resemblance to *paille-maille*, played with more hoops and more balls. It is said that the game was brought to Ireland from the south of France, and was first played on Lord Lonsdale's lawn in 1852, under the auspices of the eldest daughter of Sir Edmund Macnaghten. It came to England in 1856, or perhaps a few years earlier, and soon became popular.

In 1868 the first all-comers' meeting was held at Moreton-in-the-Marsh. In the same year the All England Croquet Club was formed, the annual contest for the championship taking place on the grounds of this club at Wimbledon.[1] But after being for ten years or so the most popular game for the country house and garden party, croquet was in its turn practically ousted by lawn tennis, until, with improved implements and a more scientific form of play, it was revived about 1894-1895. In 1896-1897 was formed the United All England Croquet Association, on the initiative of Mr Walter H. Peel. Under the name of the Croquet Association, with more than 2000 members and nearly a hundred affiliated clubs (1909), this body is the recognized ruling authority on croquet in the British

Islands. Its headquarters are at the Roehampton Club, where the championship and champion cup competitions are held each year.

The Game and its Implements.—The requisites for croquet are a level grass lawn, six hoops, two posts or pegs, balls, mallets, and hoop-clips to mark the progress of the players. The usual game is played between two sides, each having two balls, the side consisting of two players in partnership, each playing one ball, or of one player playing both balls. The essential characteristic of croquet is the scientific combination between two balls in partnership against the other two. The balls are distinguished by being coloured blue, red, black and yellow, and are played in that order, blue and black always opposing the other two.

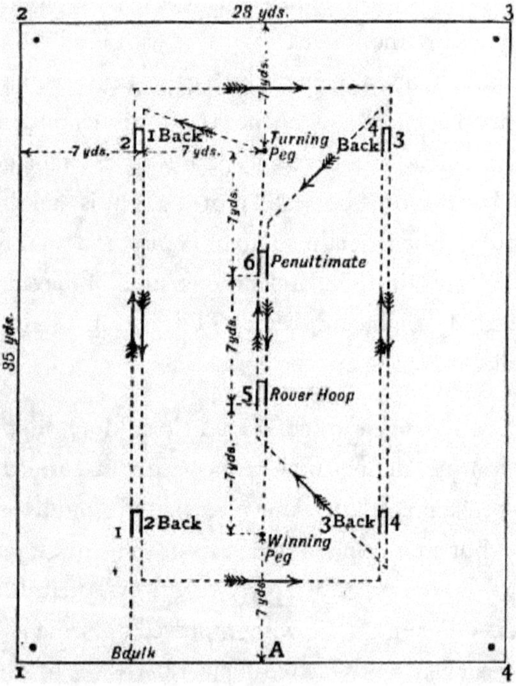

Fig. 1.—Diagram of croquet ground, showing setting of hoops and pegs, and order of play in accordance with the official Laws (1909) of the Croquet Association.

The ground for match play measures 35 yds. by 28 yds., and should be carefully marked out with white lines. In each corner a white spot is marked 1 yd. from each boundary. The hoops are made of round iron, not less than ½ in. and not more than ¾ in. in diameter, and standing 12 in. out of the ground. For match play they are 3¾ or 4 in. across, inside measurement. They are set up as in the accompanying diagram, the numbers and arrows indicating the order and direction in which they must be passed. Each hoop is run twice, and each peg struck once. The pegs may be struck from any direction.

The pegs are 1½ in. in diameter and when fixed stand 18 in. above the ground. The balls were formerly made of boxwood (earlier still of beechwood); composition balls are now in general use for tournaments. They must be 3⅝ in. in diameter and 15 oz. to 16½ oz. in weight. It will be seen that for match play the hoops are only ⅛ or at the most ⅜ in. wider than the diameter of the ball. The mallets may be of any size and weight, but the head must be made of wood (metal may be used only for weighting or strengthening purposes), and the ends must be parallel and similar. Only one mallet may be used in the course of a game, except in the case of *bona fide* damage.

The object of the player is to score the points of the game by striking his ball through each of the hoops and against each of the pegs in a fixed order; and the side wins which first succeeds in scoring all the points with both the balls of the side. A metal clip corresponding in colour with the player's ball is attached to the hoop or peg which that ball has next to make in the proper order, as a record of its progress in the game. No point is scored by passing through a hoop or hitting a peg except in the proper order. Thus, if a player has in any turn or turns driven his ball successively through hoops 1, 2, and 3, his clip is attached to hoop 4, and the next point to be made by him will be that hoop; and so on till all the points (hoops and pegs) have been scored. Each player starts in turn from any point in a "baulk" or area 3 ft. wide along the left-hand half of the "southern" boundary, marked A on the diagram, of the lawn—till 1906, from a point 1 ft. in front of the middle of hoop 1. If he fails either to make a point or to "roquet"[2] (*i.e.* drive his ball against) another ball in play, his turn is at an end and the next player in order takes his turn in like manner. If he succeeds in scoring a point, he is entitled (as in billiards) to another stroke; he may then either attempt to score another point, or he may roquet a ball. Having roqueted a ball—provided he has not already roqueted the same ball in the same turn without having scored a point in the interval—he is entitled to two further strokes: first he must "take croquet," *i.e.* he places his own ball (which from the moment of the roquet is "dead" or "in hand") in contact with the roqueted ball on any side of it, and then strikes his own ball with his mallet, being bound to move or shake both balls perceptibly. If at the beginning of a turn the striker's ball is in contact with another ball, a "roquet" is held to have been made and "croquet" must be taken at once. After taking croquet the striker is entitled to another stroke, with which he may score another point, or roquet another ball not previously roqueted in the same turn since a point was scored, or he may play for safety. Thus, by skilful alternation of making points and roqueting balls, a "break" may be made in which point after point, and even all the points in the game (for the ball in play), may be scored in a single turn, in addition to 3 or 4 points for the partner ball. The chief skill in the game perhaps consists in playing the stroke called "taking croquet" (but see below on the "rush"). Expert players can drive both balls together from one end of the ground to the other, or send one to a distance while retaining the other, or place each with accuracy in different directions as desired, the player obtaining position for scoring a point or roqueting another ball according to the strategical requirements of his position. Care has, however, to be taken in playing the croquet-stroke that both balls are absolutely moved or perceptibly shaken, and that neither of them be driven over the boundary line, for in either event the player's next stroke is forfeited and his turn brought summarily to an end.

There are three distinct methods of holding the mallet among good players. A comparatively small number still adhere to the once universal "side stroke," in which the player faces more or less at right angles to the line of aim, and strikes the ball very much like a golfer, with his hands close together on the mallet shaft. The majority use "front play," in which the player faces in the direction in which he proposes to send the ball. The essential characteristic of this stroke is that eye, hand and ball should be in the same vertical plane, and the stroke is rather a swing—the "pendulum stroke"—than a hit. There are two ways of playing it. The majority of right-handed front players swing the mallet outside the right foot, holding it with the left hand as a pivot at the top of the shaft, while the right hand (about 12 in. lower down) applies the necessary force, though it must always be borne in mind that the heavy mallet-head, weighing from 3 to 3½ ℔ or even more, does the work by itself, and the nearer the stroke is to a simple swing, like that of a pendulum, the more likely it is to be accurate. Either the right or the left foot may be in advance, and should be roughly parallel to the line of aim, the player's weight being mainly on the rear foot. Most of the best Irish and some English players swing the mallet between their feet, using a grip like that of the side player or golfer, with the hands close together, and often interlocking. It is claimed that the loss of power caused by the hampered swing—usually compensated by an extra heavy mallet—is more than counterbalanced by the greater accuracy in aim. The beginner is well advised to try all these methods, and adopt that which comes most natural to him. Skirted players, of course, are unable to use the Irish stroke; and, as one of the most meritorious features of croquet is that it is the only out-of-door game in which men and women can compete on terms of real equality, this has been put forward as a reason for barring it, if it is actually an advantage.

When a croquet ground is thoroughly smooth and level, the game gives scope for considerable skill; a great variety of strokes may be played with the mallet, each having its own well-defined effect on the behaviour of the balls, while a knowledge of angles is essential. Skilful tactics are at least as necessary as skilful execution to enable the player so to dispose the balls on the ground while making a break that they may most effectively assist him in scoring his points. The tactics of croquet are in this respect similar to those of billiards, that the player tries to make what progress he can during his own break, and to leave the balls "safe" at the end of it; he must also keep in mind the needs of the other ball of his side by leaving his own ball, or the last player's ball, or both, within easy roqueting distance or in useful positions, and that of the next player isolated. Good judgment is really more valuable than mechanical skill. Croquet is a game of combination, partners endeavouring to keep together for mutual help, and to keep their opponents apart. It is important always to leave the next player in such a position that he will be unable to score a point or roquet a ball; a break, however profitable, which does not end by doing this is often fatal. Formerly this might be done by leaving the next player's ball in such a position that either a hoop or a peg lay between it and all the other balls ("wiring"), or so near to a hoop or peg that there was no room for a proper stroke to be taken in the required direction. Under rule 36 of the *Laws of Croquet* for 1906, a ball left in such a position, provided it were within a yard of the obstacle ("close-wired"), might at the striker's option be moved one yard in any direction. This rule left to the striker whose ball was "wired" more than a yard from the hoop or peg ("distance-wired") the possibility of hitting his ball in such a way as to jump the obstacle. The jump-shot is, however, very bad for the lawn, and in 1907 a further provision was made by which the player whose ball is left "wired" from all the other balls by the stroke of an opponent may lift it and play from the "baulk" area. This practically means that "wiring" is impossible. The most that can be done is to "close-wire" the next player from two balls and leave him with a difficult shot at the third. If, however, the next player's ball has not been moved by the adversary, the adversary is entitled to wire the balls as best he can.

The following is a specimen of elementary croquet tactics. If a player is going up to hoop 5 (diagram 1) in the course of a break, he should have contrived, if possible, to have a ball waiting for him at that hoop and another at hoop 6. With the aid of the first he runs hoop 5 and sends it on to the turning peg, stopping his ball in taking croquet close to the ball at 6. The corner hoops are the difficult ones, and after running hoop 6 the assisting ball is croqueted to 1 back, the peg being struck with the aid of the ball already there, which is again struck and driven to 2 back. If the player has been able to leave the fourth ball in the centre of the ground (known as a centre ball), he hits this after taking croquet, takes croquet, going off it to the ball at 1 back, and continues the break, leaving the centre ball where it will be useful for 3 back and 4 back. A first-class player should, however, be able to make a break with 3 balls almost as easily as with 4. A useful device, especially in a losing game, is to get rid of the opponent's advanced ball if a "rover" (*i.e.* one which has run all the hoops and is for the winning peg) by croqueting it in such a way that it hits the peg and is thus out of the game. This can be done only by a ball which is itself also a rover. The opponent has then only one turn out of every three, and may be rendered practically helpless by leaving him always in a "safe" position. Inasmuch as a skilful player can cause an opponent's ball to pass through the last two or even three hoops in the course of his turn and then peg it out, it is considered prudent to leave unrun the last three hoops until the partner's ball is well advanced. There is a perennial agitation in the croquet world for a law prohibiting the player from pegging out his opponent's ball. Many good players also think it desirable that the four-ball break should be restricted or wholly forbidden, *e.g.* by barring the dead ball.

To "rush" a ball is to roquet it hard so that it proceeds for a considerable distance in a desired direction. This stroke requires absolute accuracy and often considerable force, which must be applied in such a way as to drive the player's ball evenly; otherwise it is very liable, especially if the ground be not perfectly smooth, to jump the object ball. The rush stroke is absolutely essential to good play, as it enables croquet to be taken (*e.g.*) close to the required hoop, whereas to croquet into position from a great distance and also provide a ball for use after running the hoop is extremely difficult, often impossible. To "rush" successfully, the striker's ball must lie near the object ball, preferably, though not necessarily, in the line of the rush. By means of the rush it is possible to accomplish the complete round with the assistance of one ball only. To "cut" a ball is to hit it on the edge and cause it to move at some desired angle. "Rolling croquet" is made either by hitting near the top of the player's ball which gives it "follow," or by making the mallet so hit the ball as to keep up a sustained pressure. The first impact must, however, result in a distinctly audible single tap; if a prolonged rattle or a second tap is heard the stroke is foul. The passing stroke is merely an extension of this. Here the player's ball proceeds a greater distance than the croqueted ball, but in somewhat the same direction. The "stop stroke" is made by a short, sharp tap, the mallet being withdrawn immediately after contact; the player's ball only rolls a short distance, the other going much farther. The "jump stroke" is made by striking downwards on to the ball, which can thus be made to jump over another ball, or even a hoop. "Peeling" (a term derived from Walter H. Peel, a famous advocate of the policy) is the term applied to the device of putting a partner's or an opponent's ball through the hoops with a view to ultimately pegging it out.

The laws of croquet, and even the arrangement of the hoops, have not attained complete uniformity wherever the game is played. Croquet grounds are not always of full size, and some degree of elasticity in the rules is perhaps necessary to meet local conditions. The laws by which matches for the championship and all tournaments are governed are issued annually by the Croquet Association; and though from time to time trifling amendments may be made, they have probably reached permanence in essentials.

See *The Encyclopaedia of Sport*; *The Complete Croquet Player* (London, 1896); the latest *Laws of Croquet*, published annually by the Croquet Association, and its official organ *The Croquet Gazette*. For the principles of the game and its history in England, see C. D. Locock, *Modern Croquet Tactics* (London, 1907); A. Lillie, *Croquet up to Date* (London, 1900).

Croquet in the United States: Roque.—Croquet was brought to America from England soon after its introduction into that country, and enjoyed a wide popularity as a game for boys and girls before the Civil War (see Miss Alcott's *Little Women*, cap. 12). American croquet is quite distinct from the modern English game. It is played on a lawn 60 ft. by 30, and preserves the old-fashioned English arrangement of ten hoops, including a central "cage" of two hoops. The balls, coloured red, white, blue and black, are 3¼ in. in diameter, and the hoops are from 3½ to 4 in. wide, according to the skill of the players. This game, however, is not taken seriously in the United States; the *Official Croquet Guide* of Mr Charles Jacobus emphasizes "the ease with which the game can be established," since almost every country home has a grass plot, and "no elaboration is needed." The scientific game of croquet in the United States is known as "roque." Under this title a still greater departure from the English game has been elaborated on quite independent lines from those of the English Croquet Association since 1882, in which year the National Roque Association was formed. Roque also suffered from the popularity of lawn tennis, but since 1897 it has developed almost as fast as croquet in England. A great national championship tournament is held in Norwich, Conn., every August, and the game—which is fully as scientific as modern English croquet—has numerous devotees, especially in New England.

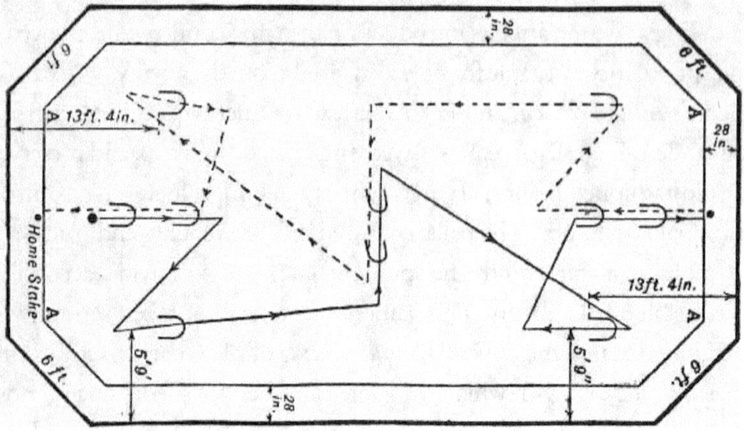

FIG. 2.—Diagram of roque ground, showing setting of arches and stakes and order of play, in accordance with the official laws (1906) of the National Roque Association.

Roque is played, not on grass, but on a prepared surface something like a cinder tennis-court. The standard ground, as adopted by the National Association in 1903, is hexagonal in shape, with ten arches (hoops) and two stakes (pegs) as shown in diagram 2. The length is 60 ft., width 30, and the "corner pieces" are 6 ft. long. An essential feature of the ground is that it is surrounded by a raised wooden border, often lined with india-rubber to facilitate the rebound of the ball, and it is permissible to play a "carom" (or rebounding shot) off this border; a skilful player can often thus hit a ball which is wired to a direct shot. A boundary line is marked 28 in. inside the border, on which a ball coming to rest outside it must be replaced. The hoops are run in the order marked on the diagram, so that the game consists of 36 points. Red and white are always partners against blue and black, and the essential features and tactics of the game are, *mutatis mutandis*, the same as in modern English croquet—*i.e.* the skilful player goes always for a break and utilizes one or both of the opponent's balls in making it. The balls are 3¼ in. in diameter, of hard rubber or composition, and the arches are 3⅜ or 3½ in. wide for first- and second-class players respectively; they are made of steel ½ in. in diameter and stand about 8 in. out of the ground. The stakes are 1 in. in diameter and only 1½ in. above the ground. The mallets are much shorter than those commonly employed in England, the majority of players using only one hand, though the two-handed "pendulum stroke," played between the legs, finds an increasingly large number of adherents, on account of the greater accuracy which it gives. The "jump shot" is a necessary part of the player's equipment, as dead wiring is allowed; it is supplemented by the carom off the border or off a stake or arch, and roque players justly claim that their game is more like billiards than any other out-of-door game.

The game of roque is opened by scoring (stringing) for lead from an imaginary line through the middle wicket (cage), the player whose ball rests nearest the southern boundary line having the choice of lead and balls. The balls are then placed on the four corner spots marked A in diagram, partner balls being diagonally opposite one another, and the starting ball having the choice of either of the upper corners. The leader, say red, usually begins by shooting at white; if he misses, a carom off the border will leave him somewhere near his partner, blue. White then shoots at red or blue, with probably a similar result. Blue is then "in," with a certain roquet and the choice of laying for red or going for an immediate break himself. The general strategy of the game corresponds to that of croquet, the most important differences being that "pegging out" is not allowed, and that on the small ground with its ten arches and two stakes the three-ball break is usually adopted, the next player or "danger ball" being wired at the earliest opportunity.

See Spalding's *Official Roque Guide*, edited by Mr Charles Jacobus (New York, 1906).

---

1 This was largely the work of W. T. Whitmore-Jones (1831-1872), generally known as W. Jones Whitmore, who subsequently formed the short-lived National Croquet Club, and was largely responsible for the first codification of the laws.
2 The words "roquet" and "croquet" are pronounced as in French, with the *t* mute.

CRORE

(Hindustani *karor*), an Anglo-Indian term for a hundred *lakhs* or ten million. It is in common use for statistics of trade and especially coinage. In the days when the rupee was worth its face value of 2s., a crore of rupees was exactly worth a million sterling, but now that the rupee is fixed at 15 to the £1, a crore is only worth £666,666.

CROSBY, HOWARD

(1826-1891), American preacher and teacher, great-grandson of Judge Joseph Crosby of Massachusetts and of Gen. William Floyd of New York, a signer of the Declaration of Independence, was born in New York City on the 27th of February 1826. He graduated in 1844 from the University of the City of New York (now New York University); became professor of Greek there in 1851, and in 1859 became professor of Greek in Rutgers College, New Brunswick, New Jersey, where two years later he was ordained pastor of the first Presbyterian church. From 1870 to 1881 he was chancellor of the University of the City of New York; from 1872 to 1881 was one of the American revisers of the English version of the New Testament; and in 1873 was moderator of the general assembly of the Presbyterian Church. He took a prominent part in politics, urged excise reform, opposed "total abstinence," was one of the founders and was the first president of the New York Society for the Prevention of Crime, and pleaded for better management of Indian affairs and for international copyright. Among his publications are *The Lands of the Moslem* (1851), *Bible Companion* (1870), *Jesus: His Life and Works* (1871), *True Temperance Reform* (1879), *True Humanity of Christ* (1880), and commentaries on the book of Joshua (1875), Nehemiah (1877) and the New Testament (1885).

His son, Ernest Howard Crosby (1856-1907), was a social reformer, and was born in New York City on the 4th of November 1856. He graduated at the University of the City of New York in 1876 and at Columbia Law School in 1878; served in the New York Assembly in 1887-1889, securing the passage of a high-licence bill; in 1889-1894 was a judge of the Mixed Tribunal at Alexandria, Egypt, resigning upon coming under the influence of Tolstoy; and died in New York City on the 3rd of January 1907. He was the first president (1894) of the Social

Reform Club of New York City, and was president in 1900-1905 of the New York Anti-Imperialist League; was a leader in settlement work and in opposition to child labour, and was a disciple of Tolstoy as to universal peace and non-resistance, and of Henry George in his belief in the "single tax" principle. His writings, many of which are in the manner of Walt Whitman, comprise *Plain Talk in Psalm and Parable* (1899), *Swords and Ploughshares* (1902), and *Broadcast* (1905), all in verse; an anti-military novel, *Captain Jinks, Hero* (1902); and essays on Tolstoy (1904 and 1905) and on Garrison (1905).

CROSS, AND CRUCIFIXION

(Lat. *crux, crucis*[1]). The meaning ordinarily attached to the word "cross" is that of a figure composed of two or more lines which intersect, or touch each other transversely. Thus, two pieces of wood, or other material, so placed in juxtaposition to one another, are understood to form a cross. It should be noted, however, that Lipsius and other writers speak of the single upright stake to which criminals were bound as a cross, and to such a stake the name of *crux simplex* has been applied. The usual conception, however, of a cross is that of a compound figure.

Punishment by crucifixion was widely employed in ancient times. It is known to have been used by nations such as those of Assyria, Egypt, Persia, by the Greeks, Carthaginians, Macedonians, and from very early times by the Romans. It has been thought, too, that crucifixion was also used by the Jews themselves, and that there is an allusion to it (Deut. xxi. 22, 23) as a punishment to be inflicted.

Two methods were followed in the infliction of the punishment of crucifixion. In both of these the criminal was first of all usually stripped naked, and bound to an upright stake, where he was so cruelly scourged with an implement, formed of strips of leather having pieces of iron, or some other hard material, at their ends, that not merely was the flesh often stripped from the bones, but even the entrails partly protruded, and the anatomy of the body was disclosed. In this pitiable state he was reclothed, and, if able to do so, was made to drag the stake to the place of execution, where he was either fastened to it, or impaled upon it, and left to die. In this method, where a single stake was employed, we have the *crux simplex* of Lipsius. The other method is that with which we are more familiar, and which is described in the New Testament account of the crucifixion of Jesus Christ. In such a case, after the scourging at the stake, the criminal was made to carry a gibbet, formed of two transverse bars of wood, to the place of execution, and he was then fastened to it by iron nails driven through the outstretched arms and through the ankles. Sometimes this was done as the cross lay on the ground, and it was then lifted into position. In other cases the criminal was made to ascend by a ladder, and was then fastened to the cross. Probably the feebleness, or state of collapse, from which the criminal must often have suffered, had much to do in deciding this. It is not quite clear which of these two plans was followed in the case of the crucifixion of Christ, but the more general opinion has been that He was nailed to the cross on the ground, and that it was then lifted into position. The contrary opinion, has, however, prevailed to some extent, and there are representations of the crucifixion which depict Him as mounting a ladder placed against the cross. Such representations may, however, have been due to a pious desire, on the part of their authors, to emphasize the voluntary offering of Himself as the Saviour of the World, rather than as being intended for actual pictures of the scene itself. It may be noted, however, that among the "Emblems of the Passion," as they are called, and which were very favourite devices in the middle ages, the ladder is not infrequently found in conjunction with the crown of thorns, nails, spear, &c.

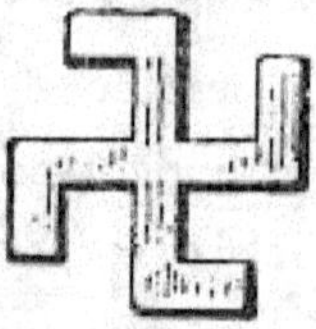

Fig. 1. Fig. 2.

From its simplicity of form, the cross has been used both as a religious symbol and as an ornament, from the dawn of man's civilization. Various objects, dating from periods long anterior to the Christian era, have been found, marked with crosses of different designs, in almost every part of the old world. India, Syria, Persia and Egypt have all yielded numberless examples, while numerous instances, dating from the later Stone Age to

Christian times, have been found in nearly every part of Europe. The use of the cross as a religious symbol in pre-Christian times, and among non-Christian peoples, may probably be regarded as almost universal, and in very many cases it was connected with some form of nature worship. Two of the forms of the pre-Christian cross which are perhaps most frequently met with are the tau cross, so named from its resemblance to the Greek capital letter Τ, and the *svastika* or *fylfot*[2] 卐, also called "*Gammadion*" owing to its form being that of four Greek capital letters *gamma* Γ placed together. The tau cross is a common Egyptian device, and is indeed often called the Egyptian cross. The *svastika* has a very wide range of distribution, and is found on all kinds of objects. It was used as a religious emblem in India and China at least ten centuries before the Christian era, and is met with on Buddhist coins and inscriptions from various parts of India. A fine sepulchral urn found at Shropham in Norfolk, and now in the British Museum, has three bands of cruciform ornaments round it. The two uppermost of these are plain circles, each of which contains a plain cross; the lowest band is formed of a series of squares, in each of which is a *svastika*. In the Vatican Museum there is an Etruscan fibula of gold which is marked with the *svastika*, but it is a device of such common occurrence on objects of pre-Christian origin, that it is hardly necessary to specify individual instances. The cross, as a device in different forms, and often enclosed in a circle, is of frequent occurrence on coins and medals of pre-Christian date in France and elsewhere. Indeed, objects marked with pre-Christian crosses are to be seen in every important museum.

The death of Christ on a cross necessarily conferred a new significance on the figure, which had hitherto been associated with a conception of religion not merely non-Christian, but in its essence often directly opposed to it. The Christians of early times were wont to trace, in things around them, hidden prophetical allusions to the truth of their faith, and such a testimony they seem to have readily recognized in the use of the cross as a religious emblem by those whose employment of it betokened a belief most repugnant to their own. The adoption by them of such forms, for example, as the tau cross and the *svastika* or *fylfot* was no doubt influenced by the idea of the occult Christian significance which they thought they recognized in those forms, and which they could use with a special meaning among themselves, without at the same time arousing the ill-feeling or shocking the sentiment of those among whom they lived.

It was not till the time of Constantine that the cross was publicly used as the symbol of the Christian religion. Till then its employment had been restricted, and private among the Christians themselves. Under Constantine it became the acknowledged symbol of Christianity, in the same way in which, long afterwards, the crescent was adopted as the symbol of the Mahommedan religion. Constantine's action was no doubt influenced by the vision which he believed he saw of the cross in the sky with the accompanying words ἐν τούτῳ νίκα, as well as by the story of the discovery of the true cross by his mother St Helena in the year 326. The legend is that, when visiting the holy places in Palestine, St Helena was guided to the site of the crucifixion by an aged Jew who had inherited traditional knowledge as to its position. After the ground had been dug to a considerable depth, three crosses were found, as well as the superscription placed over the Saviour's head on the cross, and the nails with which he had been crucified. The cross of the Lord was distinguished from the other two by the working of a miracle on a crippled woman who was stretched upon it. This finding, or "invention," of the holy cross by St Helena is commemorated by a festival on the 3rd of May, called the "Invention of the Holy Cross." The legend was widely accepted as true, and is related by writers such as St Ambrose, Rufinus, Sulpicius Severus and others, but it is discounted by the existence of an older legend, according to which the true cross was found in the reign of Tiberius, and while St James the Great was bishop of Jerusalem, by Protonice, the wife of Claudius.

In recent times an attempt has been made to reconcile the two accounts, by attributing to St Helena the rediscovery of the true cross, originally found by Protonice, and which had been buried again on the spot. A change was made in 1895 in the *Diario Romano*, when the word *Ritrovamento* was substituted for that of *Invenzione*, in the name of the festival of the 3rd of May. After St Helena's discovery a church was built upon the site, and in it she placed the greater portion of the cross. The remaining portion she conveyed to Byzantium, and thence Constantine sent a piece to Rome, where it is said to be still preserved in the church of S. Croce in Gerusalemme, which was built to receive so precious a relic. It is exposed for the veneration of the faithful on Good Friday, 3rd of May, and the third Sunday in Lent, each year.

Another festival of the holy cross is kept on the 14th of September, and is known as the "Exaltation of the Holy Cross." It seems to have originated with the dedication, in the year 335, of the churches built on the sites of the crucifixion and the holy sepulchre. The observance of this festival passed from Jerusalem to Constantinople, and thence to Rome, where it appears to have been introduced in the 7th century. By some it is thought that the feast of the Exaltation of the Cross had its origin in Constantine's vision of the cross in the sky in the year 317, but whether it originated then, or, as is more generally supposed, at the dedication of the churches at Jerusalem, there is no doubt that it was afterwards kept with much greater solemnity in consequence of the recovery of the portion of the cross St Helena had left at Jerusalem, which had been taken away in the Persian victory, and was restored to Jerusalem by Heraclitus in 627. Pope Clement VIII. (1592-1604) raised the festival of the Exaltation of the Holy Cross to the dignity, liturgically known as that of a Greater Double.

Before leaving the story of St Helena and the cross, it may be convenient to allude briefly to the superscription placed over the Saviour's head, and the nails, which it is said that she found with the cross. The earlier tradition as to the superscription is obscure, but it would seem that it ought to be considered part of the relic which Constantine sent to Rome. By some means it was entirely lost sight of until the year 1492, when it is said that it was accidentally found in a vault in the church of S. Croce in Gerusalemme at

Rome. Pope Alexander III. published a bull certifying to the truth of this rediscovery of the relic, and authenticated its character.

As regards the nails, a question has arisen whether there were three or four. In the earliest pictures of the Crucifixion the feet are shown as separately nailed to the cross, but at a later period they are crossed, and a single nail fixes them. In the former case there would be four nails, and in the latter only three. Four is the number generally accepted, and it is said that one was cast by St Helena into the sea, during a storm, in order to subdue the waves, another is said (but the legend cannot be traced far back) to have been beaten out into the iron circlet of the crown of Lombardy, while the remaining two are reputed to be preserved among the relics at Milan and Trier respectively.

The employment of the cross as the Christian symbol has been so manifold in its variety and application, and the different forms to which the figure has been adapted and elaborated are so complex, that it is only possible to deal with the outline of the subject.

We learn from Tertullian and other early Christian writers of the constant use which the Christians of those days made of the sign of the cross. Tertullian (*De Cor. Mil.* cap. iii.) says: "At each journey and progress, at each coming in and going out, at the putting on of shoes, at the bath, at meals, at the kindling of lights, at bedtime, at sitting down, whatsoever occupation engages us, we mark the brow with the sign of the cross." With so frequent an employment of the sign of the cross in their domestic life, it would be strange if we did not find that it was very frequently used in the public worship of the church. The earliest liturgical forms are comparatively late, and are without rubrics, but the allusions by different writers in early times to the ceremonial use of the sign of the cross in the public services are so numerous, and so much importance was attached to it, that we are left in no manner of doubt on the point. St Augustine, indeed, speaks of the sacraments as not duly ministered if the use of the sign of the cross were absent from their ministration (*Hom. cxviii. in S. Joan.*). Of the later liturgical use of the sign of the cross there is little need to speak, as a reference to the service books of the Greek and Latin churches will plainly indicate the frequency of, and the importance attached to, its employment. Its occasional use is retained by the Lutherans, and in the Church of England it is authoritatively used at baptism, and at the "sacring" or anointing of the sovereign at the coronation.

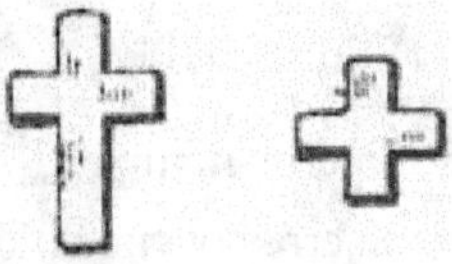

FIG. 3. FIG. 4.

Passing from the sign to the material figures of the cross, a very usual classification distinguishes three main forms: (1) the *crux immissa*, or *capitata* † (fig. 3) known also as the Latin cross, or if each limb is of the same length, + (fig. 4) as the Greek cross; (2) the *crux decussata*, formed like the letter ×, and (3) the *crux commissa* or tau cross, already mentioned. It was on a crux immissa that Christ is believed to have been crucified. The *crux decussata* is known as St Andrew's cross, from the tradition that St Andrew was put to death on a cross of that form. The *crux commissa* is often called St Anthony's cross, probably only because it resembles the crutch with which the great hermit is generally depicted.

The cross in one form or other appears, appropriately, on the flags and ensigns of many Christian countries. The English cross of St George is a plain red cross on a white ground, the Scottish cross of St Andrew is a plain diagonal white cross on a blue ground, and the Irish cross of St Patrick is a plain diagonal red cross on a white ground. These three crosses are combined in the Union Jack (see FLAG).

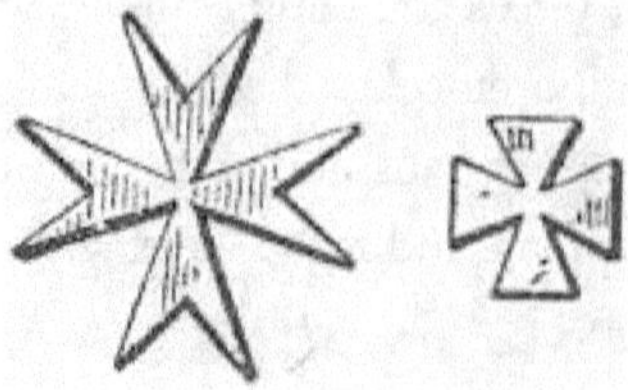

FIG. 5. FIG. 6.

The cross has also been adopted by many orders of knighthood. Perhaps the best known of these is the cross of the knights of Malta. It is a white cross of eight points on a black ground (fig. 5) and is the proper Maltese cross, a name which is often wrongly applied to the cross *patée* (fig. 6). The knights of the Garter use the cross of St George, as do those of the order of St Michael and St George, the knights of the Thistle use St Andrew's cross, and those of St Patrick the cross of St Patrick charged with a shamrock leaf. The cross of the Danish order of the Dannebrog (fig. 7) affords a good example of this use of the cross. It is in form a white cross patée, superimposed upon a red one of the same form, and is surmounted by the royal cipher and crown, and has upon its surface the royal cipher repeated, and the legend, or motto, "*Gud og Kongen*" = "God and the King." (For crosses of monastic orders see COSTUME.)

FIG. 7.—Cross of the Dannebrog.

Akin to the crosses of knightly orders are those which figure as charges on coats of arms. The science of heraldry evolved a wonderful variety of cross-forms during the period it held sway in the middle ages. The different forms of cross used in heraldry are, in fact, so numerous that it is only the larger works on that subject which attempt to record them all. For such crosses see HERALDRY.

In the middle ages the cross form, in one way or another, was predominant everywhere, and was introduced whenever opportunity offered itself for doing so. The larger churches were planned on its outline, so that the ridge line of their roofs proclaimed it far and wide. This was more particularly followed in the north of Europe, but when it was first introduced is not quite certain. All the ancient cathedral churches of England and Wales are cruciform in plan, except Llandaff.

The artistic skill and ingenuity of the medieval designer has produced cross designs of endless variety, and of singular elegance and beauty. Some of the most beautiful of these designs are the gable crosses of the old churches. Fig. 8 shows the west gable cross of Washburn church, Worcestershire; fig. 9 that of the nave of Castle Acre church, Norfolk; and fig. 10 the east gable cross of Hethersett church in that county. They may be taken as good examples of a type of cross which is often of great beauty, but it is overlooked, owing to its bad position for observation.

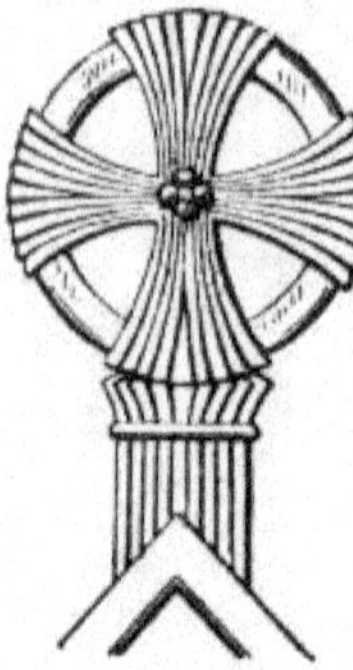

FIG. 8.

FIG. 9.

FIG. 10.

Other architectural crosses, of great beauty of design, are those which occur on the grave slabs of the middle ages. Instances of a plainer type occur in Saxon times, but it was not till after the 11th century that they were fashioned after the intricate and beautiful designs with which our ancient churches are, as a rule, so plentifully supplied. Sometimes these crosses are incised in the slab, and almost as often they are executed in low relief. The long shaft of the cross is most commonly plain, but there are a very large number of instances in which this is not so, and in which branches, with leaf designs, are thrown out at intervals the entire length of the shaft. In some cases the shaft rises from a series of steps at its base, and in such a case the name of a Calvary cross is applied to it. Fig. 11, from Stradsett church, Norfolk, and fig. 12 from Bosbury church, Herefordshire, are good examples of the designs at the head of sepulchral crosses. Often, by the side of the cross, an emblem or symbol is placed, denoting the calling in life of the person commemorated. Thus a sword is placed to indicate a knight or soldier, a chalice for a priest, and so forth; but it would be travelling beyond the scope of this article to enter into a discussion as to such symbols.

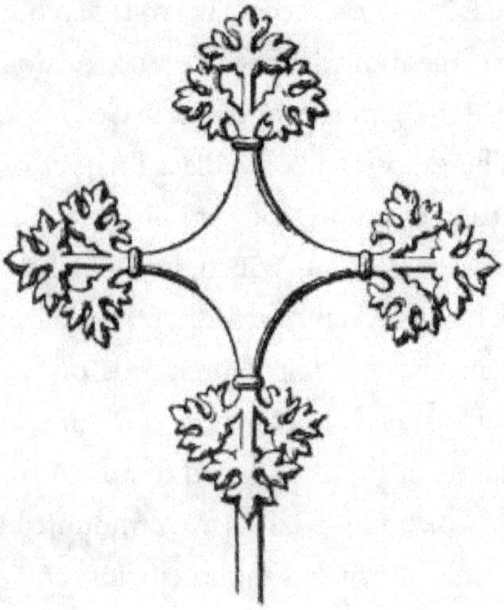

FIG. 11.

FIG. 12.

Of upright standing crosses, the Irish and Iona types are well known, and their great artistic beauty and elaboration and excellence of sculpture are universally recognized. These crosses are sometimes spoken of as "Runic Crosses"; and the interlacing knotwork design with which many of them are ornamented is also at times spoken of as "Runic." This is an erroneous application of the word, and has arisen from the fact that some of these crosses bear inscriptions in Runic characters. Standing crosses, of different kinds, were commonly set up in every suitable place during the middle ages, as the mutilated bases and shafts still remaining readily testify. Such crosses were erected in the centre of the market place, in the churchyard, on the village green, or as boundary stones, or marks to guide the traveller. Some, like the Black Friars cross at Hereford, were preaching stations, others, like the beautiful Eleanor crosses at Northampton, Geddington and Waltham, were commemorative in character. Of these latter crosses, which marked the places where the funeral procession of Queen

Eleanor halted, there were originally ten or more, erected between 1241 and 1294. They were placed at Lincoln, Northampton, Stony Stratford, Woburn, Dunstable, St Albans, Waltham and London (Cheapside and Charing Cross). The cross at Geddington differs in outline from those at Northampton and Waltham, and it is not recorded on the roll of accounts for the nine others, all of which are mentioned, but there is no real doubt that it commemorates the resting of the coffin of the queen in Geddington church on its way from Harby. These crosses, like the Black Friars cross at Hereford, are elaborate architectural erections, and very similar to them in this respect are the beautiful market crosses at Winchester, Chichester, Salisbury, Devizes, Shepton Mallet, Leighton Buzzard, &c. Of churchyard crosses, as distinguished from memorial crosses in churchyards, one only is believed to have escaped in a perfect condition the ravages of time, and the fanaticism of the past. It stands in the churchyard of Somerby, in Lincolnshire (Tennyson's birthplace), and is a tall shaft surmounted by a pedimented tabernacle, on one side of which is the crucifixion, and on the other the figure of the Virgin and Child. Churchyard crosses may have been used as occasional preaching stations, for reading the Gospel in the Palm Sunday procession, and generally for public proclamations, made usually at the conclusion of the chief Sunday morning service, much in the same way that market crosses were used on market days as places for proclamations in the towns.

Of the ecclesiastical use of the sign of the cross mention has already been made, and it is desirable to mention briefly one or two instances of the ecclesiastical use of the cross itself. From a fairly early period it has been the prerogative of an archbishop or metropolitan, to have a cross borne before him within the limits of his province. The question urged between the archbishops of Canterbury and York about the carrying of their crosses before them, in each other's province, was a fruitful source of controversy in the middle ages. The archiepiscopal cross must not be confused with the crozier or pastoral staff. The latter, which is formed with a crook at the end, is quite distinct, and is used by archbishops and bishops alike, who bear it with the left hand in processions, and when blessing the people. The archiepiscopal cross, on the contrary, is always borne before the archbishop, or during the vacancy of the archiepiscopal see before the guardian of the spiritualities *sede vacante*. The bishop of Dol in Brittany, of ordinary diocesan bishops, alone possessed the privilege of having a cross borne before him in his diocese. Good illustrations of the archiepiscopal cross occur on the monumental brasses of Archbishop Waldeby, of York (1397), at Westminster Abbey, and of Archbishop Cranley, of Dublin (1417) in New College chapel, Oxford.

The custom of carrying a cross at the head of an ecclesiastical procession can be traced back to the end of the 4th century. The cross was originally taken from the altar, and raised on a pole, and so borne before the procession. Afterwards a separate cross was provided for processions, but in poor churches, where this was not the case, the altar cross continued to be used till quite a late period. A direction to this effect occurs as late as 1829, in the *Rituel* published for the diocese of La Rochelle in that year. In England altar crosses were not very usual in the middle ages.

As a personal ornament the cross came into common use, and was usually worn suspended by a chain from the neck. A cross of this kind, of very great interest and beauty, was found about 1690, on the breast of Queen Dagmar, the wife of Waldemar II., king of Denmark (d. 1213). It is of Byzantine design and workmanship, and is of enamelled gold (fig. 13 shows both sides of it); on one side is the Crucifixion, and on the other side the half figure of our Lord in the centre, with the Virgin and St John the Evangelist on either side, and St Chrysostom and St Basil above and below. From the way in which such crosses were worn, hanging over the chest, they are called pectoral crosses. At the present day a pectoral cross forms part of the recognized insignia of a Roman Catholic bishop, and is worn by him over his robes, but this official use of the pectoral cross is not ancient, and no instance is known of it in England before the Reformation. The custom appears to have taken rise in the 16th century on the continent. It was not unusual to wear cruciform reliquaries, as objects of personal adornment, and such a reliquary was found on the body of St Cuthbert, when his tomb was opened in 1827, but it was placed under, and not over his episcopal vestments, and formed no part of his bishop's attire. The custom of wearing a pectoral cross over ecclesiastical robes has, curiously enough, been copied from the comparatively modern Roman Catholic usage by the Lutheran bishops and superintendents in Scandinavia and Prussia; and in Sweden the cross is now delivered to the new bishop, on his installation in office, by the archbishop of Upsala, together with the mitre and crozier. Within the last generation the use of a pectoral cross, worn over their robes as part of the insignia of the episcopal office, has been adopted by some bishops of the Church of England, but it has no ancient sanction or authority.

FIG. 13.—Dagmar CrosAs.

AUTHORITIES.—Mortillet, *Le Signe de la croix avant le Christianisme* (Paris, 1866); Bingham, *Antiquities of the Christian Church*; Lipsius, *De Cruce Christi*; Lady Eastlake, *History of our Lord*, vol. ii.; Cutts, *Manual of Sepulchral Slabs and Crosses*; (Anon.) *Handbook to Christian and Ecclesiastical Rome*, part ii. (London, 1897); Veldeuer, *History of the Holy Cross* (reprint, 1863).

(T. M. F.)

1 Derivatives of the Latin *crux* appear in many forms in European languages, cf.

Ger. *Kreuz*, Fr. *croix*, It. *croce*, &c.; the English form seems Norse in origin (O.N. *Krosse*, mod. *Kors*). The O.E. name was *rōd*, rood (*q.v.*).

2 The acceptance of this word as the English equivalent for this peculiar form of the cross rests only, according to the *New English Dictionary*, on a MS. of about 1500 in the Lansdowne collection, which gives details for the erection of a memorial stained-glass window, "... the fylfot in the nedermost pane under ther I knele ..."; in the sketch given with the instructions a cross occupies the space indicated. It is a question, therefore, whether "fylfot" is a name for any device suitable to "fill the foot" of any design, or the name peculiar to this particular form of cross. The word is not, as was formerly accepted, a corruption of the O. Eng. *feowerfete*, four-footed.

CROSSBILL

(Fr. *Bec-croisé*, Ger. *Kreuzschnabel*), the name given to a genus of birds, belonging to the family *Fringillidae*, or finches, from the unique peculiarity they possess among the whole class of having the horny sheaths of the bill crossing one another obliquely,[1] whence the appellation *Loxia* (λοξός, *obliquus*), conferred by Gesner on the group and continued by Linnaeus. At first sight this singular structure appears so like a deformity that writers have not been wanting to account it such,[2] ignorant of its being a piece of mechanism most beautifully adapted to the habits of the bird, enabling it to extract with the greatest ease, from fir-cones or fleshy fruits, the seeds which form its usual and almost invariable food. Its mode of using this unique instrument seems to have been first described by Townson (*Tracts on Nat. Hist.*, p. 116, London, 1799), but only partially, and it was Yarrell who, in 1829 (*Zool. Journ.*, iv. pp. 457-465, pl. xiv. figs. 1-7), explained fully the means whereby the jaws and the muscles which direct their movements become so effective in riving asunder cones or apples, while at the proper moment the scoop-like tongue is instantaneously thrust out and withdrawn, conveying the hitherto protected seed to the bird's mouth. The articulation of the mandible to the quadrate-bone is such as to allow of a very considerable amount of lateral play, and, by a particular arrangement of the muscles which move the former, it comes to pass that so soon as the bird opens its mouth the point of the mandible is brought immediately opposite to that of the maxilla (which itself is movable vertically), instead of crossing or overlapping it—the usual position when the mouth is closed. The two points thus meeting, the bill is inserted between the scales or into the pome, but on opening the mouth still more widely, the lateral motion of the mandible is once more brought to bear with great force to wrench aside the portion of the fruit attacked, and then the action of the tongue completes the operation, which is so rapidly performed as to defy scrutiny, except on very close inspection. Fortunately the birds soon become tame in confinement, and a little patience will enable an attentive observer to satisfy himself as to the process, the result of which at first seems almost as unaccountable as that of a clever conjuring trick.

The common crossbill of the Palaearctic region (*Loxia curvirostra*) is about the size of a skylark, but more stoutly built. The young (which on leaving the nest have not the tips of the bill crossed) are of a dull olive colour with indistinct dark stripes on the lower parts, and the quills of the wings and tail dusky. After the first moult the difference between the sexes is shown by the hens inclining to yellowish-green, while the cocks become diversified by orange-yellow and red, their plumage finally deepening into a rich crimson-red, varied in places by a flame-colour. Their glowing hues, are, however, speedily lost by examples which may be kept in confinement, and are replaced by a dull orange, or in some cases by a bright golden-yellow, and specimens have, though rarely, occurred in a wild state exhibiting the same tints. The cause of these changes is at present obscure, if not unknown, and it must be admitted that their sequence has been disputed by some excellent authorities, but the balance of evidence is certainly in favour of the above statement. Depending mainly for food on the seeds of conifers, the movements of crossbills are irregular beyond those of most birds, and they would seem to rove in any direction and at any season in quest of their staple sustenance. But the pips of apples are also a favourite dainty, and it is recorded by the old chronicler Matthew Paris (*Hist. Angl.* MS. fol. 252), that in 1251 the orchards of England were ravaged by birds, "pomorum grana, & non aliud de eisdem pomis comedentes," which, from his description, "Habebant autem partes rostri cancellatas, per quas poma quasi forcipi vel cultello dividebant," could be none other but crossbills. Notice of a like visitation in 1593 is recorded, but of late it has become evident that not a year passes without crossbills being observed in some part or other of England, while in certain localities in Scotland they seem to breed annually. The nest is rather rudely constructed, and the eggs, generally four in number, resemble those of the greenfinch, but are larger in size. This species ranges throughout the continent of Europe,[3] and occurs in the islands of the Mediterranean and in the fir-woods of the Atlas. In Asia it would seem to extend to Kamtschatka and Japan, keeping mainly to the forest-tracts.

Three other forms of the genus also inhabit the Old World—two of them so closely resembling the common bird that their specific validity has been often questioned. The first of these, of large stature, the parrot-crossbill (*L. pityopsittacus*), comes occasionally to Great Britain, presumably from Scandinavia, where it is known to breed. The second (*L. himalayana*), which is a good deal smaller, is only known from the Himalaya Mountains. The third, the two-barred crossbill (*L. taenioptera*), is very distinct, and its proper home seems to be the most northern forests of the Russian empire, but it has occasionally occurred in western Europe and even in England.

The New World has two birds of the genus. The first (*L. americana*), representing the common British species, but with a smaller bill, and the males easily recognizable by their more scarlet plumage, ranges from the northern limit of coniferous trees to the highlands of Mexico, or even farther. The other (*L. leucoptera*) is the equivalent of the two-barred crossbill, but smaller. It has twice occurred in England.

(A. N.)

---

1 This peculiarity is found as an accidental malformation in the crows (*Corvidae*) and other groups; it is comparable to the monstrosities seen in rabbits and other members of the order *Glires*, in which the incisor teeth grow to an inordinate length.

2 A medieval legend ascribes the conformation of bill and coloration of plumage to a divine recognition of the bird's pity, bestowed on Christ at the crucifixion.

3 Dr Malmgren found a small flock on Bear Island (lat. 74½° N.), but to this barren spot they must have been driven by stress of weather.

CROSSEN

Or KROSSEN, a town of Germany, in the kingdom of Prussia, on the Oder, here crossed by a bridge, at the influx of the Bober, 31 m. S.E. of Frankfort-on-Oder by rail. Pop. (1900) 7369. Of the churches in the town three are Protestant and one Roman Catholic. Besides the modern school (Realprogymnasium), there are a technical school for viniculture and fruit-growing and a dairy school. There are manufactories of copper and brass ware, cloth, &c., while in the surrounding country the chief industries are fruit and grape growing. There is a brisk shipping trade, mainly in wine, fruit and fish. Crossen was founded in 1005 and was important during the middle ages as a point of passage across the Oder. It attained civic rights in 1232, was for a time the capital of a Silesian duchy, which, on the death of Barbara of Brandenburg, widow of the last duke, passed to Brandenburg (1482). In May 1886 the town was devastated by a whirlwind.

CROSSING

In architecture, the term given to the intersection of the nave and transept, frequently surmounted by a tower or by a dome on pendentives.

CROSSKEY, HENRY WILLIAM

(1826-1893), English geologist and Unitarian minister, was born at Lewes in Sussex, on the 7th of December 1826. After being trained for the ministry at Manchester New College (1843-1848), he became pastor of Friargate chapel, Derby, until 1852, when he accepted charge of a Unitarian congregation in Glasgow. In 1869 he removed to Birmingham, where until the close of his life he was pastor of the Church of the Messiah. While in Glasgow his interest was awakened in geology by the perusal of A. C. Ramsay's *Geology of the Isle of Arran*, and from 1855 onwards he devoted his leisure to the pursuit of this science. He became an authority on glacial geology, and wrote much, especially in conjunction with David Robertson, on the post-tertiary fossiliferous beds of Scotland (*Trans. Geol. Soc. Glasgow*). He also prepared for the British Association a valuable series of Reports (1873-1892) on the erratic Blocks of England, Wales and Ireland. In conjunction with David Robertson and G. S. Brady he wrote the *Monograph of the Post Tertiary Entomostraca of Scotland*, &c. for the Palaeontographical Society (1874); and he edited H. Carvill Lewis' *Papers and Notes on the Glacial Geology of Great Britain and Ireland*, issued posthumously (1894). He died at Edgbaston, Birmingham, on the 1st of October 1893.

See *H. W. Crosskey: his Life and Work*, by R. A. Armstrong (with chapter on his geological work by Prof. C. Lapworth, 1895).

CROSS RIVER

A river of West Africa, over 500 m. long. It rises in 6° N, 10° 30′ E. in the mountains of Cameroon, and flows at first N.W. In 8° 48′ E., 5° 50′ N. are a series of rapids; below this point the river is navigable for shallow-draught boats. At 8° 20′ E., 6° 10′ N., its most northern point, the river turns S.W. and then S., entering the Gulf of Guinea through the Calabar estuary. The Calabar river, which rises about 5° 30′ N., 8° 30′ E., has a course parallel to, and 10 to 20 m. east of, the Cross river. Near its mouth, on its east bank, is the town of Calabar (*q.v.*). It enters the estuary in 4° 45′ N. The Cross, Calabar, Kwa and other streams farther east, which rise on the flanks of the Cameroon Mountains, form a large delta. The Calabar and Kwa rivers are wholly within the British protectorate of Southern Nigeria, as is the Cross river from its mouth to the rapids mentioned. The upper course of the river is in German territory.

CROSS-ROADS, BURIAL AT

In former times the method of disposing of executed criminals and suicides. At the cross-roads a rude cross usually stood, and this gave rise to the belief that these spots were selected as the next best burying-places to consecrated ground. The real explanation is that the ancient Teutonic peoples often built their altars at the cross-roads, and as human sacrifices, especially of criminals, formed part of the ritual, these spots came to be regarded as execution grounds. Hence after the introduction of Christianity, criminals and suicides were buried at the cross-roads during the night, in order to assimilate as far as possible their funeral to that of the pagans. An example of a cross-road execution-ground was the famous Tyburn in London, which stood on the spot where the Oxford, Edgware and London roads met.

CROSS SPRINGER

In architecture, the block from which the diagonal ribs of a vault spring or start: the top of the springer is known as the skewback (see ARCH).

CROTCH, WILLIAM

(1775-1847), English musician, was born in Green's Lane, Norwich, on the 5th of July 1775. His father was a master carpenter. The child was extraordinarily precocious, and when scarcely more than two years of age he played upon an organ of his parent's construction something like the tune of "God save the King." At the age of four he came to London and gave daily recitals on the organ in the rooms of a milliner in Piccadilly. The precocity of his musical intuition was almost equalled by a singularly early aptitude for drawing. In 1786 he went to Cambridge as assistant to Dr Randall the organist. His oratorio *The Captivity of Judah* was played at Trinity Hall, Cambridge, on the 4th of June 1789. He was then only fourteen years of age. His intention of entering the church carried him to Oxford in 1788, but the superior attractions of a musical career acquired an increasing influence over him, and in 1790 he was appointed organist of Christ Church. At the early age of twenty-two he was appointed professor of music in the university of Oxford, and there in 1799 he took his degree of doctor in that art. In 1800 and the four following years he read lectures on music at Oxford. Next he was appointed lecturer on music to the Royal Institution, and subsequently, in 1822, principal of the London Royal Academy of Music. His last years were passed at Taunton in the house of his son, the Rev. W. R. Crotch, where he died suddenly on the 29th of December 1847. He published a number of vocal and instrumental compositions, of which the best is his oratorio *Palestine*, produced in 1812. In 1831 appeared an 8vo volume containing the substance of his lectures on music, delivered at Oxford and in London. Previously, he had published three volumes of *Specimens of Various Styles of Music*. Among his didactic works is *Elements of Musical Composition and Thorough-Bass* (London, 1812). The oratorio bearing the title *The Captivity of Judah*, and produced on the occasion of the installation of the duke of Wellington as chancellor of the university of Oxford in 1834, is a totally different work from that which he wrote upon the same subject as a boy of fourteen. He arranged for the pianoforte a number of Handel's oratorios and operas, besides symphonies and quartetts of Haydn, Mozart and Beethoven. The great expectations excited by his infant precocity were not fulfilled; for he manifested no extraordinary genius for musical composition. But he was an industrious student and a sound artist, and his name remains familiar in English musical history.

CROTCHET

(from the Fr. *croche*, a hook; whence also the Anglicized "crochet," pronounced as in French, for the knitting-work done with a hook instead of on pins), properly a small hook, and so used of the hook-like *setae* or bristles found in certain worms which burrow in sand. In music, a "crotchet" is a note of half the value of a minim and double that of a quaver; it is marked by a round black head and a line without a tail or hook; the French *croche* is used of a "quaver" which has a tail, but in ancient music the *semiminima*, the modern crotchet, is marked by an open note with a hook. Derived either from an old French proverbial phrase, *il a des crochues en teste*, or from a meaning of twist or turn, as in the similar expression "crank," comes the sense of a whim, fancy or perverse idea, seen also in the adjective "crotchety" of a fussy unreasonable person.

CROTONA

Croto or Croton (Gr. Κρότων, mod. Cotrone) a Greek town on the E. coast of the territory of the Bruttii (mod. *Calabria*), on a promontory 7 m. N.W. of the Lacinian promontory. It was founded by a colony of Achaeans led by Myscellus in 710 B.C. Its name was, according to the legend, that of a local prince who afforded hospitality to Heracles, but was accidentally killed by him and buried on the spot. Like Sybaris, it soon became a city of power and wealth. It was especially celebrated for its successes in the Olympic games from 588 B.C. onwards, Milo being the most famous of its athletes. Pythagoras established himself here between 540 and 530 B.C. and formed a society of 300 disciples (among whom was Milo), who acquired considerable influence with the supreme council of 1000 by which the city was ruled. In 510 B.C. Crotona was strong enough to defeat the Sybarites, with whom it had previously been on friendly terms, and raze their city to the ground. Shortly afterwards, however, an insurrection took place, by which the disciples of Pythagoras were driven out, and a democracy established. The victory of the Locrians and Phlegians over Crotona in 480 B.C. marked the beginning of its decline. It suffered after this from the attacks of Dionysius I., who became its master for twelve years, of the Bruttii, and of Agathocles, and even more from the invasion of Pyrrhus, after which in 277 the Romans obtained possession of it. Livy states that the walls had a length of 12 m. and that about half the area within them had at that time ceased to be inhabited. After the battle of Cannae Crotona revolted from Rome, and Hannibal made it his winter quarters for three years. It was made a colony by the Romans at the end of the war (194 B.C.). After that time but little is heard of it, though Petronius mentions the corrupt morals of its inhabitants; but it continues to be mentioned down to the Gothic wars. The importance of the city was mainly due to its harbour, which, though not a good one, was the only port between Tarentum and Rhegium. The original settlement occupied the hill above it (143 ft.) and later became the acropolis. Its healthy situation was famous in antiquity, and to this was ascribed its superiority in athletics; it was the seat also of a medical school which in the days of Herodotus was considered the first in Greece. Of the exact site of the ancient city and its remains practically nothing is known; a few fragments of the productions of its art preserved in private hands at Cotrone are described by F. von Duhn in *Notizie degli scavi*, 1897, 343 seq.

(T. As.)

CROTONIC ACID

($C_4H_6O_2$). Three acids of this empirical formula are known, viz. crotonic acid, isocrotonic acid and methacrylic acid; the constitutional formulae are—

$$\begin{array}{ccc} \begin{array}{c} HC\cdot CO_2H \\ \ddot{HC}\cdot CH_3 \end{array}, & \begin{array}{c} HC\cdot CO_2H \\ CH_3\cdot \ddot{CH} \end{array}, & CH_2:C\begin{array}{l} \diagup CH_3 \\ \diagdown CO_2H \end{array}. \\ \text{Crotonic Acid.} & \text{Isocrotonic Acid.} & \text{Methacrylic Acid.} \end{array}$$

The isomerism of crotonic and isocrotonic acids is to be explained on the assumption of a different spatial arrangement of the atoms in the molecule (see STEREOCHEMISTRY).

Crotonic acid, so named from the fact that it was erroneously supposed to be a saponification product of croton oil, may be prepared by the oxidation of croton-aldehyde, $CH_3\cdot CH:CH\cdot CHO$, obtained by dehydrating aldol, or by treating acetylene successively with sulphuric acid and water; by boiling allyl cyanide with caustic potash; by the distillation of β-oxybutyric acid; by heating paraldehyde with malonic acid and acetic acid to 100° C. (T. Komnenos, *Ann.*, 1883, 218, p. 149).

$$CH_2(COOH)_2 + CH_3CHO \rightarrow CH_3CH:C(COOH)_2 \\ \rightarrow CH_3\cdot CH:CH\cdot COOH;$$

or by heating pyruvic acid with an excess of acetic anhydride and sodium acetate to 160-180° C. (B. Homolka, *Ber.*, 1885, 18, p. 987). It crystallizes in needles (from hot water) which melt at 72° C. and boil at 180-181° C. It is moderately soluble in cold water. It combines directly with bromine, and, with fuming hydrobromic acid at 100° C., it gives chiefly α-brombutyric acid. With hydriodic acid it gives only β-iodobutyric acid. Potash fusion converts it into acetic acid; nitric acid oxidizes it to acetic and oxalic acids; chromic acid mixture to acetaldehyde and acetic acid, and potassium permanganate to αβ-dioxybutyric acid.

Isocrotonic acid (Quartenylic acid) is obtained from β-chlorisocrotonic acid, formed when acetoacetic ester is treated with phosphorus pentachloride and the product poured into water, by the action of sodium amalgam (A. Geuther). It is an oil, possessing a smell like that of butyric acid. It boils at 171.9° C., with partial conversion into crotonic acid; the transformation is complete when the acid is heated to 170-180° C. in a sealed tube. Potassium permanganate oxidizes it to βγ-dioxybutyric acid.

Methacrylic acid was first obtained in the form of its ethyl ester by E. Frankland and B. F. Duppa (*Annalen*, 1865, 136, p. 12) by acting with phosphorus pentachloride on oxyisobutyric ester $(CH_3)_2 \cdot C(OH) \cdot COOC_2H_5$. It is, however, more readily obtained by boiling citra- or meso-brompyrotartaric acids with alkalis. It crystallizes in prisms, which are soluble in water, melt at 16° C., and boil at 160.5° C. When fused with an alkali, it forms propionic acid; with biomine it yields αβ-dibromisobutyric acid. Sodium amalgam reduces it to isobutyric acid. A polymeric form of methacrylic acid has been described by F. Engelhorn (*Ann.*, 1880, 200, p. 70).

CROTON OIL

(*Crotonis Oleum*), an oil prepared from the seeds of *Croton Tiglium*, a tree belonging to the natural order Euphorbiaceae, and native or cultivated in India and the Malay Islands. The tree is from 15 to 20 ft. in height, and has few and spreading branches, alternate, oval-oblong leaves, acuminate at the point, and covered when young with stellate hairs, and terminal racemes of small, downy, greenish-yellow, monoecious flowers. The male blossoms have five petals and fifteen stamens; the females have no petals but a large oblong ovary bearing three bifid styles. The fruit or capsule is obtusely three-cornered, and about the size of a hazel-nut; it contains three cells each enclosing a seed. The seeds resemble those of the castor-oil plant; they are about half an inch long, and two-fifths of an inch broad, and have a cinnamon-brown, brittle integument; between the two halves of the kernel lie the large cotyledons and radicle. The ocular distinction between the two kinds of seeds may be of great practical importance. The most obvious distinction is that the castor-oil seeds have a polished and mottled surface. The kernels contain from 50 to 60% of oil, which is obtained by pressing them, when bruised to a pulp, between hot plates. Croton oil is a transparent and viscid liquid of a brownish or pale-yellow tinge, and acrid, peculiar and persistent taste, a disagreeable odour and acid reaction. It is soluble in volatile oils, carbon disulphide, and ether, and to some extent in alcohol. It contains acetic, butyric and valeric acids, with glycerides of acids of the same series, and a volatile body, $C_5H_8O_2$, tiglic acid, metameric with angelic acid, and identical with methylcrotonic acid, $CH_3{\cdot}CH{:}C(CH_3)(CO_2H)$. The odour is due to various volatile acids, which are present to the extent of about 1%. A substance called crotonal appears to be responsible for its external, but not its internal, action. The latter is probably due to crotolinic acid, $C_9H_{14}O_2$, which has active purgative properties. The maximum dose of croton oil is two minims, one-fourth of that quantity being usually ample.

Applied to the skin, croton oil acts as a powerful irritant, inducing so much inflammation that definite pustules are formed. The destruction of the true skin gives rise to ugly scars which constitute, together with the pain caused by this application, abundant reason why croton oil should never be employed externally. Despite the pharmacopoeial liniment and the practice of a few, it may be said that this employment of croton oil is now entirely without justification or excuse.

Taken internally, even in the minute doses already detailed, croton oil very soon causes much colic and the occurrence of a fluid diarrhoea which usually recurs several times. It is characteristic of this purgative that it is a hydragogue even in minimal dose, the fluid secretions of the bowel being most markedly increased. The drug appears to act only upon the small intestine. In somewhat larger doses it produces severe gastro-enteritis. The flow of bile is somewhat increased. Such effects may all be produced, even up to the discharge of blood, by the absorption of croton oil from the skin.

The minuteness of the dose, the certainty of the action, and the large amount of fluid drained away constitute this the best drug for administration to an unconscious patient (especially in cases of apoplexy, when it is desirable to remove fluid from the body), or to insane patients who refuse to take any drug. One drop of the oil, placed on the back of the tongue, must inevitably be swallowed by reflex action. A dose should never be repeated. The characters of this drug obviously contra-indicate its use in all cases of organic disease or obstruction of the bowel, in pregnancy, or in cases of constipation in children or the aged.

CROUP

A name formerly given to diseases characterized by distress in breathing accompanied by a metallic cough and some hoarseness of speech. It is now known that these symptoms are often associated with diphtheria (*q.v.*), spasmodic laryngitis (*q.v.*), and a third disease, spasmodic croup, to which the term is now alone applied. This occurs most frequently in children above two years of age; the child goes to bed quite well, and a few hours later suddenly awakes with great difficulty in inspiration, the chest wall becomes markedly retracted, and there is a metallic cough. The child becomes cyanosed, and, to the inexperienced nurse, seems in an almost moribund condition. In the course of four or five minutes, normal respiration starts again, and the attack is over for the time being; but it may recur several times a day. The seizure may be accompanied by convulsions, and death has occurred from dyspnoea. The best treatment is to plunge the child into a warm bath, and sponge the back and chest with cold water. Subsequently this can be done two or three times a day. Should the cyanosis become very severe, respiration can be restarted by making the child sick, either with a dose of ipecacuanha wine, or by forcing one's finger down the throat. Generally the bowels should be attended to; and the throat carefully examined for enlarged tonsils or adenoids, which if present should be treated.

CROUSAZ, JEAN PIERRE DE

(1663-1750), Swiss writer, was born at Lausanne. He was a many-sided man, whose numerous works on many subjects had a great vogue in their day, but are now forgotten. He has been described as an *initiateur plutôt qu'un créateur*, chiefly because he introduced at Lausanne the philosophy of Descartes in opposition to the reigning Aristotelianism, and also as a Calvinist pendant (for he was a pastor) of the French *abbés* of the 18th century. He studied at Geneva, Leyden and Paris, before becoming (1700) professor of philosophy and mathematics at the academy of Lausanne, of which he was four times rector before 1724, when the theological disputes connected with the *Consensus*[1] led him to accept a chair of philosophy and mathematics at Groningen. In 1726 he was appointed governor to the young prince Frederick of Hesse-Cassel, and in 1735 returned to Lausanne with a good pension. In 1737 he was reinstated in his old chair, which he retained to his death. Gibbon, describing his first stay at Lausanne (1752-1755), writes in his *Autobiography*, "the logic of de Crousaz had prepared me to engage with his master Locke and his antagonist Bayle."

The most important of his works are: *Nouvel Essai de logique* (1712), *Géométrie des lignes et des surfaces rectilignes et circulaires* (1712), *Traité du beau* (1714), *Examen du traité de la liberté de penser d'Antoine Collins* (1718), *De l'éducation des enfants* (1722, dedicated to the then Princess of Wales), *Examen du pyrrhonisme ancien et moderne* (1733, an attack chiefly on Bayle), *Examen de l'essai de M. Pope sur l'homme* (1737, an attack on the Leibnitzian theory of that poem), *Logique* (6 vols., 1741), *De l'esprit humain* (1741), and *Réflexions sur l'ouvrage intitulé: La Belle Wolfienne* (1743).

(W. A. B. C.)

1 The "Consensus ecclesiarum Helveticarum reformatarum" was a document drawn up in 1675 and imposed in 1722—as a test of strict Protestant orthodoxy as to the doctrine of grace—by Bern on its subjects in Lausanne and Vaux.

CROW

(Dutch, *kraai*, Ger. *Krähe*, Fr. *corbeau*, Lat. *corvus*), a name most commonly applied in Britain to the bird properly called a rook (*Corvus frugilegus*), but perhaps originally peculiar to its congener, nowadays usually distinguished as the black or carrion-crow (*C. corone*). By ornithologists it is also used in a far wider sense, as under the title crows, or *Corvidae*, is included a vast number of birds from almost all parts of the world, and this family is probably the most highly developed of the whole class *Aves*. Leaving out of account the best known of these, as the raven, rook, daw, pie and jay, with their immediate allies, our attention will here be confined to the crows in general; and then the species of the family to which the appellation is more strictly applicable may be briefly considered. All authorities admit that the family is very extensive, and is capable of being parted into several groups, but scarcely any two agree. Especially must reserve be exercised as regards the group *Streperinae*, or piping crows, belonging to the Australian Region, and referred by some writers to the shrikes (*Laniidae*): and the jays too have been erected into a distinct family (*Garrulidae*), though it seems hardly possible to separate them even as a subfamily from the pies (*Pica* and its neighbours), which lead almost insensibly to the typical crows (*Corvinae*). Dismissing these subjects for the present, it will perhaps be most convenient to treat of the two groups which are represented by the genera *Pyrrhocorax* or choughs, and *Corvus* or true crows in the most limited sense.

Pyrrhocorax comprehends at least two very good species, which have been needlessly divided generically. The best known of them is the Cornish chough (*P. graculus*), formerly a denizen of the precipitous cliffs of the south coast of England, of Wales, of the west and north coasts of Ireland, and some of the Hebrides, but now greatly reduced in numbers, and only found in such places as are most free from the intrusion of man or of daws

(*Corvus monedula*), which last seem to be gradually dispossessing it of its sea-girt strongholds, and its present scarcity is probably in the main due to its persecution by its kindred. In Britain, indeed, it would appear to be only one of the survivors of a more ancient fauna, for in other countries where it is found it has been driven inland, and inhabits the higher mountains of Europe and North Africa. In the Himalayas a larger form occurs, which has been specifically distinguished (*P. himalayanus*), but whether justifiably so may be doubted. The general colour is a glossy black, and it has the bill and legs bright red. The remaining species (*P. alpinus*) is altogether a mountaineer, and does not affect a sea-shore life. Otherwise it frequents much the same kind of localities, but it does not occur in Britain. The alpine chough is somewhat smaller than its congener, and is easily distinguished by its shorter and bright yellow bill. Remains of both have been found in French caverns the deposits in which were formed during the "Reindeer Age." Commonly placed by systematists next to *Pyrrhocorax* is the Australian genus *Corcorax*, represented by a single species (*C. melanorhamphus*), but this assignment of the bird, which is chiefly a frequenter of woodlands, cannot be admitted without hesitation.

Coming now to what may be literally considered crows, our attention is mainly directed to the black or carrion-crow (*Corvus corone*) and the grey, hooded or Royston crow (*C. cornix*). Both these inhabit Europe, but their range and the time of their appearance are very different. The former is, speaking generally, a summer visitant to the south-western part of Europe, and the latter occupies the north-eastern portion—an irregular line drawn diagonally from about the Firth of Clyde to the head of the Adriatic roughly marking their respective distribution. But both are essentially migrants, and hence it follows that when the black crow, as summer comes to an end, retires southward, the grey crow moves downward, and in many districts replaces it during winter. Further than this, it has been incontestably proved that along or near the boundary where these two birds march they not infrequently interbreed, and it is believed that the hybrids, which sometimes wholly resemble one or other of the parents and at other times assume an intermediate plumage, pair indiscriminately among themselves or with the pure stock. Hence it has seemed to many ornithologists who have studied the subject, that these two birds, so long unhesitatingly regarded as distinct species, are only local races of one and the same dimorphic species. No structural difference—or indeed any difference except that of range (already spoken of) and colour—can be detected, and the problem they offer is one of which the solution is exceedingly interesting if not important to zoologists in general.[1] Almost omnivorous in their diet, there is little edible that comes amiss to them, and, except in South America, they are mostly omnipresent. The fish-crow of North America (*C. ossifragus*) demands a few words, since it betrays a taste for maritime habits beyond that of other species, but the crows of Europe are not averse on occasion to prey cast up by the waters. The house-crow of India (*C. splendens*) is not very nearly allied to its European namesakes, from which it can be readily distinguished by its smaller size and the lustrous tints of its darkest feathers; while its confidence in the human race has been so long encouraged by its intercourse with an unarmed and inoffensive population that it becomes a plague to the European abiding or travelling where it is abundant. Hardly a station or camp in British India is free from a crowd of feathered followers of this species, ready to dispute with the kites and the cooks the very meat at the fire.

(A. N.)

---

1 As bearing upon this question may be mentioned the fact that the crow of Australia (*C. australis*) is divisible into two forms or races, one having the irides white, the other of a dark colour. It is stated that they keep apart and do not intermix.

CROWBERRY

Or Crakeberry, the English name for a low-growing heath-like shrub, found on heaths and rocks in Scotland, Ireland and mountainous parts of England. It is known botanically as *Empetrum nigrum*, and has slender, wiry, spreading branches covered with short, narrow, stiff leaves, the margins of which are recurved so as to form a hollow cylinder concealing the hairy under face of the leaf—a device to avoid excessive loss of water from the leaf under the exposed conditions in which the plant grows. The minute flowers are succeeded by black, edible, berry-like fruits, one-fourth to one-third of an inch in diameter. The plant has a wide distribution, occurring in suitable localities throughout the north temperate zone, and on the Andes of South America.

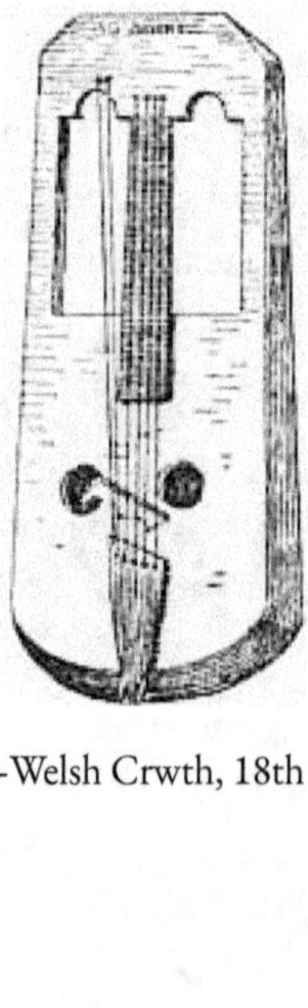

Fig. i.—Welsh Crwth, 18th century.

CROWD

CROUTH, CROWTH (Welsh *crwth*; Fr. *crout*; Ger. *Chrotta*, *Hrotta*), a medieval stringed instrument derived from the lyre, characterized by a sound-chest having a vaulted back and an open space left at each side of the strings to allow the hand to pass through in order to stop the strings on the finger-board. The Welsh crwth, which survived until the end of the 18th century, is best represented by a specimen of that date preserved in the Victoria and Albert Museum, and described and illustrated by Carl Engel.[1] The instrument consists of a rectangular sound-chest 22 in. long, 9½ in. wide and 2 in. deep; the body is scooped out of a single block, the flat belly being glued on. Right through the sound-chest on each side of the finger-board is the characteristic open space left for the hand to pass through. There are two circular sound-holes; the left foot of the flat bridge, which lies obliquely across the belly, passes through the left sound-hole and rests inside on the back of the instrument. Six catgut strings fastened to a tail-piece are wound round pegs at the top of the crwth; four of these strings lie over the sound-board and bridge, and are set in vibration by means of a bow, while the two others, used as drones and stretched across the left-hand aperture, are twanged by the thumb of the left hand. The shape and shallowness of the bridge make it impossible to sound a single string with the bow; the arrangement of the strings suggests that they were intended to be sounded in pairs. The instrument is tuned thus:

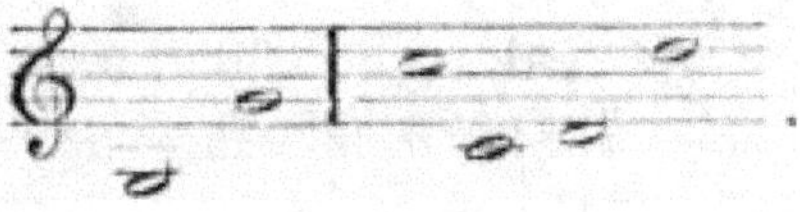

At the beginning of the 19th century, William Bingley[2] heard a Welsh peasant playing national airs on a crwth strung as follows:—.

Sir John Hawkins[3] relates that in his time there was still a Welshman living in Anglesea who understood how to play the crwth according to traditional usage. Edward Jones[4] and Daines Barrington[5] both give an account of the Welsh crwth of the 18th century which agrees substantially with Engel's; the illustration communicated by Daines Barrington shows the strings of the crwth drawn through holes at the top, and fastened on the back, as on the Persian rebab and other Oriental stringed instruments. On these somewhat scanty authentic records of the instrument, several historians of music have based an illogical claim that the crwth, or rather chrotta or rotta, mentioned by Venantius Fortunatus as a British instrument, was the Welsh crwth as it was known in the 18th century, and was the earliest bowed instrument, and therefore the ancestor of the violin. The lines of Fortunatus, who was bishop of Poictiers during the second half of the 6th century, ran thus:—[6]

> "Romanusque lyra, plaudat tibi Barbarus harpa,
> Graecus Achilliaca, chrotta Britanna canat."

The bow is not mentioned by Fortunatus, and there is no ground whatever for believing that the Welsh crwth was played with a bow in the 6th century, or indeed for several centuries after. The stringing of the Welsh crwth with the two drone strings still twanged, the form of the body without incurvations, the flat bridge which rendered bowing, even in the most highly developed specimens of the 18th century, a difficult task, together with what is known of the early history of the chrotta and rotta derived from the lyre and cithara and like them twanged by fingers or plectrum, all make the claim untenable. Carl Engel was probably the first to expose the fallacy in his work on the violin.[7]

Drawn from a plate in Auguste de Bastard's *Peintures et ornements de la bible de Charles le Chauve.*
FIG. 2.—Early Crwth, 9th century.

British lexicographers all agree in deriving the words crwth, crowd and other forms of the name, from some word meaning a bulging protuberant bellying form, while in German the etymology of the word *Chrotta* is given as *Chrota* or *Chreta*, the O.H.G. for *Kröte* = toad, *Schildkröte* = tortoise. This word *Chrotta* was undoubtedly the German equivalent term for the lyre of Hermes, having as back a tortoise-shell, χέλυς in Greek and *testudo* in Latin. Chrotta was also spelt *hrotta*, and it is easy to see how this became rotta. A thoughtful and suggestive treatment of the whole subject will be found in Engel's work, to which reference has been made. Just as the lyre and cithara, which appeared to be similar to the casual observer, and are indeed still confused at the present day, were instruments differing essentially in construction[8]; so there were, during the early middle ages, while lyre and cithara were still in transition, two types of chrotta or rotta. (1) The rotta or improved cithara had a body either rectangular with the corners rounded, or guitar-shaped with incurvations, back and sound-board being nearly or quite flat, joined as in the cithara by ribs or sides. This rotta must be reckoned among the early ancestors of the violin before the advent of the bow; it was known both as rotta and cithara, and with a neck added it became the guitar-fiddle. (2) The tortoise or lyre chrotta consisted of a protuberant, very convex back cut out of a block of wood, to which was glued a flat sound-board, at first like the lyre, without intermediary ribs. This instrument became the crwth, and there was no further development. The first step in the transition of both lyre and cithara was the incorporation of arms and cross-bar into the body, the same outline being preserved; the second step was the addition of a finger-board against which the strings were stopped, thus increasing the compass while restricting the number of strings to three or four; the third step, observed only in the rotta-cithara, consisted in the addition of a neck,[9] as in the guitar. The crwth, crowd, crouth did not undergo this third transition even when the bow was used to set the strings in vibration.

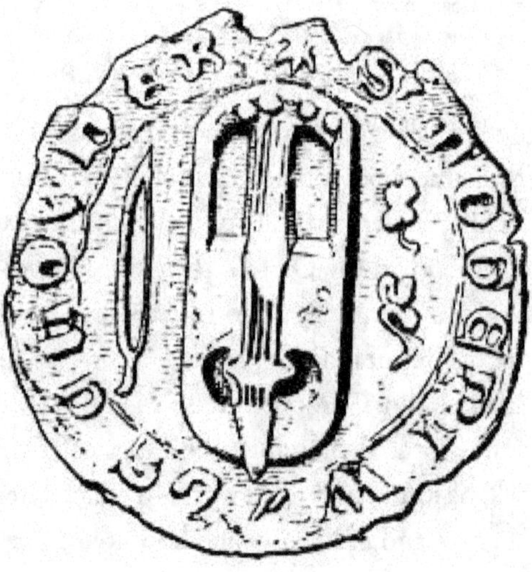

FIG. 3.—Crowd on a 14th-century Seal.

The earliest representation of the crwth yet discovered dates from the Carolingian period. In the miniatures of the Bible of Charles the Bald,[10] in the Bibliothèque Nationale, Paris, one of the musicians of King David is seen stopping strings on the finger-board with his left hand and plucking them with the right (fig. 2); this crwth has only three strings, and may be the crwth *trîthant* of Wales. A second example occurs in the Bible of St Paul,[11] another of the magnificent MSS. prepared for Charles the Bald, and preserved during the middle ages in the monastery of St Paul *extra muros* in Rome (now deposited in that of St Calixtus in Rome). Other representations are in the miniatures of the 11th, 12th and 13th centuries. To Edward Heron-Allen (*De fidiculis opuscula*, viii., 1895) is due the discovery of a representation of the Welsh crwth, showing the form still retained in the 18th cent. On the seal of Roger Wade (1316) is a crwth differing but little from the specimen in the Victoria and Albert Museum. The 14th-century instrument had four strings instead of six, and the foot of the bridge does not appear to pass through the sound-hole—a detail which may have escaped the notice of the artist who cut the seal. The original seal lies in the muniment room at Berkeley Castle in Gloucestershire attached to a defeasance of a bond between the *crowder* and his debtor Warren de l'Isle, and a cast (see fig. 3) is preserved at the British Museum. The British Museum also possesses two interesting MSS. which concern the crwth: one of these (Add. MS. 14939 ff. 4 and 27) contains an extract made by Lewis Morris in 1742 from an ancient Welsh MS. of "Instructions supposed to be wrote for the Crowd"; the other (Add. MS. 15036 ff. 65*b* and 66) consists of tracings from a 16th-century Welsh MS. copied in 1610 of a bagpipe, a harp and a *krythe*, together with the names of those who played the last at the Eisteddfod. The drawing is crude, and shows an instrument similar to Roger Wade's crowd, but having three strings instead of four.

The genealogical tree of the violin given below shows the relative positions of both kinds of rotta and chrotta.

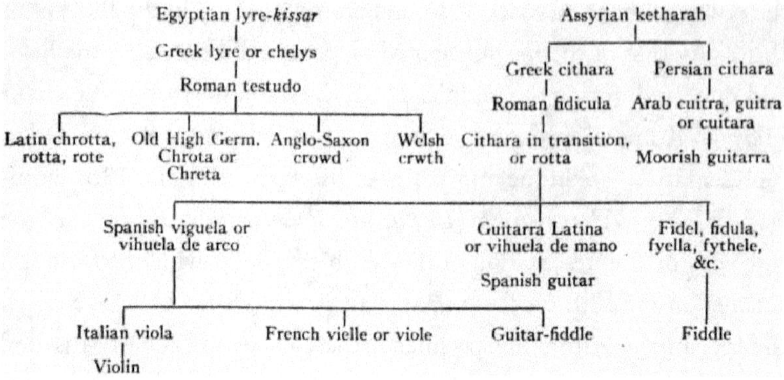

The Welsh crwth was therefore obviously not an exclusively Welsh instrument, but only a late 18th-century survival in Wales of an archaic instrument once generally popular in Europe but long obsolete. An interesting article on the subject in German by J. F. W. Wewertem will be found in *Monatshefte für Musik* (Berlin, 1881), Nos. 7-12, p. 151, &c.

(K. S.)

---

1 See *Early History of the Violin Family* (London, 1883), pp. 24-36.

2 See *A Tour round North Wales* (London, 1804), vol. ii. p. 332.

3 *History of Music* (London, 1766), vol. ii. bk. iii. ch. iii., description and illustration.

4 *Musical and Poetical Relicks of Welsh Bards* (London, 1794), illustration of crwth, also reproduced by Carl Engel; see note above.

5 *Archaeologia*, vol. iii. (London, 1775).

6 Venantius Fortunatus, Poëmata, lib. vii. cap. 8, p. 245; see Migne's *Patrologia Sacra*, vol. 88.

7 *Op. cit.* chapters "Crwth," "Chrotta," "Rotta."

8 See Kathleen Schlesinger, *Orchestral Instruments*, part ii., "The Precursors of the Violin Family" (London, 1909), pp. 14 to 23, with illustrations.

9 See also Kathleen Schlesinger, op. cit. ch. vii., "The Cithara in Transition," pp. 111-135 with illustrations.

10 See Auguste de Bastard, *Peintures et ornements des MSS. de France*, and *Peintures, ornements, &c., de la bible de Charles le Chauve*, in facsimile (Paris, 1883).

11 See J. O. Westwood, *Photographic Facsimile of the Bible of St Paul* (London, 1876).

CROWE, EYRE EVANS

(1799-1868), English journalist and historian, was born about the year 1799. He commenced his work as a writer for the London newspaper press in connexion with the *Morning Chronicle*, and he afterwards became a leading contributor to the *Examiner* and the *Daily News*. Of the latter journal he was principal editor for some time previous to his death. The department he specially cultivated was that of continental history and foreign politics. He published *Lives of Foreign Statesmen* (1830), *The Greek and the Turk* (1853), and *Reigns of Louis XVIII. and Charles X.* (1854). These were followed by his most important work, the *History of France* (5 vols., 1858-1868). It was founded upon original sources, in order to consult which the author resided for a considerable time in Paris. He died in London on the 25th of February 1868.

CROWE, SIR JOSEPH ARCHER

(1828-1896), English consular official and art critic, son of Eyre Crowe, was born in London on the 25th of October 1828. At an early age he showed considerable aptitude for painting and entered the studio of Delaroche in Paris, where his father was correspondent of the *Morning Chronicle*. During the Crimean War he was the correspondent of the *Illustrated London News*, and during the Austro-Italian War represented *The Times* in Vienna. He was British consul-general in Leipzig from 1860 to 1872, and in Düsseldorf from 1872 to 1880, when he was appointed commercial attaché in Berlin, being transferred in a like capacity to Paris in 1882. In 1883 he was secretary to the Danube Conference in London; in 1889 plenipotentiary at the Samoa Conference in Berlin; and in 1890 British envoy at the Telegraph Congress in Paris, in which year he was made K.C.M.G. During a sojourn in Italy, 1846-1847, he cemented a lifelong friendship with the Italian critic Giovanni Battista Cavalcaselle (1820-1897), and together they produced several historical works on art of classic importance, notably *Early Flemish Painters* (London, 1857); *A New History of Painting in Italy from the Second to the Sixteenth Century* (London, 1864-1871, 5 vols.). In 1895 Crowe published *Reminiscences of Thirty-Five Years of My Life*. He died at Schloss Gamburg in Bavaria on the 6th of September 1896.

Crowe and Cavalcaselle's great *History of Painting* was under revision by Crowe up to the time of his death, and then by S. A. Strong (d. 1904) and Langton Douglas, who in 1903 brought out vols. i. and ii. of Murray's new six-volume edition, the 3rd vol., edited by Langton Douglas, appearing in 1909. A reprint of the original edition, brought up to date by annotations by Edward Huttons, was published by Dent in 3 vols. in 1909.

CROW INDIANS

Or ABSAROKAS (the name for a species of hawk), a tribe of North American Indians of Siouan stock. They are now settled to the number of some 1800 on a reservation in southern Montana to the south of the Yellowstone river. Their original range included this reservation and extended eastward and southward, and no part of the country for hundreds of miles around was safe from their raids. They have ever been known as marauders and horse-stealers, and, though they have generally been cunning enough to avoid open war with the whites, they have robbed them whenever opportunity served. Physically they are tall and athletic, with very dark complexions.

CROWLAND

Or Croyland, a market-town in the S. Kesteven or Stamford parliamentary division of Lincolnshire, England; in a low fen district on the river Welland, 8 m. N.E. of Peterborough, and 4 m. from Postland station on the March-Spalding line of the Great Northern and Great Eastern railways, and Peakirk on the Great Northern. Pop. (1901) 2747. A monastery was founded here in 716 by King Æthelbald, in honour of St Guthlac of Mercia (d. 714), a young nobleman who became a hermit and lived here, and, it was said, had foretold Æthelbald's accession to the throne. The site of St Guthlac's cell, not far from the abbey, is known as Anchor (anchorite's) Church Hill. After the abbey had suffered from the Danish incursions in 870, and had been burnt in that year and in 1091, a fine Norman abbey was raised in 1113. Remains of this building appear in the ruined nave and tower arch, but the most splendid fragment is the west front, of Early English date, with Perpendicular restoration. The west tower is principally in this style. The north aisle is restored and used as the parish church. Among the abbots was Ingulphus (1085-1109), to whom was formerly attributed the *Historia Monasterii Croylandensis*. A curious triangular bridge remains, apparently of the 14th century, but referred originally to the middle of the 9th century, which spanned three streams now covered, and affords three footways which meet at an apex in the middle.

The town of Crowland grew up round the abbey. By a charter dated 716, Æthelbald granted the isle of Crowland, free from all secular services, to the abbey with a gift of money, and leave to build and enclose the town. The privileges thus obtained were confirmed by numerous royal charters extending over a period of nearly 800 years. Under Abbot Ægelric the fens were tilled, the monastery grew rich, and the town increased in size, enormous tracts of land being held by the abbey at the Domesday Survey.

The town was nearly destroyed by fire (1469-1476), but the abbey tenants were given money to rebuild it. By virtue of his office the abbot had a seat in parliament, but the town was never a parliamentary borough. Abbot Ralph Mershe in 1257 obtained a grant of a market every Wednesday, confirmed by Henry IV. in 1421, but it was afterwards moved to Thorney. The annual fair of St Bartholomew, which originally lasted twelve days, was first mentioned in Henry III.'s confirmatory charter of 1227. The dissolution of the monastery in 1539 was fatal to the progress of the town, which had prospered under the thrifty rule of the monks, and it rapidly sank into the position of an unimportant village. The abbey lands were granted by Edward VI. to Lord Clinton, from whose family they passed in 1671 to the Orby family. The inhabitants formerly carried on considerable trade in fish and wild fowl.

See R. Gough, *History and Antiquities of Croyland* (Bibl. Top. Brit. iii. No. 11) (London, 1783); W. G. Searle, *Ingulf and the Historia Croylandensis* (Camb. Antiq. Soc., No. 27); Dugdale, *Monasticon*, ii. 91 (London, 1846; Cambridge, 1894).

CROWLEY, ROBERT

(1518?-1588), English religious and social reformer, was born in Gloucestershire, and educated at Magdalen College, Oxford, of which he was successively demy and fellow. Coming to London, he set up a printing-office in Ely Rents, Holborn, where he printed many of his own writings. As a typographer, his most notable production was an edition of *Pierce Plowman* in 1550, and some of the earliest Welsh printed books came from his press. As an author, his first venture seems to have been his "Information and Petition against the Oppressors of the poor Commons of this realm," which internal evidence shows to have been addressed to the parliament of 1547. It contains a vigorous plea for a further religious reformation, but is more remarkable for its attack on the "more than Turkish tyranny" of the landlords and capitalists of that day. While repudiating communism, Crowley was a Christian Socialist, and warmly approved the efforts of Protector Somerset to stop enclosures. In his *Way to Wealth*, published in 1550, he laments the failure of the Protector's policy, and attributes it to the organized resistance of the richer classes. In the same year he published (in verse) *The Voice of the last Trumpet blown by the seventh Angel*; it is a rebuke in twelve "lessons" to twelve different classes of people; and a similar production was his *One-and-Thirty Epigrams* (1550). These, with *Pleasure and Pain* (1551), were edited for the Early English Text Society in 1872 (Extra Ser. xv.). The dozen or more other works which Crowley published are more distinctly theological: indeed, the failure of the temporal policy he advocated seems to have led Crowley to take orders, and he was ordained deacon by Ridley on the 29th of September 1551. During Mary's reign he was among the exiles at Frankfort. At Elizabeth's accession he became a popular preacher, was made archdeacon of Hereford in 1559, and prebendary of St Paul's in 1563, and was incumbent first of St Peter's the Poor in London, and then of St Giles' without Cripplegate. He refused to minister in the "conjuring garments of popery," and in 1566 was deprived and imprisoned for resisting the use of the surplice by his choir. He stated his case in "A brief Discourse against the Outward Apparel and Ministering Garments of the Popish Church," a tract "memorable," says Canon Dixon, "as the first distinct utterance of Nonconformity." He continued to preach occasionally, and in 1576 was presented to the living of St Lawrence Jewry. Nor had he abandoned his connexion with the book trade, and in 1578 he was admitted a freeman of the Stationers' Company. He died on the 18th of June 1588, and was buried in St Giles'. The most important of his works not hitherto mentioned is his continuation of Languet and Cooper's *Epitome of Chronicles* (1559).

See J. M. Cowper's *Pref. to the Select Works of Crowley* (1872); Strype's Works; Gough's *General Index to Parker Soc. Publ.*; Machyn's *Diary*; Macray's *Reg. Magdalen College*; Newcourt's *Rep. Eccles. Lond.*; Hennessy's *Nov. Rep. Eccl.* (1898); Le Neve's *Fasti Eccl. Angl.*; Pocock's Burnet; Pollard's *England under Somerset*; R. W. Dixon's *Church History*.

(A. F. P.)

CROWN

An English silver coin of the value of five shillings, hence often used to express the sum of five shillings. It was originally of gold and was first coined in the reign of Henry VIII. Edward VI. introduced silver crowns and half-crowns, and down to the reign of Charles II. crowns and half-crowns and sometimes double crowns were struck both in gold and silver. In the reign of Edward VI. also was introduced the practice of dating coins and marking them with their current value. The "Oxford crown" struck in the reign of Charles I. was designed by Rawlins (see NUMISMATICS: *Medieval*). Since the reign of Charles II. the crown has been struck in silver only. At one time during the 19th century it was proposed to abandon the issue of the crown, and from 1861 until 1887 none was struck, but since the second issue in 1887 it has been freely in circulation again.

CROWN AND CORONET

An official or symbolical ornament worn on or round the head. The crown (Lat. *corona*) at first had no regal significance. It was a garland, or wreath, of leaves or flowers, conferred on the winners in the athletic games. Afterwards it was often made of gold, and among the Romans was bestowed as a recognition of honourable service performed or distinction won, and on occasion it took such a form as to correspond with, or indicate the character of, the service rendered. The *corona obsidionalis* was formed of grass and flowers plucked on the spot and given to the general who conquered a city. The *corona civica*, made of oak leaves with acorns, was bestowed on the soldier who in battle saved the life of a Roman citizen. The mural crown (*corona muralis*) was the decoration of the soldier who was the first to scale the walls of a besieged city, and was usually a circlet of gold adorned with a series of turrets. The naval crown (*corona navalis*), decorated in like manner with a series of miniature prows of ships, was the reward of him who gained a notable victory at sea. These latter crowns form charges in English heraldry (see HERALDRY).

Many other forms of crown were used by the Romans, as the conqueror's triumphal crown of laurel, the myrtle crown, and the convivial, bridal, funeral and other crowns. Some of the emperors wore crowns on occasion, as Caligula and Domitian, at the games, and stellate or spike crowns are depicted on the heads of several of the emperors on their coins, but no idea of imperial sovereignty was indicated thereby. The Roman people, who had accepted imperial rule as a fact, were very jealous of the employment of its emblem on the part of their rulers. That emblem was the diadem, and although the diadem and crown are frequently confused with each other they were quite distinct, and it is well to bear this in mind. The diadem, which was of eastern origin, was a fillet or band of linen or silk, richly embroidered, and was worn tied round the forehead. Selden

(*Titles of Honour*, chap. viii. sect. 8) says that the diadem and crown "have been from ancient times confounded, yet the diadem strictly was a very different thing from what a crown now is or was, and it was no other then than only a fillet of silk, linen, or some such thing." It is desirable to remember the distinction, for, although diadem and crown are now used as synonymous terms, the two were originally quite distinct. The confusion between them has, perhaps, come about from the fact that the modern crown seems to be rather an evolution from the diadem than the lineal descendant of the older crowns. The linen or silk diadem was eventually exchanged for a flexible band of gold, which was worn in its place round the forehead. The further development of the crown from this was readily effected by the addition of an upper row of ornament. Thus the medieval and modern crowns may be considered as radiated diadems, and so the diadem and crown have become, as it were, merged in one another.

Among the historical crowns of Europe, the Iron Crown of Lombardy, now preserved at Monza, claims notice. It is a band of iron, enclosed in a circlet formed of six plates of gold, hinged one to the other, and richly jewelled and enamelled. It is regarded with great reverence, owing to a legend that the inner band of iron has been hammered out of one of the nails of the true cross. The crown is so small, the diameter being only 6 in., and the circlet only 2½ in. in width, that doubts have been felt as to whether it was originally intended to be worn on the head or was merely meant to be a votive crown. The legend as to the iron being that of one of the nails of the cross is rejected by Muratori and others, and cannot be traced far back. How it arose or how any credence came to be reposed in the legend, it is difficult to surmise. Another historical crown is that of Charlemagne, preserved at Vienna. It is composed of a series of four larger and four smaller plaques of gold, rounded at the tops and set together alternately. The larger plaques are richly ornamented with emeralds and sapphires, and the smaller plaques have each an enamelled figure of Our Lord, David, Solomon, and Hezekiah respectively. A jewelled cross rises from the large front plaque, and an arch bearing the name of the emperor Conrad springs across from the back of this cross to the back of the crown.

At Madrid there is preserved the crown of Svintilla, king of the Visigoths, 621-631. It is a circlet of thick gold set with pearls, sapphires and other stones. It has been given as a votive offering at some period to a church, as was often the custom. Attached to its upper rim are the chains whereby to suspend it, and from the lower rim hang letters of red-coloured glass or paste which read +SVINTILANVS REX OFFERET. Two other Visigothic crowns are also preserved with it in the Armeria Real.

FIG. I.—The Papal Tiara (without the *infulae*).

In 1858 a most remarkable discovery was made near Toledo, of eight gold crowns of the 7th century, fashioned lavishly with barbaric splendour. They are now in the Cluny Museum at Paris, having been purchased for £4000, the intrinsic value of the gold, without reckoning that of the jewels and precious stones, being not less than £600. The largest and most magnificent is the crown of Reccesvinto, king of the Visigoths from 653 to 675. It is composed of a circlet of pure gold set with pearls and precious stones in great profusion, which gives it a most sumptuous appearance.

It is 9 in. in diameter and more than ½ in. in thickness, the width of the circlet being 4 in. It has also been given as a votive offering to a church, and has the chains to hang it by attached to the upper rim, while from the lower rim depend pearls, sapphires and a series of richly jewelled letters 2 in. each in depth, which read +RECCESVINTHVS REX OFFERET. The second of these crowns in size is generally thought to be that of the queen of Reccesvinto. It has no legend, but merely a cross hanging from it. The six others are smaller, and are all most richly ornamented. They are believed to have been the crowns of Reccesvinto's children. From one of them hangs a legend which relates that they were an offering to a church, which has been identified with much probability as that of Sorbas, a small town in the province of Almeria. It has been surmised that in the disturbances which soon afterwards followed they were buried out of sight for safety, where they were eventually discovered absolutely unharmed centuries afterwards. For a detailed description of these most remarkable crowns the reader must be referred to a paper by the late Mr Albert Way (*Archaeological Journal*, xvi. 253). Mr Way, in the article alluded to, says of the custom of offering crowns to churches that frequent notices of the usage may be found in the lives of the Roman pontiffs by Anastasius. "They are usually described as having been placed over the altar, and in many instances mention is made of jewelled crosses of gold appended within such crowns as an accessory ornament.... The crowns suspended in churches suggested doubtless the sumptuous pensile luminaries, frequently designated from a very early period as *coronae*, in which the form of the royal circlet was preserved in much larger proportions, as exemplified by the remarkable *corona* still to be seen suspended in the cathedral at Aix-la-Chapelle over the crypt in which the body of Charlemagne was deposited."

Figs. 2-4 from Meyer's Konversations Lexikon.

FIG. 2.—Crown of the Holy Roman Empire.

FIG. 3.—Crown of the German Empire.

Of modern continental crowns the imperial crown of Austria (fig. 4) may be mentioned. It is composed of a circlet of gold, adorned with precious stones and pearls, heightened with fleurs-de-lys, and is raised above the circlet in the form of a cap which is opened in the middle, so that the lower part is crescent-shaped; across this opening from front to back rises an arched fillet, enriched with pearls and surmounted by an orb, on which is a cross of pearls.

The papal *tiara* (a Greek word, of Persian origin, for a form of ancient Persian popular head-dress, standing high erect, and worn encircled by a diadem by the kings), the triple crown worn by the popes, has taken various forms since the 9th century. It is important to remember that the tiaras in old Italian pictures are inventions of the artists and not copied from actual examples. In its present shape, dating substantially from the Renaissance, it is a peaked head-covering not unlike a closed mitre (*q.v.*), round which are placed one above the other three circlets or open crowns.[1] Two bands, or *infulae*, as they are called, hang from it as in the case of a mitre. The tiara is the crown of the pope as a temporal sovereign (see TIARA).

Pictorial representations in early manuscripts, and the rude effigies on their coins, are not very helpful in deciding as to the form of crown worn by the Anglo-Saxon and Danish kings of England before the Norman Conquest. In some cases it would appear as if the diadem studded with pearls had been worn, and in others something more of the character of a crown. We reach surer ground after the Conquest, for then the great seals, monumental effigies, and coins become more and more serviceable in determining the forms the crown took.

Fig. 5.

Fig. 6.

Fig. 7.

Fig. 8.

Fig. 9.

Fig. 10.

Royal Crowns. William I. to Henry IV.

Fig. 11.

Fig. 12.

Fig. 13.

Fig. 14.

Fig. 15.

Royal Crowns. Henry V. to Charles I.

The crown of William the Conqueror and his immediate successors seems to have been a plain circlet with four uprights, which terminated in trefoils (fig. 5), but Henry I. enriched the circlet with pearls or gems (fig. 6), and on his great seal the trefoils have something of the character of fleurs-de-lys. The effigy of Richard I. at Fontevrault shows a development of the crown; the trefoil heads are expanded, and are chased and jewelled. The crown of John is shown on his effigy at Worcester, though unfortunately it is rather badly mutilated. It shows, however, that the upper ornament was of fleurons set with jewels. Fig. 7 shows generally this development of the crown in a restored form. The crown on the effigy of Henry III. at Westminster had a beaded row below the circlet, which is narrow and plain, and from it rises a series of plain trefoils with slightly raised points between them. The tomb was opened in 1774, and on the king's head was found an imitation crown of tin or latten gilt, with trefoils rising from its upper edge. This, although only made of base metal for the king's burial, may nevertheless be taken as exhibiting the form of the royal crown at the time, and it may be usefully compared with that on the effigy of the king, which was made in Edward I.'s reign (fig. 8). Edward I. used a crown of very similar design. In the crown of Edward II. we have perhaps the most graceful and elegant of all the forms which the English medieval crown assumed (fig. 9), and it seems to have continued without any marked alteration during the reigns of Edward III. and Richard II. The crown on the head of the effigy of Henry IV. at Canterbury evidently represents one of great magnificence, both of design and ornament. What is perhaps lost of the grace of form of the crown of Edward II. is made up for by a profusion of adornment and ornamentation unsurpassed at any later period (fig. 10). The circlet is much wider and is richly chased and jewelled, and from it rise eight large leaves, the intervening spaces being filled with fleurs-de-lys of definite outline. It will be noted that this crown is, like its predecessors, what is known as an open crown, without any arches rising from the circlet, but in the accounts of the coronation of Henry IV. by Froissart and Waurin it is distinctly stated that the crown was arched in the form of a cross. This is the earliest mention of an arched crown, which is not represented on the great seal till that of Edward IV. in 1461. The crown, as shown on Henry IV.'s effigy, very probably represents the celebrated "Harry crown" which was afterwards broken up and employed as surety for the loan required by Henry V. when he was about to embark on his expedition to France. Fig. 11 shows the crown of Henry V. The crown of Henry VI. seems to have had three arches, and there is the same number shown on the crown of Henry VII., which ensigns the hawthorn bush badge of that king. The crown of Edward IV. (fig. 12) shows two arches, and a crown similarly arched appears on the great seal of Richard III. Crowns, both open and arched, are represented in sculpture and paintings until the end of the reign of Edward IV., and the royal arms are occasionally ensigned by an open crown as late as the reign of Henry VIII. The crown of Henry VII. on his effigy in Westminster Abbey shows a circlet surmounted by four crosses and four fleurs-de-lys alternately, and has two arches rising from it. A similar crown appears on the great seal of Henry VIII. The crown of Henry VII. (fig. 13), which ensigns the royal arms above the south door of King's College chapel, Cambridge, has the motto of the order of the Garter round the circlet. Fig. 14 shows the form of crown used by Edward VI., but a tendency (not shown in the illustration) began of flattening the arches of the crown, and on some of the coins of Elizabeth the arches are not merely flattened, but are depressed in the centre, much after the character of the arches of the crown on many of the silver coins of the 19th century prior to 1887. The crowns of James I. and Charles I. had four arches, springing from the alternate crosses and fleurs-de-lys of the circlet (fig. 15). The crown which strangely enough surmounts the shield with the arms of the Commonwealth on the coins of Oliver Cromwell (as distinguished from those of the Commonwealth itself, which have no crown) is a royal crown with alternate crosses and fleurs-de-lys round the circlet, and is surmounted by three arches, which, though somewhat flattened, are not bent. On them rests the orb and cross. The crown used by Charles II. (fig. 16) shows the arches depressed in the centre, a feature of the royal crown which seems to have been continued henceforward till 1887, when the pointed form of the arches was resumed, in consonance with an idea that such a form indicated an imperial rather than a regal crown, Queen Victoria having been proclaimed empress of India in 1877. In the foregoing account the changes of the form of the crowns of the kings have been briefly noticed. Those crowns were the personal crowns, worn by the different kings on various state occasions, but they were all crowned before the Commonwealth with the ancient crown of St Edward, and the queens consort with that of Queen Edith. There were, in fact, two sets of regalia, the one used for the coronations and kept at Westminster, and the other that used on other occasions by the kings and kept in the Tower. The crowns of this latter set were the personal crowns made to fit the different wearers, and are those which have been briefly described. The crown of St Edward, with which the sovereigns were crowned, had a narrow circlet from which rose alternately four crosses and four fleurs-de-lys, and from the crosses sprang two arches, which at their crossing supported an orb and cross. These arches must have been a later addition, and possibly were first added for the coronation of Henry IV. (*vide supra*). Queen Edith's crown had a plain circlet with, so far as can be determined, four crosses of pearls or gems on it, and a large cross patée rising from it in front, and arches of jewels or pearls terminating in a large pearl at the top. A valuation of these ancient crowns was made at the time of the Commonwealth prior to their destruction. From this valuation we learn that St Edward's crown was of gold filigree or "wirework" as it is called, and was set with stones, and was valued at £248. Queen Edith's crown was found to be only of silver-gilt, with counterfeit pearls, sapphires and other stones, and was only valued at £16. At the Restoration an endeavour was made to reproduce as well as possible the old crowns and regalia according to their ancient form, and a new crown of St Edward was made on the lines of the old one for the coronation of Charles II. The framework of this crown, bereft of its jewels, is in the possession of Lady Amherst of Hackney. The crowns of James II., William III. and Anne generally resembled it in form (fig. 16). The later crowns of the Georges and William IV. are represented in general form in fig. 17. Although the marginal note in the coronation order of Queen Victoria indicates "K. Edward's crown" as that with which the late queen was to be crowned, it was actually the state or imperial crown worn by the sovereign when leaving the church after the ceremony that was used. It had been altered for the coronation, and the arches were formed of oak leaves (fig. 18). Fig. 19 shows Queen Victoria's crown with raised arches and without the inner cap of estate, which since the reign of

Henry VII. has been degraded into forming a lining to the crowns of the sovereigns and the coronets of the peers. Fig. 20 shows the coronation crown of King Edward VII. The crown of Scotland, preserved with the Scottish regalia at Edinburgh, is believed to be composed of the original circlet worn by King Robert the Bruce. James V. made additions to it in 1535, and in general characteristics it much resembles an English crown of that date.

FIG. 16.

FIG. 17.

FIG. 19.

Recent Forms of the English Crown.

FIG. 18.

FIG. 20.

Coronation Crowns of Queen Victoria and King Edward VII.

The kings of arms in England, Scotland and Ireland wear crowns, the ornamentation of which round the upper rim of the circlet is composed of a row of acanthus or oak leaves. Round the circlet is the singularly inappropriate text from Psalm li., "*Miserere mei Deus secundum magnam misericordiam tuam.*" The form of these crowns seems to have been settled in the reign of Charles II. Before that period they varied at different times, according to representations given of them in grants of arms, &c.

This brings us to the crowns of lesser dignity, known for that reason as coronets, and worn by the five orders of peers.

FIG. 21.

FIG. 22.

FIG. 23.
Coronets of Dukes, Marquesses and Earls.

The use of crowns by dukes originated in 1362, when Edward III. created his sons Lionel and John dukes of Clarence and Lancaster respectively. This was done by investing them with a sword, a cap of maintenance or estate, and with a circlet of gold set with precious stones, which was imposed on the head. Previous to this dukes had been invested at their creation by the girding on of a sword only. In 1387 Richard II. created Richard de Vere marquess of Dublin, and invested him by girding on a sword, and by placing a golden circlet on his head. The golden circlet was confined to dukes and marquesses till 1444, when Henry VI. created Henry Beauchamp, earl of Warwick, premier earl, and the letters patent effecting this concede that the earl and his heirs shall wear a golden circlet on the head on feast days, even in the royal presence. As to the form of these circlets we have no clear knowledge. The dignity of a viscount was first created by Henry VI. in 1439, but nothing is said of any insignia pertaining to that dignity. It is believed that a circlet of gold with an upper rim of pearls was first conferred on a viscount by James I., who conceded it to Robert Cecil, Viscount Cranborne. However, in 1625-1626 it is definitely recorded that the viscounts carried their coronets in their hands in the coronation procession from Westminster Hall to the Abbey church. The use of a coronet by the barons dates from the coronation of Charles II., and by letters patent of the 7th of August 1661 their coronet is described as a circle of gold with six pearls on it.

Fig. 24. Fig. 25.

Coronets of Viscounts and Barons.

At the present day the coronet of a duke (fig. 21) is formed of a circlet of gold, from which rise eight strawberry leaves. The coronet of a marquess (fig. 22) differs from that of a duke in having only four strawberry leaves, the intervening spaces being occupied by four low points which are surmounted by pearls. The coronet of an earl (fig. 23) differs again by having eight tall rays on each of which is set a pearl, the intervening spaces being occupied by strawberry leaves one-fourth of the height of the rays. The coronet of a viscount (fig. 24) has sixteen small pearls fixed to the golden circlet, and the coronet of a baron (fig. 25) has six large pearls similarly arranged.

Authorities.—L. G. Wickham Legg, *English Coronation Records* (London, 1901); *The Ancestor*, Nos. i. and ii. (London, 1902); Stothard, *The Monumental Effigies of Great Britain* (London, 1817).

(T. M. F.)

1 A coloured drawing, done in the first half of the 18th century, of the magnificent tiara made by the celebrated goldsmith, Caradosso, for Julius II., is in the Print-Room, British Museum. It was re-fashioned by Pius VI., but went with other treasure as part of the indemnity to Napoleon. The splendid emerald at the summit, which was engraved with the arms of Gregory XIII., was restored by Napoleon and now adorns another papal tiara at Rome. In this drawing the three crowns (a feature introduced at the beginning of the 14th century) are represented by three bands of **X**-shaped ornament in enamelled gold.

CROWN DEBT

In English law, a debt due to the crown. By various statutes—the first dating from the reign of Henry VIII. (1541)—the crown has priority for its debts before all other creditors. At common law the crown always had a lien on the lands and goods of debtors by record, which could be enforced even when they had passed into the hands of other persons. The difficulty of ascertaining whether lands were subject to a crown lien or not was often very great, and a remedy was provided by the Judgments Act 1839, and the Crown Suits Act 1865. Now by the Land Charges Act 1900, no debt due to the crown operates as a charge on land until a writ of execution for the purpose of enforcing it has been registered under the Land Charges Registration and Searches Act 1888. By the Act of 1541 specialty debts were put practically on the same footing as debts by record. Simple contract debts due to the crown also become specialty debts, and the rights of the crown are enforced by a summary process called an *extent* (see WRIT).

CROWNE, JOHN

(d. *c.* 1703), British dramatist, was a native of Nova Scotia. His father "Colonel" William Crowne, accompanied the earl of Arundel on a diplomatic mission to Vienna in 1637, and wrote an account of his journey. He emigrated to Nova Scotia where he received a grant of land from Cromwell, but the French took possession of his property, and the home government did nothing to uphold his rights. When the son came to England his poverty compelled him to act as gentleman usher to an Independent lady of quality, and his enemies asserted that his father had been an Independent minister. He began his literary career with a romance, *Pandion and Amphigenia, or the History of the coy Lady of Thessalia* (1665). In 1671 he produced a romantic play, *Juliana, or the Princess of Poland*, which has, in spite of its title, no pretensions to rank as an historical drama. The earl of Rochester procured for him, apparently with the sole object of annoying Dryden by infringing on his rights as poet-laureate, a commission to supply a masque for performance at court. *Calisto* gained him the favour of Charles II., but Rochester proved a fickle patron, and his favour was completely alienated by the success of Crowne's heroic play in two parts, *The Destruction of Jerusalem by Titus Vespasian* (1677). This piece contained a thinly disguised satire on the Puritan party in the description of the Pharisees, and about 1683 he produced a distinctly political play, *The City Politiques*, satirizing the Whig party and containing characters which were readily recognized as portraits of Titus Oates and others. This made him many enemies, and he petitioned the king for a small place that would release him from the necessity of writing for the stage. The king exacted one more comedy, which should, he suggested, be based on the *No pued esser* of Moreto. This had already been unsuccessfully adapted, as Crowne discovered later, by Sir Thomas St Serfe, but in Crowne's hands it developed into *Sir Courtly*

Nice, It Cannot Be (1685), a comedy which kept its place as a stock piece for nearly a century. Unfortunately Charles II. died before the play was completed, and Crowne was disappointed of his reward. He continued to write plays, and it is stated that he was still living in 1703, but nothing is known of his later life.

Crowne was a fertile writer of plays with an historical setting, in which heroic love was, in the fashion of the French romances, made the leading motive. The prosaic level of his style saved him as a rule from the rant to be found in so many contemporary heroic plays, but these pieces are of no particular interest. He was much more successful in comedy of the kind that depicts "humours."

The History of Charles the Eighth of France, or The Invasion of Naples by the French (1672) was dedicated to Rochester. In *Timon*, generally supposed to have been written by the earl, a line from this piece—"whilst sporting waves smil'd on the rising sun"—was held up to ridicule. *The Ambitious Statesman, or The Loyal Favourite* (1679), one of the most extravagant of his heroic efforts, deals with the history of Bernard d'Armagnac, Constable of France, after the battle of Agincourt; *Thyestes, A Tragedy* (1681), spares none of the horrors of the Senecan tragedy, although an incongruous love story is interpolated; *Darius, King of Persia* (1688), *Regulus* (acted 1692, pr. 1694) and *Caligula* (1698) complete the list of his tragedies. *The Country Wit: A Comedy* (acted 1675, pr. 1693), derived in part from Molière's *Le Sicilien, ou l'amour peintre*, is remembered for the leading character, Sir Mannerly Shallow; *The English Frier; or The Town Sparks* (acted 1689, pr. 1690), perhaps suggested by Molière's *Tartuffe*, ridicules the court Catholics, and in Father Finical caricatures Father Petre; and *The Married Beau; or The Curious Impertinent* (1694), is based on the *Curioso Impertinente* in Don Quixote. He also produced a version of Racine's *Andromaque*, an adaptation from Shakespeare's Henry VI., and an unsuccessful comedy, *Justice Busy*.

See *The Dramatic Works of John Crowne* (4 vols., 1873), edited by James Maidment and W. H. Logan for the *Dramatists of the Restoration*.

CROWN LAND

In the United Kingdom, land belonging to the crown, the hereditary revenues of which were surrendered to parliament in the reign of George III.

In Anglo-Saxon times the property of the king consisted of (*a*) his private estate, (*b*) the demesne of the crown, comprising palaces, &c., and (*c*) rights over the folkland of the kingdom. By the time of the Norman Conquest the three became merged into the estate of the crown, that is, land annexed to the crown, held by the king as king. The king, also, ceased to hold as a private owner,[1] but he had full power of disposal by grant of the crown lands, which were increased from time to time by confiscation, escheat, forfeiture, &c. The history of the crown lands to the reign of William III. was one of continuous alienation to favourites. Their wholesale distribution by William III. necessitated the intervention of parliament, and in the reign of Queen Anne an act was passed limiting the right of alienation of crown lands to a period of not more than thirty-one years or three lives. The revenue from the crown lands was also made to constitute part of the civil list. At the beginning of his reign George III. surrendered his interest in the crown lands in return for a fixed "civil list" (*q.v.*). The control and management of the crown lands is now regulated by the Crown Lands Act 1829 and various amending acts. Under these acts their management is entrusted to the commissioners of Woods, Forests and Land Revenues, who have certain statutory powers as to leasing, selling, exchanging, &c.

In theory, also, state lands in the British colonies are supposed to be vested in the crown, and they are called crown lands; actually, however, the various colonial legislatures have full control over them and power of disposal. The term "crown-lands," in Austria, is applied to the various provinces into which that country is divided. (See AUSTRIA.)

1 The duchy of Lancaster, which was the private property of Henry IV. before he ascended the throne, was assured to him and his heirs by a special act of parliament. In the first year of Henry VII. it was united to the crown, but as a separate property.

CROWN POINT

A village of Essex county, New York, U.S.A., in a township of the same name, about 90 m. N.E. of Albany and about 10 m. N. of Ticonderoga, on the W. shore of Lake Champlain. Pop. of the township (1890) 3135; (1900) 2112; (1905) 1890; (1910) 1690; of the village, about 1000. The village is served by the Delaware & Hudson Railway and by the Champlain Canal. Among the manufactures are lumber and woodenware. Graphite has been found in the western part of the township, and spar is mined. In 1609 Champlain fought near here the engagement with the Iroquois Indians which marked the beginning of the long enmity between the Five (later Six) Nations and the French. Subsequently Dutch and English traders trafficked in the vicinity, the latter maintaining here for many years a regular trading-post. In 1731 the French built here Fort Frédéric, the first military post at Crown Point, and the place was subsequently for many years of considerable strategic importance, owing to its situation on Lake Champlain, which with Lake George furnished a comparatively easy route from Canada to New York. Twice during the French and Indian War, in 1755 and again in 1756, English and colonial expeditions were sent against it in vain; it remained in French hands until 1759, when, after Lord Jeffrey Amherst's occupation of Ticonderoga, the garrison joined that of the latter place and retreated to Canada. Crown Point was then occupied by Amherst, who during the winter of 1759-1760 began the construction, about a quarter of a mile from the old Fort Frédéric, of a large fort, which was garrisoned but was never completed; the ruins of this fort (not of Fort Frédéric) still remain. At the outbreak of the War of Independence, on the 11th of May 1775, the fort, whose garrison then consisted of only a dozen men, was captured by Colonel Seth Warner and a force of "Green Mountain Boys," sent from Ticonderoga by Ethan

Allen; and it remained in American hands save for a brief period in 1777, when it was occupied by a detachment of Burgoyne's invading army.

CROWTHER, SAMUEL ADJAI

(1809?-1891), African missionary-bishop, was born at Ochugu in the Yoruba country, West Africa, and was sold into slavery in 1821. Next year he was rescued, with many other captives, by H.M. ship "Myrmidon," and was landed at Sierra Leone. Educated there in a missionary school, he was baptized on the 11th of December 1825. In time he became a teacher at Furah Bay, and afterwards an energetic missionary on the Niger. He came to England in 1842, entered the Church Missionary College at Islington, and in June 1843 was ordained by Bishop Blomfield. Returning to Africa, he laboured with great success amongst his own people and afterwards at Abeokuta. Here he devoted himself to the preparation of school-books, and the translation of the Bible and Prayer-Book into Yoruba and other dialects. He also established a trade in cotton, and improved the native agriculture. In 1857 he commenced the third expedition up the Niger, and after labouring with varied success, returned to England and was consecrated, on St Peter's Day 1864, first bishop of the Niger territories. Before long a commencement was made of the missions to the delta of the Niger, and between 1866 and 1884 congregations of Christians were formed at Bonny, Brass and New Calabar, but the progress made was slow and subject to many impediments. In 1888 the tide of persecution turned, and several chiefs embraced Christianity, and on Crowther's return from another visit to England, the large iron church known as "St Stephen's cathedral" was opened. Crowther died of paralysis on the 31st of December 1891, having displayed as a missionary for many years untiring industry, great practical wisdom, and deep piety.

CROYDON

A municipal, county and parliamentary borough of Surrey, England, suburban to London, 10 m. S. of London Bridge. Pop. (1891) 102,695; (1901) 133,895. The borough embraces a great residential district. Several railway stations give it communication with all parts of the metropolis, the principal railways serving it being the London, Brighton & South Coast and the South-Eastern & Chatham. It stands near the sources of the river Wandle, under Banstead Downs, and is a place of great antiquity. The original site, farther west than the present town, is mentioned in Domesday Book. The derivation indicated is from the O. Fr. *croie dune*, chalk hill. The supposition that here was the Roman station of *Noviomagus* is rejected. The site is remarkable for the number of springs which issue from the soil. One of these, called the "Bourne," bursts forth a short way above the town at irregular intervals of one to ten years or more; and after running a torrent for two or three months, as quickly vanishes. Until its course was diverted it caused destructive floods. This phenomenon seems to arise from rains which, falling on the chalk hills, sink into the porous soil and reappear after a time from crevices at lower levels. The manor of Croydon was presented by William the Conqueror to Archbishop Lanfranc, who is believed to have founded the archiepiscopal palace there, which was the occasional residence of his successors till about 1750, and of which the chapel and hall remain. Addington Park, 3½ m. from Croydon, was purchased for the residence, in 1807, of the archbishop of Canterbury, but was sold in consequence of Archbishop Temple's decision to reside at the palace, Canterbury. The neighbouring church, which is Norman and Early English, contains several memorials of archbishops. Near the park a group of tumuli and a circular encampment are seen. Croydon is a suffragan bishopric in the diocese of Canterbury. The parish church of St John the Baptist appears to have been built in the 14th and 15th centuries, but to have contained remains of an older building. The church was restored or rebuilt in the 16th century, and again restored by Sir Gilbert Scott in 1857-1859. It was destroyed by fire, with the exception of the tower, on the 5th of January 1867, and was at once rebuilt by Scott on the old lines. In 1596 Archbishop Whitgift founded the hospital or almshouse which bears his name, and remains in its picturesque brick buildings surrounding two quadrangles. His grammar school was housed in new buildings in 1871, and is a flourishing day school. The principal public building of Croydon is that erected by the corporation for municipal business; it included court-rooms and the public library. At Addiscombe in the neighbourhood was formerly a mansion dating from 1702, and acquired by the East India Company in 1809 for a Military College, which on the abolition of the Company became the Royal Military College for the East Indian Army, and was closed in 1862. Croydon was formed into a municipal borough in 1883, a parliamentary borough, returning one member, in 1885, and a county borough in 1888. The corporation consists of a mayor, 12 aldermen and 36 councillors. Area, 9012 acres.

CROZAT, PIERRE

(1661-1740), French art collector, was born at Toulouse, one of a family who were prominent French financiers and collectors. He became treasurer to the king in Paris, and gradually acquired a magnificent collection of pictures and *objets d'art*. Between 1729 and 1742 a finely illustrated work was published in two volumes, known as the *Cabinet Crozat*, including the finest pictures in French collections. Most of his own treasures descended to his nephews, Louis François (d. 1750), Joseph Antoine (d. 1750), and Louis Antoine (d. 1770), and were augmented by them, being dispersed after their deaths; the collection of Louis Antoine Crozat went to St Petersburg.

CROZET ISLANDS

An uninhabited group in the Indian Ocean, in 46°-47° S. and 51° E. They are mountainous, with summits from 4000 to 5000 ft. high, and are disposed in two divisions—Penguin or Inaccessible, Hog, Possession and East Islands; and the Twelve Apostles. Like Kerguelen, and other clusters in these southern waters, they appear to be of igneous formation; but owing to the bleak climate and their inaccessible character they are seldom visited, and have never been explored since their discovery in 1772 by Marion-Dufresne, after one of whose officers they are named. Possession, the highest, has a snowy peak said to exceed 5000 ft. Hog Island takes its name from the animals which were here let loose by an English captain many years ago, but have since disappeared. Rabbits burrow in the heaps of scoria on the slopes of the mountains.

CROZIER, WILLIAM

(1855-), American artillerist and inventor, born at Carrollton, Carroll county, Ohio, on the 19th of February 1855, was the son of Robert Crozier (1827-1895), chief justice of Kansas in 1863-1866, and a United States senator from that state from December 1873 to February 1874. He graduated at West Point in 1876, was appointed a 2nd lieutenant in the 4th Artillery, and served on the Western frontier for three years against the Sioux and Bannock Indians. From 1879 to 1884 he was instructor in mathematics at West Point, and was superintendent of the Watertown (Massachusetts) Arsenal from 1884 to 1887. In 1888 he was sent by the war department to study recent developments in artillery in Europe, and upon his return he was placed in full charge of the construction of gun carriages for the army, and with General Adelbert R. Buffington (1837-), the chief of ordnance, he invented the Buffington-Crozier disappearing gun carriage (1896). He also invented a wire-wound gun, and perfected many appliances connected with heavy and field ordnance. In 1890 he attained the rank of captain. During the Spanish-American War he was inspector-general for the Atlantic and Gulf coast defences. In 1899 he was one of the American delegates to the Peace Conference at the Hague. He later served in the Philippine Islands on the staffs of Generals John C. Bates and Theodore Schwan, and in 1900 was chief of ordnance on the staff of General A. R. Chaffee during the Pekin Relief Expedition. In November 1901 he was appointed brigadier-general and succeeded General Buffington as chief of ordnance of the United States army. His *Notes on the Construction of Ordnance*, published by the war department, are used as text-books in the schools for officers, and he is also the author of other important publications on military subjects.

CROZIER

Or pastoral staff, one of the insignia of a bishop, and probably derived from the *lituus* of the Roman augurs. It is crook-headed, and borne by bishops and archbishops alike (see PASTORAL STAFF). The word "crozier" or "crosier" represents the O. Fr. *crocier*, Med. Lat. *crociarius*, the bearer of the episcopal crook (Med. Lat. *crocea*, *croccia*, &c., Fr. *croc*). The English representative of *crocea* was *crose*, later *crosse*, which, becoming confused with "cross" (*q.v.*), was replaced by "crozier-staff" or "crozier's staff," and then, at the beginning of the 16th century, by "crozier" (see J. T. Taylor, *Archaeologia*, Iii., "On the Use of the Terms Crosier, Pastoral Staff and Cross").

CRUCIAL

(from Lat. *crux*, a cross), that which has the form of a cross, as the "crucial ligaments" of the knee-joint, which cross each other, connecting the femur and the tibia. From Francis Bacon's expression *instantia crucis* (taken, as he says, from the finger-post or *crux* at cross-roads) for a phenomenon which decides between two causes which have each similar analogies in its favour, comes the use of "crucial" for that which decides between two alternatives, hence, generally, as a synonym for "critical." The word is also used, with a reference to the use of a "crucible," of something which tests and tries.

CRUCIFERAE

Or Crucifer family, a natural order of flowering plants, which derives its name from the cruciform arrangement of the four petals of the flower. It is an order of herbaceous plants, many of which, such as wallflower, stock, mustard, cabbage, radish and others, are well-known garden or field-plants. Many of the plants are annuals; among these are some of the commonest weeds of cultivation, shepherd's purse (*Capsella Bursa-pastoris*), charlock (*Brassica Sinapis*), and such common plants as hedge mustard (*Sisymbrium officinale*), Jack-by-the-hedge (*S. Alliaria* or *Alliaria officinalis*). Others are biennials producing a number of leaves on a very short stem in the first year, and in the second sending up a flowering shoot at the expense of the nourishment stored in the thick tap-root during the previous season. Under cultivation this root becomes much enlarged, as in turnip, swede and others. Wallflower (*Cheiranthus Cheiri*) (fig. 1) is a perennial. The leaves when borne on an elongated stem are arranged alternately and have no stipules. The flowers are arranged in racemes without bracts; during the life of the flower its stalk continues to grow so that the open flowers of an inflorescence stand on a level (that is, are corymbose). The flowers are regular, with four free sepals arranged in two pairs at right angles, four petals arranged crosswise in one series, and two sets of stamens, an outer with two members and an inner with four, in two pairs placed in the middle line of the flower and at right angles to the outer series. The four inner stamens are longer than the two outer; and the stamens are hence collectively described as tetradynamous. The pistil, which is above the rest of the members of the flower, consists of two carpels joined at their edges to form the ovary, which becomes two-celled by subsequent ingrowth of a septum from these united edges; a row of ovules springs from each edge. The fruit is a pod or siliqua splitting by two valves from below upwards and leaving the placentas with the seeds attached to the *replum* or framework of the septum. The seeds are filled with the large embryo, the two cotyledons of which are variously folded. In germination the cotyledons come above ground and form the first green leaves of the plant.

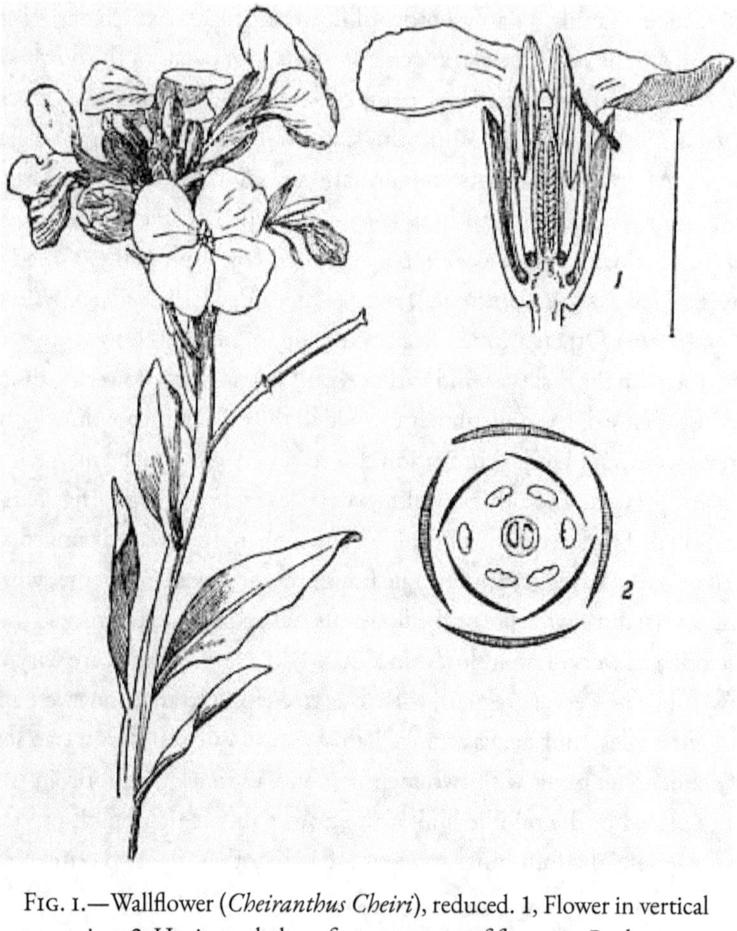

FIG. I.—Wallflower (*Cheiranthus Cheiri*), reduced. 1, Flower in vertical section. 2, Horizontal plan of arrangement of flower in *Barbarea*.

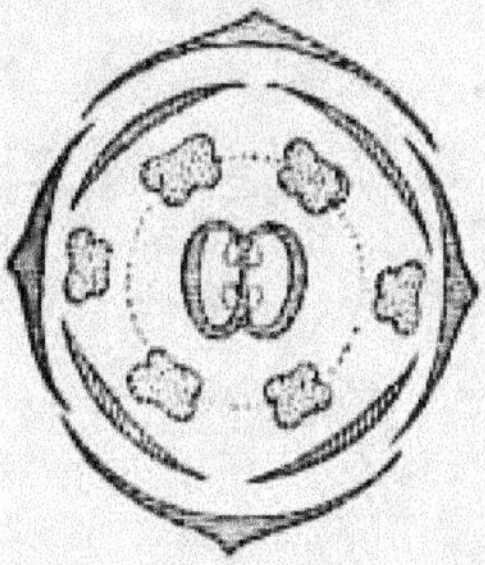

FIG. 2.—*Cruciferae*. Floral Diagram (*Brassica*).

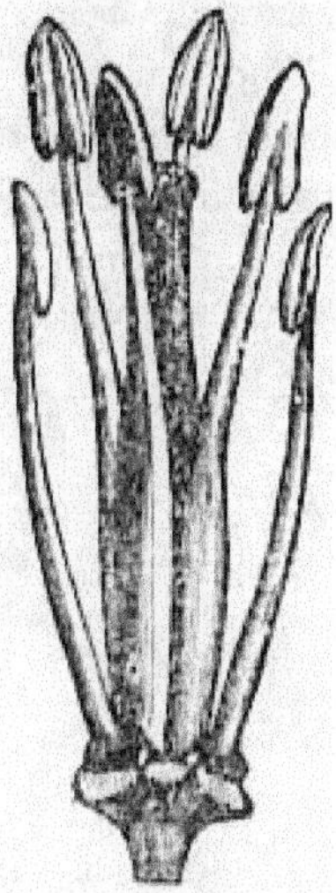

FIG. 3.—*Cardamine pratensis*. Flower with Perianth removed. (After Baillon.)

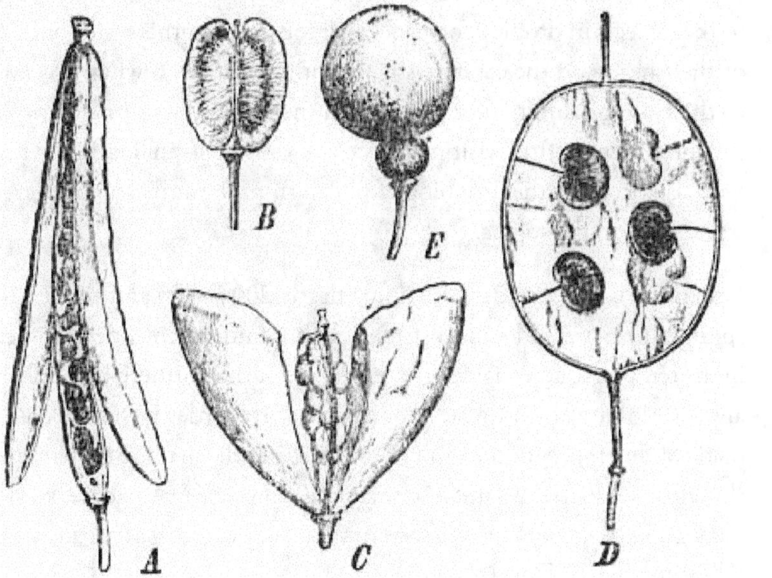

FIG. 4.—Cruciferous Fruits. (After Baillon.)

A, Cheiranthus Cheiri.
B, Lepidium sativum.
C, Capsella Bursa-pastoris. D, Lunaria biennis, showing the septum after the carpels have fallen away.
E, Crambe maritima.

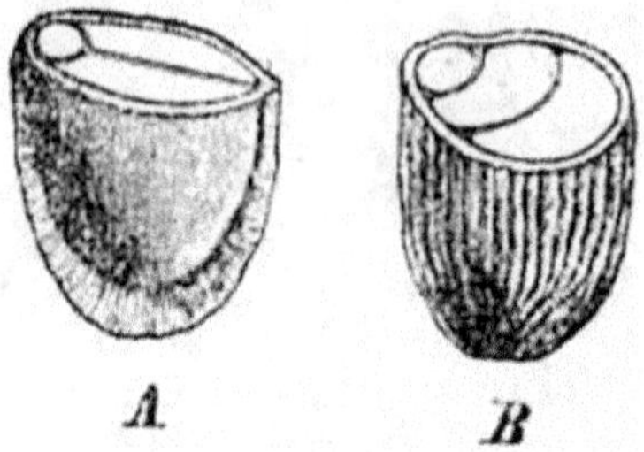

FIG. 5.—Seeds of Cruciferae cut across to show the radicle and cotyledons. (After Baillon.)

Pollination is effected by aid of insects. The petals are generally white or yellow, more rarely lilac or some other colour, and between the bases of the stamens are honey-glands. Some or all of the anthers become twisted so that insects in probing for honey will touch the anthers with one side of their head and the capitate stigma with the other. Owing, however, to the close proximity of stigma and anthers, very slight irregularity in the movements of the visiting insect will cause self-pollination, which may also occur by the dropping of pollen from the anthers of the larger stamens on to the stigma.

Cruciferae is a large order containing nearly 200 genera and about 1200 species. It has a world-wide distribution, but finds its chief development in the temperate and frigid zones, especially of the northern hemisphere, and as Alpine plants. In the subdivision of the order into tribes use is made of differences in the form of the fruit and the manner of folding of the embryo. When the fruit is several times longer than broad it is known as a siliqua, as in stock or wallflower; when about as long as broad, a silicula, as in shepherd's purse.

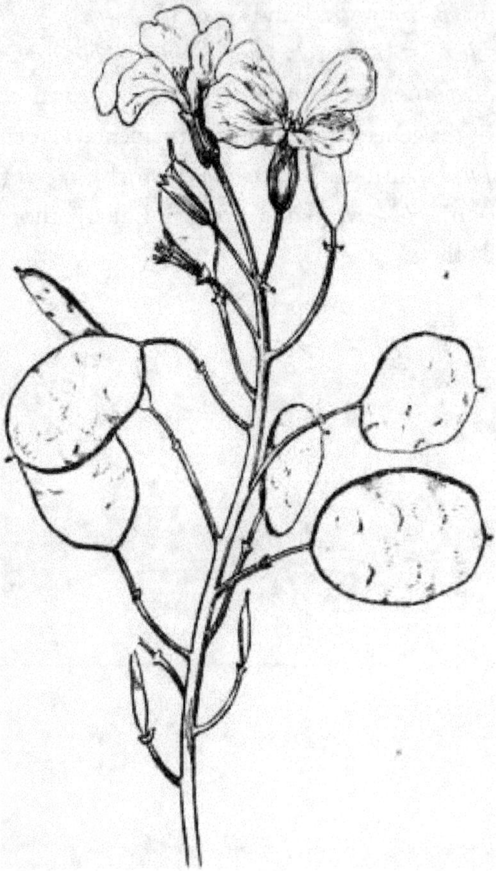

FIG. 6.—Honesty (*Lunaria biennis*), showing Flower and Fruit. Reduced.

The order is well represented in Britain—among others by *Nasturtium* (*N. officinale*, water-cress), *Arabis* (rock-cress), *Cardamine* (bitter-cress), *Sisymbrium* (hedge mustard, &c.; *S. Irio* is London rocket, so-called because it sprang up after the fire of 1666), *Brassica* (cabbage and mustard), *Diplotaxis* (rocket), *Cochlearia* (scurvy-grass), *Capsella* (shepherd's purse), *Lepidium* (cress), *Thlaspi* (penny-cress), *Cakile* (sea rocket), *Raphanus* (radish), and others. Of economic importance are species of *Brassica*, including mustard (*B. nigra*), white mustard, used when young in salads (*B. alba*), cabbage (*q.v.*) and its numerous forms derived from *B.*

oleracea, turnip (*B. campestris*), and swede (*B. Napus*), *Raphanus sativus* (radish), *Cochlearia Armoracia* (horse-radish), *Nasturtium officinale* (water-cress), *Lepidium sativum* (garden cress). *Isatis* affords a blue dye, woad. Many of the genera are known as ornamental garden plants; such are *Cheiranthus* (wallflower), *Matthiola* (stock), *Iberis* (candy-tuft), *Alyssum* (Alison), *Hesperis* (dame's violet), Lunaria (honesty) (fig. 6), *Aubrietia* and others.

CRUDEN, ALEXANDER

(1701-1770), author of the well-known concordance (*q.v.*) to the English Bible, was born at Aberdeen on the 31st of May 1701. He was educated at the grammar school, Aberdeen, and studied at Marischal College, intending to enter the ministry. He took the degree of master of arts, but soon after began to show signs of insanity owing to a disappointment in love. After a term of confinement he recovered and removed to London. In 1722 he had an engagement as private tutor to the son of a country squire living at Eton Hall, Southgate, and also held a similar post at Ware. Years afterwards, in an application for the title of bookseller to the queen, he stated that he had been for some years corrector for the press in Wild Court. This probably refers to this time. In 1729 he was employed by the 10th earl of Derby as a reader and secretary, but was discharged on the 7th of July for his ignorance of French pronunciation. He then lodged in a house in Soho frequented exclusively by Frenchmen, and took lessons in the language in the hope of getting back his post with the earl, but when he went to Knowsley in Lancashire, the earl would not see him. He returned to London and opened a bookseller's shop in the Royal Exchange. In April 1735 he obtained the title of bookseller to the queen by recommendation of the lord mayor and most of the Whig aldermen. The post was an unremunerative sinecure. In 1737 he finished his concordance, which, he says, was the work of several years. It was presented to the queen on the 3rd of November 1737, a fortnight before her death.

Although Cruden's biblical labours have made his name a household word among English-speaking people, he was disappointed in his hopes of immediate profit, and his mind again became unhinged. In spite of his earnest and self-denying piety, and his exceptional intellectual powers, he developed idiosyncrasies, and his life was marred by a harmless but ridiculous egotism, which so nearly bordered on insanity that his friends sometimes thought it necessary to have him confined. He paid unwelcome addresses to a widow, and was confined in a madhouse in Bethnal Green. On his release he published a pamphlet dedicated to Lord H. (probably Harrington, secretary of state) entitled *The London Citizen exceedingly injured, or a British Inquisition Displayed*. He also published an account of his trial, dedicated to the king. In December 1740 he writes to Sir H. Sloane saying he has been employed since July as Latin usher in a boarding-school at Enfield. He then found work as a proof-reader, and several editions of Greek and Latin classics are said to have owed their accuracy to his care. He superintended the printing of one of Matthew Henry's commentaries, and in 1750 printed a small *Compendium of the Holy Bible* (an abstract of the contents of each chapter), and also reprinted a larger edition of the *Concordance*.

About this time he adopted the title of "Alexander the Corrector," and assumed the office of correcting the morals of the nation, especially with regard to swearing and Sunday observance. For this office he believed himself divinely commissioned, but he petitioned parliament for a formal appointment in this capacity. In April 1755 he printed a letter to the speaker and other members of the House of Commons, and about the same time an "Address to the King and Parliament." He was in the habit of carrying a sponge, with which he effaced all inscriptions which he thought contrary to good morals. In September 1753, through being involved in a street brawl, he was confined in an asylum in Chelsea for seventeen days at the instance of his sister, Mrs Wild. He brought an unsuccessful action against his friends, and seriously proposed that they should go into confinement as an atonement. He published an account of this second restraint in "The Adventures of Alexander the Corrector." He made attempts to present to the king in person an account of his trial, and to obtain the honour of knighthood, one of his predicted honours. In 1754 he was nominated as parliamentary candidate for the city of London, but did not go to the poll. In 1755 he paid unwelcome addresses to the daughter of Sir Thomas Abney, of Newington (1640-1722), and then published his letters and the history of his repulse in the third part of his "Adventures." In June and July 1755 he visited Oxford and Cambridge. He was treated with the respect due to his learning by officials and residents in both universities, but experienced some boisterous fooling at the hands of the undergraduates. At Cambridge he was knighted with mock ceremonies. There he appointed "deputy correctors" to represent him in the university. He also visited Eton, Windsor, Tonbridge and Westminster schools, where he appointed four boys to be his deputies. (An *Admonition to Cambridge* is preserved among letters from J. Neville of Emmanuel to Dr Cox Macro, in the British Museum.) *The Corrector's Earnest Address to the Inhabitants of Great Britain*, published in 1756, was occasioned by the earthquake at Lisbon. In 1762 he saved an ignorant seaman, Richard Potter, from the gallows, and in 1763 published a pamphlet recording the history of the case. Against John Wilkes, whom he hated, he wrote a small pamphlet, and used to delete with his sponge the number 45 wherever he found it, this being the offensive number of the *North Briton*. In 1769 he lectured in Aberdeen as "Corrector," and distributed copies of the fourth commandment and various religious tracts. The wit that made his eccentricities palatable is illustrated by the story of how he gave to a conceited young minister whose appearance displeased him *A Mother's Catechism dedicated to the young and ignorant*. The *Scripture Dictionary*, compiled about this time, was printed in Aberdeen in two volumes shortly after his death. Alexander Chalmers, who in his boyhood heard Cruden lecture in Aberdeen and wrote his biography, says that a verbal index to Milton, which accompanied the edition of Thomas Newton, bishop of Bristol, in 1769, was Cruden's.

The second edition of the Bible *Concordance* was published in 1761, and presented to the king in person on the 21st of December. The third appeared in 1769. Both contain a pleasing portrait of the author. He is said to have gained £800 by these two editions. He returned to London from Aberdeen, and died suddenly while praying in his lodgings in Camden Passage, Islington, on the 1st of November 1770. He was buried in the ground of a Protestant dissenting congregation in Dead Man's Place, Southwark. He bequeathed a portion of his savings for a £5 bursary at Aberdeen, which preserves his name on the list of benefactors of the university.

(D. Mn.)

CRUDEN

A village and parish on the E. coast of Aberdeenshire, Scotland. Pop. of parish (1901) 3444. It is situated at the head of Cruden Bay, 29¾ m. N.N.E. of Aberdeen by the Great North of Scotland railway company's branch line from Ellon to Boddam. The golf-course of 18 holes is one of the best in Scotland, and there is a sandy beach, with good bathing. There is some good fishing at Port Erroll, also called Ward of Cruden. Prehistoric remains have been found in the parish, and near Ardendraught, not far from the shore, Malcolm II. is said to have defeated Canute in 1014. The Water of Cruden, which rises a few miles to the west, flows through the village into the North Sea. Slains Castle, a seat of the earl of Erroll, lies to the north of Cruden, but must not be confounded with the old castle of Slains, about 5 m. to the south-west, near the point where, according to tradition, the "St Catherine" of the Spanish Armada foundered in 1588. The Bullers of Buchan are within 2 m. walk of Cruden.

CRUELTY

(through the O. Fr. *crualté*, mod. *cruauté*, from the Lat. *crudelitas*), the intentional infliction of pain or suffering. It is only necessary to deal here with the legal relations involved. Statutory provision for the prevention of cruelty to those who are unable to protect themselves has been particularly marked in the 19th century. The increase of legislation for the protection of children, lunatics and animals is a proof of the growing humanitarianism of the age. There was at one time a tendency among jurists to question whether, for instance, the prevention of cruelty to animals was not a recognition of a certain quasi-right in animals, or whether it was merely that such exhibitions as bull- and bear-baiting, cock-fights, &c., were demoralizing to the public generally. The true fact seems to be that the first introduction of such legislation was undoubtedly due to the desire for the promotion of humanity, but that the principle, for the recognition of which the time was not yet ripe, had to be excused in the eyes of the public by the plea that cruelty had a demoralizing effect upon spectators (see A. V. Dicey, *Law and Opinion in England*, p. 188; T. E. Holland, *Jurisprudence*, 10th ed., p. 372).

Cruelty to Animals.—The English common law has never taken cognizance of the commission of acts of cruelty upon animals, and direct legislation upon the subject, dating from the 19th century, was due in a great measure to public agitation, supported by the Royal Society for the Prevention of Cruelty to Animals (founded in 1824). Various acts were passed in 1822 (known as Martin's Act), 1835 and 1837, and these were amended and consolidated by the Cruelty to Animals Acts 1849 and 1854, which, with the Wild Animals in Captivity Protection Act 1900, are the main acts upon the subject. There are also, in addition, many other acts that impose certain liabilities in respect of animals and indirectly prevent cruelty. The Cruelty to Animals Acts 1849 and 1854 render liable to prosecution and fine practically any act of cruelty to an animal; such acts as dubbing a cock, cropping the ears of a dog or dishorning cattle, are offences. The latter practice, however, is allowed both in Scotland and Ireland, the courts having held that the advantages to be obtained from dishorning outweigh the pain caused by the operation. The word "animal" is defined as meaning "any domestic animal" of whatever kind or species, and whether a quadruped or not. The act of 1849 also forbids bull- and bear-baiting, or fighting between any kinds of animals; requires the provision of food and water to animals impounded; lays down regulations as to the treatment of animals sent for slaughter, and imposes a penalty for improperly conveying animals. The Wild Animals in Captivity Protection Act 1900 extends to wild animals in captivity that protection which the acts of 1849 and 1854 conferred on domestic animals, making exception of any act done or any omission in the preparation of animals for the food of man or for sport. The word "animal" in the act includes bird, beast, fish or reptile. The Dogs Act 1865 rendered owners of dogs liable for injuries to cattle and sheep; the Dogs Act 1906 extended the owner's liability for injury done to any cattle by a dog, and further, where a dog is proved to have injured cattle or chased sheep it may be treated as a dangerous dog and must be kept under proper control or be destroyed. The Drugging of Animals Act 1876 imposes a penalty on giving poisonous drugs to any domestic animal unlawfully. The Cruelty to Animals Act 1876 was passed for the purpose of regulating the practice of vivisection (*q.v.*). The Ground Game Act 1880, prohibits night shooting, or the use of spring traps above ground or poison. The Injured Animals Act 1907 enables police constables to cause any animal when mortally or seriously injured to be slaughtered. The Diseases of Animals Act 1894 and orders under it are for the purpose of securing animals from unnecessary suffering, as well as from disease. Finally, the Wild Birds Protection Acts 1880 to 1904, with various game acts (see GAME LAWS), extend the protection of the law to wild birds. The acts establish a close time for wild birds and impose penalties for shooting or taking them within that time; prohibit the exposing or offering for sale within certain dates any wild bird recently killed or taken unless bought or received from some person residing out of the United Kingdom; the taking or destroying of wild birds' eggs, the setting of pole traps, and the taking of a wild bird by means of a hook or other similar instrument.

For the law relating to the prevention of cruelty to children see CHILDREN, LAW RELATING TO; for cruelty in the sense of such conduct as entitles a husband or wife to judicial separation see DIVORCE.

(T. A. I.)

CRUIKSHANK, GEORGE

(1792-1878), English artist, caricaturist and illustrator, was born in London on the 27th of September 1792. By natural disposition and collateral circumstances he may be accepted as the type of the born humoristic artist predestined for this special form of art. His grandfather had taken up the arts, and his father, Isaac Cruikshank, followed the painter's profession. Amidst these surroundings the children were born and brought up, their first playthings the materials of the arts their father practised. George followed the family traditions with amazing facility, easily surpassing his compeers as an etcher. When the father died, about 1811, George, still in his teens, was already a successful and popular artist. All his acquisitions were native gifts, and of home-growth; outside training, or the serious apprenticeship to art, were dispensed with, under the necessity of working for immediate profit. This lack of academic training the artist at times found cause to regret, and at some intervals he made exertions to cultivate the knowledge obtainable by studying from the antique and drawing from life at the schools. From boyhood he was accustomed to turn his artistic talents to ready account, disposing of designs and etchings to the printsellers, and helping his father in forwarding his plates. Before he was twenty his spirited style and talent had secured popular recognition; the contemporary of Gillray, Rowlandson, Alken, Heath, Dighton, and the established caricaturists of that generation, he developed great proficiency as an etcher. Gillray's matured and trained skill had some influence upon his executive powers, and when the older caricaturist passed away in 1815, George Cruikshank had already taken his place as a satirist. Prolific and dexterous beyond his competitors, for a generation he delineated Tories, Whigs and Radicals with fine impartiality. Satirical capital came to him from every public event,—wars abroad, the enemies of England (for he was always fervidly patriotic), the camp, the court, the senate, the Church;

low life, high life; the humours of the people, the follies of the great. In this wonderful gallery the student may grasp the popular side of most questions which for the time being engaged public attention. George Cruikshank's technical and manipulative skill as an etcher was such that Ruskin and the best judges have placed his productions in the foremost rank; in this respect his works have been compared favourably with the masterpieces of etching. He died at 263 Hampstead Road on the 1st of February 1878. His remains rest in St Paul's cathedral.

A vast number of Cruikshank's spirited cartoons were published as separate caricatures, all coloured by hand; others formed series, or were contributed to satirical magazines, the *Satirist*, *Town Talk*, *The Scourge* (1811-1816) and the like ephemeral publications. In conjunction with William Hone's scathing tracts, G. Cruikshank produced political satires to illustrate the series of facetiae and miscellanies, like *The Political House that Jack Built* (1819).

Of a more genially humoristic order are his well-known book illustrations, now so deservedly esteemed for their inimitable fun and frolic, among other qualities, such as the weird and terrible, in which he excelled. Early in this series came *The Humorist* (1819-1821) and *Life in Paris* (1822). The well-known series of *Life in London*, conjointly produced by the brothers I. R. and G. Cruikshank, has enjoyed a prolonged reputation, and is still sought after by collectors. Grimm's *Collection of German Popular Stories* (1824-1826), in two series, with 22 inimitable etchings, are in themselves sufficient to account for G. Cruikshank's reputation. To the first fourteen volumes (1837-1843) of *Bentley's Miscellany* Cruikshank contributed 126 of his best plates, etched on steel, including the famous illustrations to *Oliver Twist*, *Jack Sheppard*, *Guy Fawkes* and *The Ingoldsby Legends*. For W. Harrison Ainsworth, Cruikshank illustrated *Rookwood* (1836) and *The Tower of London* (1840); the first six volumes of *Ainsworth's Magazine* (1842-1844) were illustrated by him with several of his finest suites of etchings. For C. Lever's *Arthur O'Leary* he supplied 10 full-page etchings (1844), and 20 spirited graphic etchings for Maxwell's lurid *History of the*

Irish Rebellion in 1798 (1845). Of his own speculations, mention must be made of *George Cruikshank's Omnibus* (1841) and *George Cruikshank's Table Book* (1845), as well as his *Comic Almanack* (1835-1853). *The Life of Sir John Falstaff* contained 20 full-page etchings (1857-1858). These are a few leading items amongst the thousands of illustrations emanating from that fertile imagination. As an enthusiastic teetotal advocate, G. Cruikshank produced a long series of pictures and illustrations, pictorial pamphlets and tracts; the best known of these are *The Bottle*, 8 plates (1847), with its sequel, *The Drunkard's Children*, 8 plates (1848), with the ambitious work, *The Worship of Bacchus*, published by subscription after the artist's oil painting, now in the National Gallery, London, to which it was presented by his numerous admirers.

See *Cruikshank's Water-Colours*, with introduction by Joseph Grego (London, 1903).

(J. Go.*)

CRUNDEN, JOHN

(d. 1828), English architectural and mobiliary designer. Most of his early inspiration was drawn from Chippendale and his school, but he fell later under the influence of a bastard classicism. He produced a very large number of designs which were published in numerous volumes; among the most ambitious were ornamental centres for ceilings in which he introduced cupids with bows and arrows, Fame sounding her trumpet, and such like motives. Sport and natural history supplied him with many other themes, and one of his ceilings is a hunting scene representing a "kill." His principal works were *Designs for Ceilings*; *Convenient and Ornamental Architecture*; *The Carpenter's Companion for Chinese Railings, Gates*, &c. (1770); *The Joiner and Cabinet-maker's Darling*, or *Sixty Designs for Gothic, Chinese, Mosaic and Ornamental Frets* (1765); and *The Chimney Piece Maker's Daily Assistant* (1776). Much of his work was either absurd or valueless.

CRUSADES

The name given to the series of wars for delivering the Holy Land from the Mahommedans, so-called from the cross worn as a badge by the crusaders. By analogy the term "crusade" is also given to any campaign undertaken in the same spirit.

1. *The Meaning of the Crusades.*—The Crusades may be regarded partly as the *decumanus fluctus* in the surge of religious revival, which had begun in western Europe during the 10th, and had mounted high during the 11th century; partly as a chapter, and a most important chapter, in the history of the interaction of East and West. Contemporaries regarded them in the former of these two aspects, as "holy wars" and "pilgrims' progresses" towards Christ's Sepulchre; the reflective eye of history must perhaps regard them more exclusively from the latter point of view. Considered as holy wars the Crusades must be interpreted by the ideas of an age which was dominated by the spirit of otherworldliness, and accordingly ruled by the clerical power which represented the other world. They are a *novum salutis genus*—a new path to Heaven, to tread which counted "for full and complete satisfaction" *pro omni poenitentia* and gave "forgiveness of sins" (*peccaminum remissio*)[1]; they are, again, the "foreign policy" of the papacy, directing its faithful subjects to the great war of Christianity against the infidel. As such a *novum salutis genus*, the Crusades connect themselves with the history of the penitentiary system; as the foreign policy of the Church they belong to that clerical purification and direction of feudal society and its instincts, which appears in the institution of "God's Truce" and in chivalry itself. The penitentiary system, according to which the priest enforced a code of moral law in the confessional by the sanction of penance—penance which must be performed as a condition of admission to the sacrament of the Eucharist—had been from early times a great instrument in the civilization of the raw Germanic races. Penance might

consist in fasting; it might consist in flagellation; it might consist in pilgrimage. The penitentiary pilgrimage, which seems to have been practised as early as A.D. 700, was twice blessed; not only was it an act of atonement in itself, like fasting and flagellation; it also gained for the pilgrim the merit of having stood on holy ground. Under the influence of the Cluniac revival, which began in the 10th century, pilgrimages became increasingly frequent; and the goal of pilgrimage was often Jerusalem. Pilgrims who were travelling to Jerusalem joined themselves in companies for security, and marched under arms; the pilgrims of 1064, who were headed by the archbishop of Mainz, numbered some 7000 men. When the First Crusade finally came, what was it but a penitentiary pilgrimage under arms—with the one additional object of conquering the goal of pilgrimage? That the Pilgrims' Progress should thus have turned into a Holy War is a fact readily explicable, when we turn to consider the attempts made by the Church, during the 11th century, to purify, or at any rate to direct, the feudal instinct for private war (*Fehde*). Since the close of the 10th century diocesan councils in France had been busily acting as legislatures, and enacting "forms of peace" for the maintenance of God's Peace or Truce (*Pax Dei* or *Treuga Dei*). In each diocese there had arisen a judicature (*judices pacis*) to decide when the form had been broken; and an executive, or *communitas pacis*, had been formed to enforce the decisions of the judicature. But it was an easier thing to consecrate the fighting instinct than to curb it; and the institution of chivalry represents such a clerical consecration, for ideal ends and noble purposes, of the martial impulses which the Church had hitherto endeavoured to check. In the same way the Crusades themselves may be regarded as a stage in the clerical reformation of the fighting laymen. As chivalry directed the layman to defend what was right, so the preaching of the Crusades directed him to attack what was wrong—the possession by "infidels" of the Sepulchre of Christ. The Crusades are the offensive side of chivalry: chivalry is their parent—as it is also their child. The knight who joined the Crusades might thus still indulge the bellicose side of his genius—under the aegis and at the bidding of the Church; and in so doing he would also attain what the spiritual side of his nature ardently sought—a perfect salvation and remission of sins. He might butcher all day, till he waded ankle-deep in blood, and then at nightfall kneel, sobbing for very joy, at the altar of the Sepulchre—for was he not red from the winepress of the Lord? One can readily understand the popularity of the Crusades, when one reflects that they permitted men to get to the other world by fighting hard on earth, and allowed them to gain the fruits of asceticism by the ways of hedonism. Nor was the Church merely able, through the Crusades, to direct the martial instincts of a feudal society; it was also able to pursue the object of its own immediate policy, and to attempt the universal diffusion of Christianity, even at the edge of the sword, over the whole of the known world.

Thus was renewed, on a greater scale, that ancient feud of East and West, which has never died. For a thousand years, from the Hegira in 622 to the siege of Vienna in 1683, the peril of a Mahommedan conquest of Europe was almost continually present. From this point of view, the Crusades appear as a reaction of the West against the pressure of the East—a reaction which carried the West into the East, and founded a Latin and Christian kingdom on the shores of Asia. They protected Europe from the new revival of Mahommedanism under the Turks; they gave it a time of rest in which the Western civilization of the middle ages developed. But the relation of East and West during the Crusades was not merely hostile or negative. The Latin kingdom of Jerusalem was the meeting-place of two civilizations: on its soil the East learned from the West, and—perhaps still more—the West learned from the East. The culture developed in the West during the 13th century was not only permitted to develop by the protection of the Crusades, it grew upon materials which the Crusades enabled it to import from the East. Yet the debt of Europe to the Crusades in this last respect has perhaps been unduly emphasized. Sicily was still more the meeting-place of East and West than the kingdom of Jerusalem; and the Arabs of Spain gave more to the culture of Europe than the Arabs of Syria.

2. *Historical Causes of the Crusades.*—Within fifteen years of the Hegira Jerusalem fell before the arms of Omar (637), and it continued to remain

in the hands of Mahommedan rulers till the end of the First Crusade. For centuries, however, a lively intercourse was maintained between the Latin Church in Jerusalem, which the clemency of the Arab conquerors tolerated, and the Christians of the West. Charlemagne in particular was closely connected with Jerusalem: the patriarch sent him the keys of the city and a standard in 800; and in 807 Harun al-Rashid recognized this symbolical cession, and acknowledged Charlemagne as protector of Jerusalem and owner of the church of the Sepulchre. Charlemagne founded a hospital and a library in the Holy City; and later legend, when it made him the first of crusaders and the conqueror of the Holy Land, was not without some basis of fact. The connexion lasted during the 9th century; kings like Alfred of England and Louis of Germany sent contributions to Jerusalem, while the Church of Jerusalem acquired estates in the West. During the 10th century this intercourse still continued; but in the 11th century interruptions began to come. The fanaticism of the caliph Hakim destroyed the church of the Sepulchre and ended the Frankish protectorate (1010); and the patronage of the Holy Places, a source of strife between the Greek and the Latin Churches as late as the beginning of the Crimean War, passed to the Byzantine empire in 1021. This latter change in itself made pilgrimages from the West increasingly difficult: the Byzantines, especially after the schism of 1054, did not seek to smooth the way of the pilgrim, and Victor II. had to complain to the empress Theodora of the exactions practised by her officials. But still worse for the Latins was the capture of Jerusalem by the Seljukian Turks in 1071. Without being intolerant, the Turks were a rougher and ruder race than the Arabs of Egypt whom they displaced; while the wars between the Fatimites of Egypt and the Abbasids of Bagdad, whose cause was represented by the Seljuks, made Syria (one of the natural battle-grounds of history) into a troubled and unquiet region. The native Christians suffered; the pilgrims of the West found their way made still more difficult, and that at a time when greater numbers than ever were thronging to the East. Western Christians could not but feel hampered and checked in their natural movement towards the fountain-head of their religion, and it was natural that they should ultimately endeavour to clear the way. In much the same way, at a later date and in a lesser sphere, the closing of the trade-routes by the advance of the Ottoman Turks led traders to endeavour to find new channels, and issued in the rounding of the Cape of Good Hope and the discovery of America. Nor, indeed, must it be forgotten that the search for new and more direct connexions with the routes of Oriental trade is one of the motives underlying the Crusades themselves, and leading to what may be called the 13th-century discovery of Asia.

It was thus natural, for these reasons, that the conquest of the Holy Land should gradually become an object for the ambition of Western Christianity—an object which the papacy, eager to realize its dream of a universal Church subject to its sway, would naturally cherish and attempt to advance. Two causes combined to make this object still more natural and more definite. On the one hand, the reconquest of lost territories from the Mahommedans by Christian powers had been proceeding steadily for more than a hundred years before the First Crusade; on the other hand, the position of the Eastern empire after 1071 was a clear and definite summons to the Christian West, and proved, in the event, the immediate occasion of the holy war. As early as 970 the recovery of the territories lost to Mahommedanism in the East had been begun by emperors like Nicephoras Phocas and John Zimisces: they had pushed their conquests, if only for a time, as far as Antioch and Edessa, and the temporary occupation of Jerusalem is attributed to the East Roman arms. At the opposite end of the Mediterranean, in Spain, the Omayyad caliphate was verging to its fall: the long Spanish crusade against the Moor had begun; and in 1018 Roger de Toeni was already leading Normans into Catalonia to the aid of the native Spaniard. In the centre of the Mediterranean the fight between Christian and Mahommedan had been long, but was finally inclining in favour of the Christian. The Arabs had begun the conquest of Sicily from the East Roman empire in 827, and they had attacked the mainland of Italy as early as 840. The popes had put themselves at the head of Italian resistance: in 848 Leo IV. is already promising a sure and certain hope of salvation to those who die in defence of the cross; and by 916, with the capture of the Arab fortress on the Garigliano, Italy was safe. Then came the reconquest of the Mediterranean islands near Italy. The Pisans conquered Sardinia at the instigation of Benedict VIII. about 1016; and, in a thirty years' war which lasted from 1060 to 1090, the Normans, under a banner blessed by Pope Alexander II., wrested Sicily from the Arabs. The Norman conquest of Sicily may with justice be called a crusade before the Crusades; and it cannot but have given some impulse to that later attempt to wrest Syria from the Mahommedans, in which the virtual leader was Bohemund, a scion of the same house which had conquered Sicily. But while the Christians of the West were thus winning fresh ground from the Mahommedans, in the course of the 11th century, the East Roman empire had now to bear the brunt of a Mahommedan revival under the Seljuks—a revival which, while it crushed for a time the Greeks, only acted as a new incentive to the Latins to carry their arms to the East. The Seljukian Turks, first the mercenaries and then the masters of the caliph, had given new life to the decadent caliphate of Bagdad. Under the rule of their sultans, who assumed the rôle of mayors of the palace in Bagdad about the middle of the 11th century, they pushed westwards towards the caliphate of Egypt and the East Roman empire. While they wrested Jerusalem from the former (1071), in the same year they inflicted a crushing defeat on the Eastern emperor at Manzikert. The result of the defeat was the loss of almost the whole of Asia Minor; the dominions of the Turks extended to the sea of Marmora. An appeal for assistance, such as was often to be heard again in succeeding centuries, was sent by Michael VII. of Constantinople to Gregory VII. in 1073. Gregory listened to the appeal; he projected—not, indeed, as has often been said, a crusade,[2] but a great expedition, which should recover Asia Minor for the Eastern empire, in return for a union of the Eastern with the Western Church. In 1074 Gregory actually assembled a considerable army; but his disagreement with Robert Guiscard, followed by the outbreak of the war of investitures, hindered the realization of his plans, and the only result was a precedent and a suggestion for the events of 1095. The appeal of

Michael VII. was re-echoed by Alexius Comnenus himself. Brave and sage as he was, he could hardly cope at one and the same time with the hostility of the Normans on the west, of the Petchenegs (Patzinaks) on the north, and of the Seljuks on the east and south. Already in 1087 and 1088 he had appealed to Baldwin of Flanders, verbally and by letter,[3] for troops; and Baldwin had answered the appeal. The same appeal was made, more than once, to Urban II.; and the answer was the First Crusade. The First Crusade was not, indeed, what Alexius had asked or expected to receive. He had appealed for reinforcements to recover Asia Minor; he received hundreds of thousands of troops, independent of him, and intending to conquer Jerusalem for themselves, though they might incidentally recover Asia Minor for the Eastern empire on their way. Alexius may almost be compared to a magician, who has uttered a charm to summon a ministering spirit, and is surrounded on the instant by legions of demons. In truth the appeal of Alexius had set free forces in the West which were independent of, and even ultimately hostile to, the interests of the Eastern empire.

The primary force, which thus transmuted an appeal for reinforcements into a holy war for the conquest of Palestine, was the Church. The creative thought of the middle ages is clerical thought. It is the Church which creates the Carolingian empire, because the clergy thinks in terms of empire. It is the Church which creates the First Crusade, because the clergy believes in penitentiary pilgrimages, and the war against the Seljuks can be turned into a pilgrimage to the Sepulchre; because, again, it wishes to direct the fighting instinct of the laity, and the consecrating name of Jerusalem provides an unimpeachable channel; above all, because the papacy desires a perfect and universal Church, and a perfect and universal Church must rule in the Holy Land. But it would be a mistake to regard the Crusades (as it would be a mistake to regard the Carolingian empire) as a *pure* creation of the Church, or as *merely* due to the policy of a theocracy directing men to the holy war which is the only war possible for a theocracy. It would be almost truer, though only half the truth, to say that the clergy gave the name of Crusade to sanctify interests and ambitions which, while set on other ends than those of the Church, happened to coincide in their choice of means. There was, for instance, the ambition of the adventurer prince, the younger son, eager to carve a principality in the far East, of whom Bohemund is the type; there was the interest of Italian towns, anxious to acquire the products of the East more directly and cheaply, by erecting their own emporia in the eastern Mediterranean. The former was the driving force which made the First Crusade successful, where later Crusades, without its stimulus, for the most part failed; the latter was the one staunch ally which alone enabled Baldwin I. and Baldwin II. to create the kingdom of Jerusalem. So far as the Crusades led to permanent material results in the East, they did so in virtue of these two forces. Unregulated enthusiasm might of itself have achieved little or nothing; enthusiasm caught and guided by the astute Norman, and the no less astute Venetian or Genoese, could not but achieve tangible results. The principality or the emporium, it is true, would supply motives to the prince and the merchant only; and it may be urged that to the mass of the crusaders the religious motive was all in all. In this way we may return to the view that the First Crusade, at any rate, was *un fait ecclésiastique*. It is indeed true that to thousands the hope of acquiring spiritual merit must have been a great motive; it is also true, as the records of crusading sermons show, that there was a strong element of "revivalism" in the Crusades, and that thousands were hurried into taking the cross by a gust of that uncontrollable enthusiasm which is excited by revivalist meetings to-day. But it must also be admitted that there were motives of this world to attract the masses to the Crusades. Famine and pestilence at home drove men to emigrate hopefully to the golden East. In 1094 there was pestilence from Flanders to Bohemia: in 1095 there was famine in Lorraine. *Francigenis occidentalibus facile persuaderi poterat sua rura relinquere; nam Gallias per annos aliquot nunc seditio civilis, nunc fames, nunc mortalitas nimis afflixerat.*[4] No wonder that a stream of emigration set towards the East, such as would in modern times flow towards a newly discovered gold-field—a stream carrying in its turbid waters much refuse, tramps and bankrupts, camp-followers and hucksters, fugitive monks and escaped villeins, and marked by the same motley grouping, the same fever of life, the same alternations of affluence and beggary, which mark the rush for a gold-field to-day.

Such were the forces set in movement by Urban II., when, after holding a synod at Piacenza (March, 1095), and receiving there fresh appeals from Alexius, he moved to Clermont, in the S.E. of France, and there on the 26th of November delivered the great speech which was followed by the First Crusade. In this speech he appealed, indeed, for help for the Greeks, *auxilio ... saepe acclamato indigis* (Fulcher i. c. i.); but the gist of his speech was the need of Jerusalem. Let the truce of God be observed at home; and let the arms of Christians be directed to the winning of Jerusalem in an expedition which should count for full and complete penance. Like Gregory, Urban had thus sought for aid for the Eastern empire; unlike Gregory, who had only mentioned the Holy Sepulchre in a single letter, and then casually, he had struck the note of Jerusalem. The instant cries of *Deus vult* which answered the note showed that Urban had struck aright. Thousands at once took the cross; the first was Bishop Adhemar of Puy, whom Urban named his legate and made leader of the First Crusade (for the holy war, according to Urban's original conception, must needs be led by a clerk). Fixing the 15th of August 1096 as the time for the departure of the crusaders, and Constantinople as the general rendezvous, Urban returned from France to Italy. It is noticeable that it was on French soil that the seed had been sown.[5] Preached on French soil by a pope of French descent, the Crusades began—and they continued—as essentially a French (or perhaps better Norman-French) enterprise; and the kingdom which they established in the East was essentially a French kingdom, in its speech and its customs, its virtues and its vices. It was natural that France should be the home of the Crusades. She was already the home of the Cluniac movement, the centre from which radiated the truce of God, the chosen place of chivalry; she could supply a host of feudal nobles, somewhat loosely tied to their place in society, and ready to break loose for a great enterprise; she had suffered from battle and murder, pestilence and famine, from which any escape was welcome. To the Normans particularly the Crusades had an intimate appeal. They appealed to the old Norse instinct for wandering—an instinct which, as it had long before sent the Norseman eastward to find his El Dorado of Micklegarth, could now find a natural outlet in the expedition to Jerusalem: they appealed to the Norman religiosity, which had made them a people of pilgrims, the allies of the papacy, and, in England and Sicily, crusaders before the Crusades: finally, they appealed to that desire to gain fresh territory, upon which Malaterra remarks as characteristic of Norman princes.[6] No wonder, then, that the crusading armies were recruited in France, or that they were led by men of the stock of the d'Hautevilles. Meanwhile newly-conquered England had its own problems to solve; and Germany, torn by civil war, and not naturally quick to kindle, could only deride the "delirium" of the crusader.[7]

3. *Course of the First Crusade.*—The First Crusade falls naturally into two parts. One of these may be called the Crusade of the people: the other may be termed the Crusade of the princes. Of these the people's Crusade—prior in order of time, if only secondary in point of importance—may naturally be studied first. The sermon of Urban II. at Clermont became the staple for wandering preachers, among whom Peter the Hermit distinguished himself by his fiery zeal.[8] Riding on an ass from place to place through France and along the Rhine, he carried away by his eloquence thousands of the poor. Some three or four months before the term fixed by Urban II., in April and May 1096, five divisions of *pauperes* had already collected. Three of these, led by Fulcher of Orleans, Gottschalk and William the Carpenter respectively, failed to reach even Constantinople. The armies of Fulcher and Gottschalk were destroyed by the Hungarians in just revenge for their excesses (June); the third, after joining in a wild *Judenhetze* in the towns of the valley of the Rhine, during which some 10,000 Jews perished as the first-fruits of crusading zeal, was scattered to the winds in Hungary (August). Two other divisions, however, reached Constantinople in safety. The first of these, under Walter the Penniless,

passed through Hungary in May, and reached Constantinople, where it halted to wait for the Hermit, in the middle of July. The second, led by Peter himself, passed safely through Hungary, but suffered severely in Bulgaria, and only attained Constantinople with sadly diminished numbers at the end of July. These two divisions (which in spite of good treatment by Alexius began to commit excesses against the Greeks) united and crossed the Bosporus in August, Peter himself remaining in Constantinople. By the end of October they had perished utterly at the hands of the Seljuks; a heap of whitening bones also remained to testify to the later crusaders, when they passed in the spring of 1097, of the fate of the people's Crusade.

Meanwhile the knights had already begun to assemble in March 1096. In small bands, and by divers ways, they streamed gradually southward and eastward, in a steady flow, throughout 1096. But three large divisions, under three considerable leaders, were pre-eminent among the rest. Godfrey of Bouillon, with his brother Baldwin, led the crusaders of Lorraine along "the road of Charles the Great," through Hungary, to Constantinople, where he arrived on the 23rd of December. Raymund of Toulouse (the first prince to join the crusading movement) along with Bishop Adhemar, the papal commissary, led the Provençals down the coast of Illyria, and then due east to Constantinople, arriving towards the end of April 1097. Bohemund of Otranto, the destined leader of the Crusade, with his nephew Tancred, led a fine force of Normans by sea to Durazzo, and thence by land to Constantinople, which he reached about the same time as Raymund. To the same great rendezvous other leaders also gathered, some of higher rank than Godfrey or Raymund or Bohemund, but none destined to exercise an equal influence on the fate of the Crusade. Hugh of Vermandois, younger brother of Philip I. of France, had reached Constantinople in November 1096, in a species of honourable captivity, and had done Alexius homage; Robert of Normandy and Stephen of Blois, to whom Urban II. had given St Peter's banner at Lucca, only arrived—the last of the crusaders—in May 1097 (their original companion in arms,

Count Robert of Flanders, having left them to winter at Bari, and crossed to Constantinople before the end of 1096).

Thus was gathered at Constantinople, in the spring of 1097, a great host, which Fulcher computes at 600,000 men (I. c. iv.), Urban II. at 300,000, and which was probably some 150,000 strong.[9] Before we follow this host into Asia, we may pause to inquire into the various factors which would determine its course, or condition its activity. On the Western side, and among the crusaders themselves, there were two factors of importance, already mentioned above—the aims of the adventurer prince, and the interests of the Italian merchant; while on the Eastern side there are again two—the policy of the Greeks, and the condition of the Mahommedan East. We have already seen that among the princes who joined the First Crusade there were some who were rather *politiques* than *dévots*, and who aimed at the acquisition of temporal profit as well as of spiritual merit. Of these the type—and, it may almost be said, the inspirer of the rest—was Bohemund. From the first he had an Eastern principality in his mind's eye; and if we may judge from the follower of Bohemund who wrote the *Gesta Francorum*, there had already been some talk at Constantinople of Antioch as the seat of this principality. Bohemund's policy seems to have inspired Baldwin, the brother of Godfrey of Bouillon to emulation; on the one hand he strove to thwart the endeavours of Tancred, the nephew of Bohemund, to begin the foundation of the Eastern principality for his uncle by conquering Cilicia, and, on the other, he founded a principality for himself in Edessa. Raymond of Provence, the third and last of the great *politiques* of the First Crusade, was, like Baldwin, envious of Bohemund; and jealousy drove him first to attempt to wrest Antioch from Bohemund, and then to found a principality of Tripoli to the south of Antioch, which would check the growth of his power. The political motives of these three princes, and the interaction of their different policies, was thus a great factor in determining the course and the results of the First Crusade. The influence of the Italian towns did not make itself greatly felt till after the end of the First Crusade, when it made possible the foundation of a kingdom in Jerusalem, in addition to the three principalities established by Bohemund, Baldwin and Raymond; but during the course of the Crusade itself the Italian ships which hugged the shores of Syria were able to supply the crusaders with provisions and munition of war, and to render help in the sieges of Antioch and Jerusalem.[10] Sea-power had thus some influence in determining the victory of the crusaders.

In the East the conditions were, on the whole, favourable to the crusaders. The one difficulty—and it was serious—was the attitude adopted by Alexius. Confronted by crusaders where he had asked for auxiliaries, Alexius had two alternative policies presented to his choice. He might, in the first place, have frankly admitted that the crusaders were independent allies, and treating them as equals, he might have waged war in concert with them, and divided the conquests achieved in the war. A boundary line might have been drawn somewhere to the N.W. of Antioch; and the crusaders might have been left to acquire what they could to the south and east of that line. Unhappily, clinging to the conviction that all the lands which the crusaders would traverse were the "lost provinces" of his empire, he induced the crusaders to do him homage, so that, whatever they conquered, they would conquer in his name, and whatever they held, they would hold by his grant and as his vassals. Thus Hugh of Vermandois became the man of Alexius in November 1096; Godfrey of Bouillon was induced, not without difficulty, to do homage in January 1097; and in April and May the other leaders, including Bohemund and the obstinate Raymond himself, followed his example. The policy of Alexius was destined to produce evil results, both for the Eastern empire and for the crusading movement. The West had already its grievances against the East: the Greek emperors had taken advantage of their protectorate of the Holy Places to lay charges on the pilgrims, against which the Papacy had already been forced to remonstrate; nor were the Italian towns, with the exception of favoured Venice, disposed to be friendly to the great monopolist city of Constantinople. The old dissension of the Eastern and Western Churches had blazed out afresh in 1054; and the policy of Alexius only added new rancours to an old grudge, which culminated in the Latin conquest of Constantinople in 1204. On the other hand, the success of the crusading movement was imperilled, both now and afterwards, by the jealousy of the Comneni. Always hostile to the principality, which Bohemund established in spite of his oath, they helped by their hostility to cause the loss of Edessa in 1144, and thus to hasten the disintegration of the Latin kingdom of Jerusalem. Yet one must remember, in justice to Alexius, the gravity of the problem by which he was confronted; nor was the conduct of the crusaders themselves such that he could readily make them his brethren in arms.

The condition of Asia Minor and Syria in 1097 was almost altogether such as to favour the success of the crusaders. The Seljukian sultans had only achieved a military occupation of the country which they had conquered. There were Seljukian garrisons in towns like Nicaea and Antioch, ready to offer an obstinate resistance to the crusaders; and here and there in the country there were Seljukian armies, either cantoned or nomadic. But the inhabitants of the towns were often hostile to the garrisons, and over wide tracts of country there were no forces at all. Accordingly, when the crusaders had captured the town at Nicaea, and defeated the Seljukian field-army at Dorylaeum their way lay clear before them through Asia Minor. Not only so, but they could count, at the very least, on a benevolent neutrality from the native population; while from the Armenian principalities in the S.E. of Asia Minor, which survived unsubdued in the general deluge of Seljukian conquest, they could expect active assistance (the hope of which will explain the north-easterly line of march which they followed after leaving Heraclea). But the purely military character of the Seljukian occupation helped the crusaders in yet another way. Strong generals were needed in the separate divisions of the empire, and these, as has always been the case in Eastern empires, made themselves independent in their spheres of command, because there was no organization to keep them together under a single control. On the death of Malik Shah, the last of the great Seljukian emperors (1092), the empire dissolved. A new sultan, Barkiyāroq or Barkiarok, ruled in Bagdad (1094-1104); but in Asia Minor Kilij Arslan held sway as the independent sultan of Konia (Iconium), while the whole of Syria was also practically independent. Not only was Syria thus weakened by being detached from the body of the Seljukian empire; it was divided by dissensions within, and assailed by the Fatimite caliph of Egypt from without. In 1095 two brothers, Ridwan and Dekak, ruled in Aleppo and Damascus respectively; but they were at war with one another, and Yagi-sian, the ruler of Antioch, was a party to their dissensions. Ridwan and Yagi-sian were only stopped in an attack on Damascus by news of the approach of the crusaders, which led the latter to throw himself hastily into Antioch, in the autumn of 1097. Meanwhile the Fatimites were not slow to take advantage of these dissensions. A great religious difference divided the Fatimite caliph of Cairo, the head of the Shiite sect, from the Abbasid caliph of Bagdad, who was the head of the Sunnites. The difference may be compared to the dissension between the Greek and the Latin Churches; but it had perhaps more of the nature of a political difference. In any case, it hampered the Mahommedans as much as the jealousy between Alexius and the Latins hampered the progress of the Crusade. The crusading princes were well enough aware of the gulf which divided the caliph of Cairo from the Sunnite princes of Syria; and they sought by envoys to put themselves into connexion with him, hoping by his aid to gain Jerusalem (which was then ruled for the Turks by Sokman, the son of the amir Ortok).[11] But the caliph preferred to act for himself, and took advantage of the wars of the Syrian princes, and of the terror inspired by the advance of the crusaders to conquer Jerusalem (August 1098). But though the leaders of the First Crusade did not succeed in utilizing the dissensions of the Mahommedans as fully as they desired, it still remains true that these dissensions very largely explain their success. It was the disunion of the Syrian amirs, and the division between the Abbasids and the Fatimites, that made possible the conquest of the Holy City and the foundation of the kingdom of Jerusalem. When a power arose in Mosul, about 1130, which was able to unify Syria—when, again, in the hands of Saladin, unified Syria was in turn united to Egypt—the cause of Latin Christianity in the East was doomed.

We are now in a position to follow the history of the First Crusade. By the beginning of May 1097 the crusaders were crossing the Bosporus, and entering the dominions of Kilij Arslan. Their first operation was the siege of Nicaea, defended by a Seljuk garrison, but eventually captured, with the aid of Alexius, after a month's siege (June 18). Alexius took possession of the town; and though he rewarded the crusading princes richly, some discontent was excited by his action. After the capture of Nicaea, the field-army of Kilij Arslan had to be met. In a long and obstinate encounter, it was defeated at Dorylaeum (July 1); and the crusaders marched unmolested in a south-easterly direction to Heraclea. Here Tancred, followed by Baldwin, turned into Cilicia, and began to take possession of the Cilician towns, and especially of Tarsus—thus beginning, it would seem, the creation of the Norman principality of Antioch. The main army turned to the N.E., in the direction of Caesarea (in order to bring itself into touch with the Armenian princes of this district), and then marched southward again to Antioch. At Marash, half way between Caesarea and Antioch, Baldwin, who had meanwhile wrested Tarsus from Tancred, rejoined the ranks; but he soon left the main body again, and struck eastward towards Edessa, to found a principality there. At the end of October the crusaders came into position before Antioch, which was held by Yagi-sian, and began the siege of the city, which lasted from October 21, 1097, to June 3, 1098. The great figure in the siege was naturally Bohemund (who had also been the hero of Dorylaeum). He repelled attempts at relief made by Dekak (Dec. 31, 1097) and Ridwan (Feb. 9, 1098); he put the besiegers in touch with the Genoese ships lying in the harbour of St Simeon, the port of Antioch (March 1098)—a move which at once served to remedy the want of provisions from which the crusaders suffered, and secured materials for the building of castles, with which Bohemund sought—in the Norman fashion—to overawe the besieged city. But it was finally by the treachery of one of Yagi-sian's commanders, the amir Firuz, that Bohemund was able to effect its capture. The other leaders had, however, to promise him possession of the city, before he would bring his negotiations with Firuz to a conclusion; and the matter was so long protracted that an army of relief under Kerbogha of Mosul was only at a distance of three days' march, when the city was taken (June 3, 1098). The besiegers were no sooner in the city, than they were besieged in their turn by Kerbogha; and the twenty-five days which followed were the worst period of stress and strain which the crusaders had to encounter. Under the pressure of this strain "spiritualistic" phenomena began to appear. It was in the ranks of the Provençals, where the religiosity of Count Raymund seems to have extended to his followers, that these phenomena appeared; and they culminated in the discovery of the Holy Lance, which had pierced the side of the Saviour. The excitement communicated itself to the whole army; and the nervous strength which it gave enabled the crusaders to meet and defeat Kerbogha in the open (June 28), but not before many of their number, including even Count Stephen of Blois, had deserted and fled.

With the discovery of the Lance, which became as it were a Provençal asset, Count Raymund assumes a new importance. Mingled with the religiosity of his nature there was much obstinacy and self-seeking; and when Kerbogha was finally repelled, he began to dispute the possession of Antioch with Bohemund, pleading in excuse his oath to Alexius. The struggle lasted for some months, and helped to delay the further progress of the crusaders. Raymund, indeed, left Antioch in November, and moved S.E. to Marra; but his men still held two positions in Antioch, from which they were not dislodged by Bohemund till January 1099. Expelled from Antioch, the obstinate Raymund endeavoured to recompense himself in the south (where indeed he subsequently created the county of Tripoli); and from February to May 1099 he occupied himself with the siege of Arca, to the N.E. of Tripoli. It was during the siege of Arca that Peter Bartholomew, to whom the vision of the Holy Lance had first appeared, was subjected, with no definite result, to the ordeal of fire—the hard-headed Normans doubting the genuine character of any

Provençal vision, the more when, as in this case, it turned to the political advantage of the Provençals. The siege was long protracted; the mass of the pilgrims were anxious to proceed to Jerusalem, and, as the altered tone of the author of the *Gesta* sufficiently indicates, thoroughly weary of the obstinate political bickerings of Raymund and Bohemund. Here Godfrey of Bouillon finally came to the front, and placing himself at the head of the discontented pilgrims, he forced Raymund to accept the offers of the amir of Tripoli, to desist from the siege, and to march to Jerusalem (in the middle of May 1099). Bohemund remained in Antioch: the other leaders pressed forward, and following the coast route, arrived before Jerusalem in the beginning of June. After a little more than a month's siege, the city was finally captured (July 15). The slaughter was terrible; the blood of the conquered ran down the streets, until men splashed in blood as they rode. At nightfall, "sobbing for excess of joy," the crusaders came to the Sepulchre from their treading of the winepress, and put their blood-stained hands together in prayer. So, on that day of July, the First Crusade came to an end.

It remained to determine the future government of Jerusalem; and here the eternal problem of the relations of Church and State emerged. It might seem natural that the Holy City, conquered in a holy war by an army of which the pope had made a churchman, Bishop Adhemar, the leader, should be left to the government of the Church. But Adhemar had died in August 1098 (whence, in large part, the confusion and bickerings which followed in the end of 1098 and the beginning of 1099); nor were there any churchmen left of sufficient dignity or weight to secure the triumph of the ecclesiastical cause. In the meeting of the crusaders on the 22nd of July, some few voices were raised in support of the view that a "spiritual vicar" should first be chosen in the place of the late patriarch of Jerusalem (who had just died in Cyprus), before the election of any lay ruler was taken in hand. But the voices were not heard; and the princes proceeded at once to elect a lay ruler. Raymund of Provence refused to accept their nomination, nominally on the pious ground that he did not wish to reign where Christ had suffered on the cross; though one may suspect that the establishment of a principality in Tripoli—in which he had been interrupted by the pressure of the pilgrims—was still the first object of his ambition. The refusal of Raymund meant the choice of Godfrey of Bouillon, who had, as we have seen, become prominent since the siege of Arca; and Godfrey accordingly became—not king, but "advocate of the Holy Sepulchre," while a few days afterwards Arnulf, the chaplain of Robert of Normandy, and one of the sceptics in the matter of the Holy Lance, became "vicar" of the vacant patriarchate. Godfrey's first business was to repel an Egyptian attack, which he accomplished successfully at Ascalon, with the aid of the other crusaders (August 12). At the end of August the other crusaders returned,[12] and Godfrey was left with a small army of 2000 men, and the support of Tancred, now prince of Galilee, to rule in some four isolated districts—Jaffa, Jerusalem, Ramlah and Haifa. At the end of the year came Bohemund and Godfrey's brother Baldwin (now count of Edessa) on a pilgrimage to Jerusalem. The result of Bohemund's visit was new trouble for Godfrey. Bohemund procured the election of Dagobert, the archbishop of Pisa, to the vacant patriarchate, disliking Arnulf, and perhaps hoping to find in the new patriarch a political supporter. Bohemund and Godfrey together became Dagobert's vassals; and in the spring Godfrey even seems to have entered into an agreement with the patriarch to cede Jerusalem and Jaffa into his hands, in the event of acquiring other lands or towns, especially Cairo, or dying without direct heirs. When Godfrey died in July 1100 (after successful forays against the Mahommedans which took him as far as Damascus), it might seem as if a theocracy were after all to be established in Jerusalem, in spite of the events of 1099.

4. *The Latin Kingdom of Jerusalem under the First Three Kings,*[13] *1100-1143.*—The theocracy, however, was not destined to be established. Godfrey had died without direct heirs; but in far Edessa there was his brother Baldwin, ready to take his place. Dagobert had at first consented to the dying Godfrey's wish that Baldwin should be his successor; but

when Godfrey died he saw an opportunity too precious to be missed, and opposed Baldwin, counting on the support of Bohemund, to whom he sent an appeal for assistance.[14] But a party in Jerusalem, headed by the late "vicar" Arnulf, opposed itself to the hierarchical pretensions of Dagobert and the Norman influence by which they were backed; and this party, representing the Lotharingian laity, carried the day. Baldwin was summoned from Edessa; and when he arrived, towards the end of the year, he was crowned king by Dagobert himself. Thus was founded, on Christmas day 1100, the Latin kingdom of Jerusalem; and thus was the possibility of a theocracy finally annihilated. A feudal kingdom of Frankish seigneurs was to be planted on the soil of Palestine, instead of a *dominium temporale* of the patriarch like that of the pope in central Italy. Nor were any great difficulties with the Church to hamper the growth of this kingdom. For two years, indeed, a struggle raged between Baldwin I. and Dagobert: Baldwin accused the patriarch of treachery, and attempted to force him to contribute to the defence of the kingdom. But in 1102 the struggle ceased with the deposition of the patriarch and the victory of the king; and though it was renewed for a time by the patriarch Stephen in the reign of Baldwin II. (1128-1130), the new struggle was of short duration, and was soon ended by Stephen's death.

The establishment of a kingdom in Jerusalem in 1100 was a blow, not only to the Church but to the Normans of Antioch. At the end of 1099 any contemporary observer must have believed that the capital of Latin Christianity in the East was destined to be Antioch. Antioch lay in one of the most fertile regions of the East; Bohemund was almost, if not quite, the greatest genius of his generation; and when he visited Jerusalem at the end of 1099, he led an army of 25,000 men—and those men, at any rate in large part, Normans. What could Godfrey avail against such a force? Yet the principality of Godfrey was destined to higher things than that of Bohemund. Jerusalem, like Rome, had the shadow of a mighty name to lend prestige to its ruler; and as residence in Rome was one great reason of the strength of the medieval papacy, so was residence in Jerusalem a reason for the ultimate supremacy of the Lotharingian kings. Jerusalem attracted the flow of pilgrims from the West as Antioch never could; and though the great majority of the pilgrims were only birds of passage, there were always many who stayed in the East. There was thus a steady immigration into the kingdom, to strengthen its armies and recruit with new blood the vigour of its inhabitants. Still more important perhaps was the fact that the ports of the kingdom attracted the Italian towns; and it was therefore to the kingdom that they lent the strength of their armies and the skill of their siege-artillery—in return, it is true, for concessions of privileges so considerable as to weaken the resources of the kingdom they helped to create. While Jerusalem possessed these advantages, Antioch was not without its defects. It had to meet—or perhaps it would be more true to say, it brought upon itself—the hostility of strong Mahommedan powers in the vicinity. As early as 1100 Bohemund was captured in battle by Danishmend of Sivas; and it was his captivity, depriving the patriarch as it did of Norman assistance, which allowed the uncontested accession of Baldwin I. Again, in 1104, the Normans, while attempting to capture Harran, were badly defeated on the river Balikh, near Rakka; and this defeat may be said to have been fatal to the chance of a great Norman principality.[15] But the hostility of Alexius, aided and abetted by the jealousy of Raymund of Toulouse, was almost equally fatal. Alexius claimed Antioch; was it not the old possession of his empire, and had not Bohemund done him homage? Raymund was ready to defend the claims of Alexius; was not Bohemund a successful rival? Thus it came about that Alexius and Raymund became allies; and by the aid of Alexius Raymund established, from 1102 onwards, the principality which, with the capture of Tripoli in 1109, became the principality of Tripoli, and barred the advance of Antioch to the south. Meanwhile the armies of Alexius not only prevented any farther advance to the N.W., but conquered the Cilician towns (1104). No wonder that Bohemund flung himself in revenge on the Eastern empire in 1108—only, however, to meet with a humiliating defeat at Durazzo.

Thus it was that Baldwin waxed while Bohemund waned. The growth of Baldwin's kingdom, as it was suggested above, owed more to the interests of Italian traders than it did to crusading zeal. In 1100, indeed, it might appear that a new Crusade from the West, which the capture of Antioch in 1098 had begun, and the conquest of Jerusalem in 1099 had finally set in motion, was destined to achieve great things for the nascent kingdom. Thousands had joined this new Crusade, which should deal the final blow to Mahommedanism: among the rest came the first of the troubadours, William IX., Count of Poitiers, to gather copy for his muse, and even some, like Stephen of Blois and Hugh of Vermandois, who had joined the First Crusade, but had failed to reach Jerusalem. The new crusaders cherished high plans; they would free Bohemund and capture Bagdad. But each of the three sections of their army was routed in turn in Asia Minor by the princes of Sivas, Aleppo and Harran, in the middle of 1101; and only a few escaped to report the crushing disaster. Baldwin I. had thus no assistance to expect from the West, save that of the Italian towns. From an early date Italian ships had followed the crusaders. There were Genoese ships in St Simeon's harbour in the spring of 1098 and at Jaffa in 1099; in 1099 Dagobert, the archbishop of Pisa, led a fleet from his city to the Holy Land; and in 1100 there came to Jaffa a Venetian fleet of 200 sail, whose leaders promised Venetian assistance in return for freedom from tolls and a third of each town they helped to conquer. But it was the Genoese who helped Baldwin I. most. The Venetians already enjoyed, since 1080, a favoured position in Constantinople, and had the less reason to find a new emporium in the East; while Pisa connected itself, through Dagobert, with Antioch[16] rather than with Jerusalem, and was further, in 1111, invested by Alexius with privileges, which made an outlet in the Holy Land no longer necessary. But the Genoese, who had helped with provisions and siege-tackle in the capture of Antioch and of Jerusalem, had both a stronger claim on the crusaders, and a greater interest in acquiring an eastern emporium. An alliance was accordingly struck in 1101 (Fulcher II. c. vii.), by which the Genoese promised their assistance, in return for a third of all booty, a quarter in each town captured, and a grant of freedom from tolls. In this way Baldwin I. was able to take Arsuf and Caesarea in 1101 and Acre in 1104. But Genoese aid was given to others beside Baldwin (it enabled Raymund to capture Byblus in 1104, and his successor, William, to win Tripoli in 1109); while, on the other hand, Baldwin enjoyed other aid besides that of the Genoese. In 1110, for example, he was enabled to capture Sidon by the aid of Sigurd of Norway, the Jorsalafari, who came to the Holy Land with a fleet of 55 ships, starting in 1107, and in a three years' "wandering," after the old Norse fashion, fighting the Moors in Spain, and fraternizing with the Normans in Sicily. At a later date, in the reign of Baldwin II., Venice also gave her aid to the kings of Jerusalem. Irritated by the concessions made by Alexius to the Pisans in 1111, and furious at the revocation of her own privileges by John Comnenus in 1118, the republic naturally sought a new outlet in the Holy Land. A Venetian fleet of 120 sail came in 1123, and after aiding in the repulse of an attack, which the Egyptians had taken advantage of Baldwin II.'s captivity to deliver, they helped the regent Eustace to capture Tyre (1124), in return for considerable privileges—freedom from toils throughout the kingdom, a quarter in Jerusalem, baths and ovens in Acre, and in Tyre one-third of the city and its suburbs, with their own court of justice and their own church. After thus gaining a new footing in Tyre, the Venetians could afford to attack the islands of the Aegean as they returned, in revenge for the loss of their privileges in Constantinople; but the hostility between Venice and the Eastern empire was soon afterwards appeased, when John Comnenus restored the old privileges of the Venetians. The Venetians, however, maintained their position in Palestine; and their quarters remained, along with those of the Genoese, as privileged commercial franchises in an otherwise feudal state.

In this way the kingdom of Jerusalem expanded until it came to embrace a territory stretching along the coast from Beirut (captured in 1110[17]) to el-Arish on the confines of Egypt—a territory whose strength lay not in Judaea, like the ancient kingdom of David, but, somewhat paradoxically (though commercial motives explain the paradox), in Phoenicia and the land of the Philistines. With all its length, the territory had but little breadth: towards the north it was bounded by the amirate of Damascus; in the centre, it spread little, if at all, beyond the Jordan; and it was only in the south that it had any real extension. Here there were two considerable annexes. To the south of the Dead Sea stretched a tongue of land, reaching to Aila, at the head of the eastern arm of the Red Sea. This had been won by Baldwin I., by way of revenge for the attacks of the Egyptians on his kingdom; and here, as early as 1116, he had built the fort of Monreal, half way between Aila and the Dead Sea. To the east of the Dead Sea, again, lay a second strip of territory, in which the great fortress was Krak (Kerak) of the Desert, planted somewhere about 1140 by the royal butler, Paganus, in the reign of Fulk of Jerusalem. These extensions in the south and east had also, it is easy to see, a commercial motive. They gave the kingdom a connexion of its own with the Red Sea and its shipping; and they enabled the Franks to control the routes of the caravans, especially the route from Damascus to Egypt and the Red Sea. Thus, it would appear, the whole of the expansion of the Latin kingdom (which may be said to have attained its height in 1131, at the death of Baldwin II.) may be shown to have been dictated, at any rate in large part, by economic motives; and thus, too, it would seem that two of the most powerful motives which sway the mind of man—the religious motive and the desire for gain—conspired to elevate the kingdom of Jerusalem (at once the country of Christ, and a natural centre of trade) to a position of supremacy in Latin Syria. During this process of growth the kingdom stood in relation to two sects of powers—the three Frankish principalities in northern Syria, and the Mahommedan powers both of the Euphrates and the Nile—whose action affected its growth and character.

Of the three Frankish principalities, Edessa, founded in 1098 by Baldwin I. himself, was a natural fief of Jerusalem. Baldwin de Burgh, the future Baldwin II., ruled in Edessa as the vassal of Baldwin I. from 1100 to 1118; and thereafter the county was held in succession by the two Joscelins of Tell-bashir until the conquest of Edessa by Zengi in 1144. Lying to the east of the Euphrates, at once in close contact with the Armenians, and in near proximity to the great route of trade which came up the Euphrates to Rakka, and thence diverged to Antioch and Damascus, the county of Edessa had an eventful if brief life. The county of Tripoli, the second of these principalities, had also come under the aegis of Jerusalem at an early date. Founded by Raymund of Toulouse, between 1102 and 1105, with the favour of Alexius and the alliance of the Genoese, it did not acquire its capital of Tripoli till 1109. Even before the conquest of Tripoli, there had been dissensions between William, the nephew and successor of Raymund, and Bertrand, Raymund's eldest son, which it had needed the interference of Baldwin I. to compose; and it was only by the aid of the king that the town of Tripoli had been taken. At an early date therefore the county of Tripoli had already come under the influence of the kingdom. Meanwhile the principality of Antioch, ruled by Tancred, after the departure of Bohemund (1104-1112), and then by Roger his kinsman (1112-1119), was, during the reign of Baldwin I., busily engaged in disputes both with its Christian neighbours at Edessa and Tripoli, and with the Mahommedan princes of Mardin and Mosul. On the death of Roger in 1119, the principality came under the regency of Baldwin II. of Jerusalem, until 1126, when Bohemund II. came of age. Bohemund had married a daughter of Baldwin; and on his death in 1130 Baldwin II. had once more become the guardian of Antioch. From his reign therefore Antioch may be regarded as a dependency of Jerusalem; and thus the end of Baldwin's reign (1131) may be said to mark the time when the Latin kingdom of Jerusalem stands complete, with its own boundaries stretching from Beirut in the north to el-Arish and Aila in the south, and with the three Frankish powers of the north admitting its suzerainty.

The Latin power thus established and organized in the East had to face in the north a number of Mahommedan amirs, in the south the caliph of Egypt. The disunion between the Mahommedans of northern Syria and the Fatimites of Egypt, and the political disintegration of the former, were both favourable to the success of the Franks; but they had nevertheless to maintain their ground vigorously both in the north and the south against almost incessant attacks. The hostility of the decadent caliphate of Cairo was the less dangerous; and though Baldwin I. had at the beginning of his reign to meet annual attacks from Egypt, by the end he had pushed his power to the Red Sea, and in the very year of his death (1118) he had penetrated along the north coast of Egypt as far as Farama (Pelusium). The plan of conquering Egypt had indeed presented itself to the Franks from the first, as it continued to attract them to the end; and it is significant that Godfrey himself, in 1100, promised Jerusalem to the patriarch, "as soon as he should have conquered some other great city, and especially Cairo." But the real menace to the Latin kingdom lay in northern Syria; and here a power was eventually destined to rise, which outstripped the kings of Jerusalem in the race for Cairo, and then—with the northern and southern boundaries of Jerusalem in its control—was able to crush the kingdom as it were between the two arms of a vice. Until 1127, however, the Mahommedans of northern Syria were disunited among themselves. The beginning of the 12th century was the age of the atabegs (regents or stadtholders). The atabegs formed a number of dynasties, which displaced the descendants of the Seljukian amirs in their various principalities. These dynasties were founded by emancipated mamelukes, who had held high office at court and in camp under powerful amirs, and who, on their death, first became stadtholders for their descendants, and then usurped the throne of their masters. There was an atabeg dynasty in Damascus founded by Tughtigin (1103-1128): there was another to the N.E., that of the Ortokids, represented by Sokman, who established himself at Kaifa in Diarbekr about 1101, and by his brother Ilghazi, who received Mardin from Sokman about 1108, and added to it Aleppo in 1117.[18] But the greatest of the atabegs were those of Mosul on the Tigris—Maudud, who died in 1113; Aksunkur, his successor; and finally, greatest of all, Zengi himself, who ruled in Mosul from 1127 onwards.

Before the accession of Zengi, there had been constant fighting, which had led, however, to no definite result, between the various Mahommedan princes and the Franks of northern Syria. The constant pressure of Tancred of Antioch and Baldwin de Burgh of Edessa led to a series of retaliations between 1110 and 1115; Edessa was attacked in 1110, 1111, 1112 and 1114; and in 1113 Maudud of Mosul had even penetrated as far as the vicinity of Acre and Jerusalem.[19] But the dissensions of the Mahommedans made their attacks unavailing; in 1115, for instance, we find Antioch actually aided by Ilghazi and Tughtigin against Aksunkur of Mosul. Again, in the reign of Baldwin II., there was steady fighting in the north; Roger of Antioch was defeated by Ilghazi at Balat in 1119, and Baldwin II. himself was captured by Balak, the successor of Ilghazi, in 1123, but on the whole the Franks held the upper hand. Baldwin conquered part of the territory of Aleppo (in 1121 and the following years), and extorted a tribute from Damascus (1126). But when Zengi established himself in Mosul in 1127, the tide gradually began to turn. He created for himself a great and united principality, comprising not only Mosul, but also Aleppo,[20] Harran, Nisibin and other districts; and in 1130, Alice, the widow of Bohemund II., sought his alliance in order to maintain herself in power at Antioch. In the beginning of the reign of Fulk of Jerusalem (1131-1143) the progress of Zengi was steady. He conquered in 1135 several fortresses in the east of the principality of Antioch, and in this year and the next pressed the count of Tripoli hard; while in 1137 he defeated Fulk at Barin, and forced the king to capitulate and surrender the town. If Fulk had been left alone to wage the struggle against Zengi, and if Zengi had enjoyed a clear field against the Franks, the fall of the kingdom of Jerusalem might have come far sooner than it did.[21] But there were two powers which aided Fulk, and impeded the progress of Zengi—the amirate of Damascus and the emperors of Constantinople. The position of Damascus is a position of crucial importance from 1130 to 1154. Lying between Mosul and Jerusalem, and important both strategically and from its position on the great route of commerce from the Euphrates to Egypt, Damascus became the arbiter of Syrian politics. During the greater part of the period between 1130 and 1154 the policy of Damascus was guided by the vizier Muin-eddin Anar, who ruled on behalf of the descendants of the atabeg

Tughtigin. He saw the importance of finding an ally against the ambition of Zengi, who had already attacked Damascus in 1130. The natural ally was Jerusalem. As early as 1133 the alliance of the two powers had been concluded; and in 1140 the alliance was solemnly renewed between Fulk and the vizier. Henceforth this alliance was a dominant factor in politics. One of the great mistakes made by the Franks was the breach of the alliance in 1147—a breach which was widened by the attack directed against Damascus during the Second Crusade; and the conquest of Damascus by Nureddin in 1154 was ultimately fatal to the Latin kingdom, removing as it did the one possible ally of the Franks, and opening the way to Egypt for the atabegs of Mosul.

The alliance of the emperors of Constantinople was of far more dubious value to the kings of Jerusalem. We have already seen that it was the theory of the Eastern emperors—a theory which logically followed from the homage of the crusaders to Alexius—that the conquests of the crusaders belonged to their empire, and were held by the crusading princes as fiefs. We have seen that the action of Bohemund at Antioch was the negation of this theory, and that Alexius in consequence helped Raymund to establish himself in Tripoli as a thorn in the side of Bohemund, and sent an army and a fleet which wrested from the Normans the towns of Cilicia (1104). The defeat of Bohemund at Durazzo in 1108 had resulted in a treaty, which made Antioch a fief of Alexius; but Tancred (who in 1107 had recovered Cilicia from the Greeks) refused to fulfil the terms of the treaty, and Alexius (who attempted—but in vain—to induce Baldwin I. to join an alliance against Tancred in 1112) was forced to leave Antioch independent. Thus, although Alexius had been able, in the wake of the crusading armies, to recover a large belt of land round the whole coast of Asia Minor,—the interior remaining subject to the sultans of Konia (Iconium) and the princes of Sivas,—he left the territories to the east of the western boundary of Cilicia in the hands of the Latins when he died in 1118. Not for 20 years after his death did the Eastern empire make any attempt to gain Cilicia or wrest homage from Antioch. But in 1137

John Comnenus appeared, instigated by the opportunity of dissensions in Antioch, and received its long-denied homage, as well as that of Tripoli; while in the following year he entered into hostilities with Zengi, without, however, achieving any considerable result. In 1142 he returned again, anxious to create a principality in Cilicia and Antioch for his younger son Manuel. The people of Antioch refused to submit; a projected visit to Jerusalem, during which John was to unite with Fulk in a great alliance against the Moslem, fell through; and in the spring of 1143 the emperor died in Cilicia, with nothing accomplished. On the whole, the interference of the Comneni, if it checked Zengi for the moment in 1138, may be said to have ultimately weakened and distracted the Franks, and to have helped to cause the loss of Edessa (1144), which marks the turning-point in the history of the kingdom of Jerusalem.

5. *Organization of the Kingdom.*—Before we turn to describe the Second Crusade, which the loss of Edessa provoked, and to trace the fall of the kingdom, which the Second Crusade rather hastened than hindered, we may pause at this point to consider the organization of the Frankish colonies in Syria. The first question which arises is that of the relation of the kingdom of Jerusalem to the three counties or principalities of Antioch, Tripoli and Edessa, which acknowledged their dependence upon it. The degree of this dependence was always a matter of dispute. The rights of the king of Jerusalem chiefly appear when there is a vacancy or a minority in one of the principalities, or when there is dissension either inside one of the principalities or between two of the princes. On the death of one of the princes without heirs of full age, the kings of Jerusalem were entitled to act as regents, as Baldwin II. did twice at Antioch, in 1119 and 1130; but the kings regarded this right of regency as a burden rather than a privilege, and it is indeed characteristic of the relation of the king to the three princes, that it imposes upon him duties without any corresponding rights. It is his duty to act as regent; it is his duty to compose the dissensions in the principality of Antioch, and to repress the violences of the prince towards his patriarch (1154); it is his duty to reconcile

Antioch with Edessa, when the two fall to fighting. The princes on their side acted independently: if they joined the king with their armies, it was as equals doing a favour; and they sometimes refused to join until they were coerced. They made their own treaties with the Mahommedans, or attacked them in spite of the king's treaties; they dated their documents by the year of their own reign, and they had each their separate laws or assizes. There was, in a word, co-ordination rather than subordination; nor did the kings ever attempt to embark on a policy of centralization.

The relation of the king to his own barons within his immediate kingdom of Jerusalem is not unlike the relation of the king to the three princes. In Norman England the king insisted on his rights; in Frankish Jerusalem the barons insisted on his duties. The circumstances of the foundation of the kingdom explain its characteristics. As the crusaders advanced to Jerusalem, says Raymund of Agiles (c. xxxiii.), it was their rule that the first-comer had the right to each castle or town, provided that he hoisted his standard and planted a garrison there. The feudal nobility was thus the first to establish itself, and the king only came after its institution—the reverse of Norman England, where the king first conquered the country, and then plotted it out among his nobles. The predominance of the nobility in this way became as characteristic of feudalism in the Latin kingdom of Jerusalem as the supremacy of the crown was of contemporary feudalism in England; and that predominance expressed itself in the position and powers of the high court, in which the ultimate sovereignty resided. The kingdom of Jerusalem consisted of a society of peers, in which the king might be *primus*, but in which he was none the less subject to a punctilious law, regulating his position equally with that of every member of the society. In such a society the election of the head by the members may seem natural; and in the case of Godfrey and the first two Baldwins this was the case. But the conception of the equality of the king and his peers in the long run led to hereditary monarchy; for if the king held his kingdom as a fief, like other nobles, the laws of descent which applied to a fief applied to the kingdom, and those laws demanded heredity. Yet the high court, which decided all problems of descent, would naturally intervene if a problem of descent arose, as it frequently did, in the kingdom; and thus the barons had the right of deciding between different claimants, and also of formally "approving" each new successor to the throne. The conception of the kingdom as a fief not only subjected it to the jurisdiction of the high court; it involved the more disastrous result that the kingdom, like other fiefs, might be carried by an heiress to her husband; and the proximate causes of the collapse of the kingdom in 1187 depend on this fact and the dissensions which it occasioned.

Thus conceived as the holder of a great fief, the king had only the rights of *suzerain* over the four great baronies and the twelve minor fiefs of his kingdom. He had not those rights of sovereign which the Norman kings of England inherited from their Anglo-Saxon predecessors, or the Capetian kings of France from the Carolings; nor was he able therefore to come into direct touch with each of his subjects, which William I., in virtue of his sovereign rights, was able to attain by the Salisbury oath of 1086. Amalric I. indeed, by his *assise sur la ligèce*, attempted to reach the vassals of his vassals; he admitted arrière-vassaux to the *haute cour*, and encouraged them to carry their cases to it in the first instance. But this is the only attempt at that policy of *immédiatisation* which in contemporary England was carried to far greater lengths; and even this attempt was unsuccessful. No alliance was actually formed between the king and the mesne nobility against the immediate baronage. The body of the tenants-in-chief continued to limit the power of the crown: their consent was necessary to legislation, and grants of fiefs could not be made without their permission. Nor was the crown only limited in this way. The *duties* of the king towards his tenants are prominent in the *assises*. The king's oath to his men binds him to respect and maintain their rights, which are as prominent as are his duties; and if the men feel that the royal oath has not been kept, they may lawfully refuse military service (*gager le roi*), and may even rise in authorized and legal rebellion. The system of military service and the organization of justice corresponded to the part which the monarchy was thus constrained to play. The vassal was bound to pay military service, not, as in western Europe, for a limited period of forty days, but for the whole year—the Holy Land being, as it were, in a perpetual state of siege. On the other hand, the vassal was not bound to render service, unless he were *paid* for his service; and it was only famine, or Saracen devastation, which freed the king from the obligation of paying his men. The king was also bound to insure the horses of his men by a system called the *restor*: if a vassal lost his horse otherwise than by his own fault, it must be replaced by the treasury (which was termed, as it also was in Norman Sicily, the *secretum*).[22] But the king had another force in addition to the feudal levy—a paid force of *soudoyers*,[23] holding fiefs, not of land, but of pay (*fiefs de soudée*). Along with this paid cavalry went another branch of the army, the Turcopuli, a body of light cavalry, recruited from the Syrians and Mahommedans, and using the tactics of the Arabs; while an infantry was found among the Armenians, the best soldiers of the East, and the Maronites, who furnished the kingdom with archers. To all these various forces must be added the knights and native levies of the great orders, whose masters were practically independent sovereigns like the princes of Antioch and Tripoli;[24] and with these the total levy of the kingdom may be reckoned at some 25,000 men. But the strength of the kingdom lay less perhaps in the army than in the magnificent fortresses which the nobility, and especially the two orders, had built; and the most visible relic of the crusades to-day is the towering ruins of a fortress like Krak (Kerak) des Chevaliers, the fortress of the Knights of St John in the principality of Tripoli. These fortresses, garrisoned not by the king, as in Norman England, but by their possessors, would only strengthen the power of the feudatories, and help to dissipate the kingdom into a number of local units.

In the organization of its system of justice the kingdom showed its most characteristic features. Two great central courts sat in Jerusalem to do justice—the high court of the nobles, and the court of burgesses for the rest of the Franks. (1) The high court was the supreme source of justice for the military class; and in its composition and procedure the same limitation of the crown, which appears in regard to military service, is again evident. The high court is not a *curia regis*, but a *curia baronum*, in which the theory of *judicium parium* is fully realized. If the king presides in the court, the motive of its action is none the less the preservation of the rights of the nobles, and not, as in England, the extension of the rights of the crown. It is a court of the king's peers: it tries cases of dispute between the king and his peers—with regard, for instance, to military service—and it settles the descent of the title of king. (2) The court of burgesses was almost equally sovereign within its sphere. While the body of the noblesse formed the high court, the court of the burgesses was composed of twelve legists (probably named by the king) under the presidency of the *vicomte*—a knight also named by the king, who was a great financial as well as a judicial officer. The province of the court included all acts and contracts between burgesses, and extended to criminal cases in which burgesses were involved. Like the high court, the court of burgesses had also its assizes[25]—a body of unwritten legal custom. The independent position of the burgesses, who thus assumed a position of equality by the side of the feudal class, is one of the peculiarities of the kingdom of Jerusalem. It may be explained by reference to the peculiar conditions of the kingdom. Burgesses and nobles, however different in status, were both of the same Frankish stock, and both occupied the same superior position with regard to the native Syrians. The commercial motive, again, had been one of the great motives of the crusade; and the class which was impelled by that motive would be both large and, in view of the quality of the Eastern goods in which it dealt, exceptionally prosperous. Finally, when one remembers how, during the First Crusade, the *pedites* had marched side by side with the *principes*, and how, from the beginning of 1099, they had practically risen in revolt against the selfish ambitions of princes like Count Raymund, it becomes easy to understand the independent position which the burgesses assumed in the organization of the kingdom. Burgesses could buy and possess property in towns, which knights were forbidden to acquire; and though they could not intermarry with the feudal classes, it was easy and regular for a burgess to thrive to knighthood. Like the nobles, again, the burgesses had the right of confirming royal grants and of taking part in legislation; and they may be said to have formed—socially, politically and judicially—an independent and powerful estate. Yet (with the exception of Antioch, Tripoli and Acre in the course of the 13th century) the Frankish towns never developed a communal government: the domain of their development was private law and commercial life.

Locally, the consideration of the system of justice administered in the kingdom involves some account of three things—the organization of the fiefs, the position of the Italian traders in their quarters, and the privileges of the Church. Each fief was organized like the kingdom. In each there was a court for the noblesse, and a court (or courts) for the bourgeoisie. There were some thirty-seven *cours de bourgeoisie* (several of the fiefs having more than one), each of which was under the presidency of a *vicomte*, while all were independent of the court of burgesses at Jerusalem. Of the feudal courts there were some twenty-two. Each of these followed the procedure and the law of the high court; but each was independent of the high court, and formed a sovereign court without any appeal. On the other hand, the revolution wrought by Amalric I. in the status of the *arrière-vassaux*, which made them members of the high court, allowed them to carry their cases to Jerusalem in the first instance, if they desired. Apart from this, the characteristic of seignorial justice is its independence and its freedom from the central court; though, when we reflect that the central court is a court of seigneurs, this characteristic is seen to be the logical result of the whole system. Midway between the seignorial *cours de bourgeoisie* and the privileged jurisdictions of the Italian quarter, there were two kinds of courts of a commercial character—the *cours de la fonde* in towns where trade was busy, and the *cours de la chaîne* in the sea-ports. The former courts, under their bailiffs, gradually absorbed the separate courts which the Syrians had at first been permitted to enjoy under their own *reïs*; and the bailiff with his 6 assessors (4 Syrians and 2 Franks) thus came to judge both commercial cases and cases in which Syrians were involved. The *cours de la chaîne*, whose institution is assigned to Amalric I. (1162-1174), had a civil jurisdiction in admiralty cases, and, like the *cours de la fonde*, they were composed of a bailiff and his assessors. Distinct from all these courts, if similar in its sphere, is the court which the Italian quarter generally enjoyed in each town under its own consuls—a court privileged to try all but the graver cases, like murder, theft and forgery. The court was part of the general immunity which made these quarters *imperia in imperio*: their exemptions from tolls and from financial contributions is parallel to their judicial privileges. Regulated by their mother-town, both in their trade and their government, these Italian quarters outlasted the collapse of the kingdom, and continued to exist under Mahommedan rulers. The Church had its separate courts, as in the West; but their province was perhaps greater than elsewhere. The church courts could not indeed decide cases of perjury; but, on the other hand, they tried all matters in which clerical property was concerned, and all cases of dispute between husband and wife. In other spheres the immunities and exemptions of the Church offered a far more serious problem, and especially in the sphere of finance. Perhaps the supreme defect of the kingdom of Jerusalem was its want of any financial basis. It is true that the king had a revenue, collected by the vicomte and paid into the *secretum* or treasury—a revenue composed of tolls on the caravans and customs from the ports, of the profits of monopolies and the proceeds of justice, of poll-taxes on Jews and Mahommedans, and of the tributes paid by Mahommedan powers. But his expenditure was large: he had to pay his feudatories; and he had to provide fiefs in money and kind to those who had not fiefs of land. The contributions sent to the Holy Land by the monarchs of western Europe, as commutations in lieu of personal participation in crusades, might help; the fatal policy of razzias against the neighbouring Mahommedan powers might procure temporary resources; but what was really necessary was a wide measure of native taxation, such as was once, and once only, attempted in 1183. To any such measure the privileges of the Italian quarters, and still more those of the Church, were inimical. In spite of provisions somewhat parallel to those of the English statute of mortmain, the clergy continued to acquire fresh lands at the same time that they refused to contribute to the defence of the kingdom, and rigorously exacted the full quota of tithe from every source which they could tap, and even from booty captured in war. The richest proprietor in the Holy Land,[26] but practically immune from any charges on its property, the Church helped, unconsciously, to ruin the kingdom which it should have supported above all others. It refused to throw its weight into the scale, and to strengthen the hands of the king against an over-mighty nobility. On the other hand, it must be admitted that the Church did not, after the first struggle between Dagobert and Baldwin I., actively oppose by any hierarchical pretensions the authority of the crown. The assizes may speak of patriarch and king as conjoint seigneurs in Jerusalem; but as a matter of fact the king could secure the nomination of his own patriarch, and after Dagobert the patriarchs are, with the temporary exception of Stephen in 1128, the confidants and supporters of the kings. It was the two great orders of the Templars and the Hospitallers which were, in reality, most dangerous to the kingdom. Honeycombed as it was by immunities—of seigneurs, of Italian quarters, of the clergy—the kingdom was most seriously impaired by these overweening immunists, who, half-lay and half-clerical, took advantage of their ambiguous position to escape from the duties of either character. They built up great estates, especially in the principality of Tripoli; they quarrelled with one another, until their dissensions prevented any vigorous action; they struggled against the claims of the clergy to tithes and to rights of jurisdiction; they negotiated with the Mahommedans as separate powers; they conducted themselves towards the kings as independent sovereigns. Yet their aid was as necessary as their influence was noxious. Continually recruited from the West, they retained the vigour which the native Franks of Palestine gradually lost; and their corporate strength gave a weight to their arms which made them indispensable.

In describing the organization of the kingdom, we have also been describing the causes of its fall. It fell because it had not the financial or political strength to survive. "Les vices du gouvernement avaient été plus puissants que les vertus des gouvernants." But the vices were not only vices of the government: they were also vices, partly inevitable, partly moral, in the governing race itself. The climate was no doubt responsible for much. The Franks of northern Europe attempted to live a life that suited a northern climate under a southern sun. They rode incessantly to battle over burning sands, in full armour—chain mail, long shield and heavy casque—as if they were on their native French soil. The ruling population was already spread too thin for the work which it had to do; and exhausted by its efforts, it gradually became extinct. A constant immigration from the West, bringing new blood and recruiting the stock, could alone have maintained its vigour; and such immigration never came. Little driblets of men might indeed be added to the numbers of the Franks; but the great bodies of crusaders either perished in Asia Minor, as in 1101 and 1147, or found themselves thwarted and distrusted by the native Franks. It was indeed one of the misfortunes of the kingdom that its inhabitants could never welcome the reinforcements which came to their aid.[27] The barons suspected the crusaders of ulterior motives, and of designing to get new principalities for themselves. In any case the native Frank, accustomed to commercial intercourse and diplomatic negotiations with the Mahommedans, could hardly share the unreasoning passion to make a dash for the "infidel." As with the barons, so with the burgesses: they profited too much by their intercourse with the Mahommedans to abandon readily the way of peaceful commerce, and they were far more ready to hinder than to help any martial enterprise. Left to itself, the native population lost physical and moral vigour. The barons alternated between the extravagances of Western chivalry and the attractions of Eastern luxury: they returned from the field to divans with frescoed walls and floors of mosaic, Persian rugs and embroidered silk hangings. Their houses, at any rate those in the towns, had thus the characteristics of Moorish villas; and in them they lived a Moorish life. Their sideboards were covered with the copper and silver work of Eastern smiths and the confectioneries of Damascus. They dressed in flowing robes of silk, and their women wore oriental gauzes covered with sequins. Into these divans where figures of this kind moved to the music of Saracen instruments, there entered an inevitable voluptuousness and corruption of manners. The hardships of war and the excesses of peace shortened the lives of the men; the kingdom of Jerusalem had eleven kings within a century. While the men died, the women, living in comparative indolence, lived longer lives. They became regents to their young children; and the experience of all medieval minorities reiterates the lesson—woe to the land where the king is a child and the regent a woman. Still worse was the frequent remarriage of widowed princesses and heiresses. By the assizes of the high court, the widow, on the death of her husband, took half of the estate for herself, and half in guardianship for her children. *Liberae ire cum terra*, widows carried their estates or titles to three or four husbands; and as in 15th-century England, the influence of the heiress was fatal to the peace of the country. At Antioch, for instance, after the death of Bohemund II. in 1130, his widow Alice headed a party in favour of the marriage of the heiress Constance to Manuel of Constantinople, and did not scruple to enter into negotiations with Zengi of Mosul. Her policy failed; and Constance successively married Raymund of Antioch and Raynald of Chatillon. The result was the renewed enmity of the Greek empire, while the French adventurers who won the prize ruined the prospects of the Franks by their conduct. In the kingdom matters were almost worse. There was hardly any regular succession to the throne; and Jerusalem, as Stubbs writes, "suffered from the weakness of hereditary right and the jealousies of the elective system" at one and the same time. With the frequent remarriages of the heiresses of the kingdom, relationships grew confused and family quarrels frequent; and when Sibylla carried the crown to Guy de Lusignan, a newcomer disliked by all the relatives of the crown, she sealed the fate of the kingdom.

It may be doubted—though it seems a harsh verdict to pass on a kingdom founded by religious zeal on holy soil—whether the kingdom possessed that moral basis which alone can give a right of survival to any institution or organization. The crusading states had been founded by adventurers who thirsted for gain; and the primitive appetite did not lose its edge with the progress of time. We cannot be certain, indeed, how far the Frankish lords oppressed their Syrian tenants: the stories of such oppression have been discredited; while if we may trust the evidence of a Mahommedan traveller, Ibn Jubair, the lot of the Mahommedan who lived on Frankish manors was better than it had been under their native lords.[28] But the habits of the Franks were none the less habits of lawless greed: they swooped down from their castles, as Raynald of Chatillon did from Krak of the Desert, to capture Saracens and hold them to ransom or to plunder caravans. The lust of unlawful gain had infected the Frankish blood, as it seems to have infected England during the Hundred Years' War; and in either case nemesis infallibly came. The Moslems might have endured a state of "infidels"; they could not endure a state of brigands.

6. *The History of the Kingdom and the Crusades from the Loss of Edessa in 1144 to the Fall of Jerusalem in 1187.*—The years 1143-1144 are in many ways the turning point in the history of the Latin East. In 1143 began the reign of the first native king;[29] and about this date may be placed the final organization of the kingdom, witnessed by the completion of its body of customary law. At the same date, however, the decline of the kingdom also begins; the fall of Edessa is the beginning of the end. In 1143 John Comnenus and Fulk had just died, and Zengi, seeing his way clear, threw himself on the great Christian outpost, against which the tides of Mahommedan attack had so often vainly surged, and finally entered on Christmas Day 1144. Two years later Zengi died; but he left an able successor in his son, Nureddin, and an attempt to recover Edessa was successfully repelled in November 1146. Not only so, but in the spring of 1147 the Franks were unwise enough to allow the hope of gaining two small towns to induce them to break the vital alliance with Damascus. Thus, in itself, the position of affairs in the Holy Land in 1147 was certainly ominous; and the kingdom might well seem dependent for its safety on such aid as it might receive from the West.

Early in 1145 news had come from Antioch to Eugenius III. of the fall of

Edessa, and at the end of the year he had sent an encyclical to France—the natural soil, as we have seen, of crusading zeal. The response was instantaneous: the king of France himself, who bore on his conscience the burden of an unpunished massacre by his troops at Vitry in 1142,[30] took the crusading vow on the Christmas day of 1145. But the greatest success was attained when St Bernard—no great believer in pilgrimages, and naturally disposed to doubt the policy of a second Crusade—was induced by the pope to become the preacher of the new movement. To the crusading king of France St Bernard added the king of Germany, when, in Christmas week of 1146, he induced Conrad III. to take the vow by his sermon in the cathedral of Spires. Thus was begun the Second Crusade,[31] under auspices still more favourable than those which attended the beginning of the First, seeing that kings now took the place of knights, while the new crusaders would no longer be penetrating into the wilds, but would find a friendly basis of operations ready to their hands in Frankish Syria. But the more favourable the auspices, the greater proved the failure. Already at the final meeting at Étampes, in 1147, difficulties arose. Manuel Comnenus demanded that all conquests made by the crusaders should be his fiefs; and the question was debated whether the crusaders should follow the land route through Hungary, along the old road of Charlemagne, or should go by sea to the Holy Land. In this question the envoys of Manuel and of Roger of Sicily, who were engaged in hostilities with one another, took opposite sides. Conrad, related by marriage to Manuel, decided in favour of the land route, which Manuel desired because it brought the Crusade more under his direction, and because, if the route by sea were followed, Roger of Sicily might be able to divert the crusading ships against Constantinople. As it was, a struggle raged between Roger and Manuel during the whole progress of the Crusade, which greatly contributed towards its failure, preventing, as it did, any assistance from the Eastern empire. Nor was there any real unity among the crusaders themselves. The crusaders of northern Germany never went to the Holy Land at all; they were allowed the crusaders' privileges for attacking the Wends to the east of the Elbe—a fact which at once attests the cleavage between northern and southern Germany (intensified of late years by the war of investitures), and anticipates the age of the Teutonic knights and their long Crusade on the Baltic. The crusaders of the Low Countries and of England took the sea route, and attacked and captured Lisbon on their way, thus helping to found the kingdom of Portugal, and achieving the one real success which was gained by the Second Crusade.[32] Among the great army of crusaders who actually marched to Jerusalem there was little real unity. Conrad and Louis VII. started separately, and at different times, in order to avoid dissensions between their armies; and when they reached Asia Minor (after encountering some difficulties in Greek territory) they still acted separately. Eager to win the first spoils, the German crusaders, who were in advance of the French, attempted a raid into the sultanate of Iconium; but after a stern fight at Dorylaeum they were forced to retreat (October 1147), and for the most part perished by the way. Louis VII., who now appeared, was induced by this failure to take the long and circuitous route by the west coast of Asia Minor; but even so he had lost the majority of his troops when he reached the Holy Land in 1148. Here he joined Conrad (who had come by sea from Constantinople) and Baldwin III., and after some deliberation the three sovereigns resolved to attack Damascus. The attack was impolitic: Damascus was the one ally which could help the Franks to stem the advance of Nureddin. It proved as futile as it was impolitic; for the vizier of Damascus, Muin-eddin-Anar, was able to sow dissension between the native Franks and the crusaders; and by bribes and promises of tribute he succeeded in inducing the former to make the siege an absolute failure, at the end of only four days (July 28th, 1148). The Second Crusade now collapsed. Conrad returned to Constantinople in the autumn of 1148, and Louis VII. returned by sea to France in the spring of 1149. The only effects of this great movement were effects prejudicial to the ends towards which it was directed. The position of the Franks in the Holy Land was not improved by the attack on Damascus; while the ignominious failure of a Crusade led by two kings brought the whole crusading movement into discredit in western Europe, and it was utterly in vain that Suger and St Bernard attempted to gather a fresh Crusade in 1150.

The result of the failure of the Second Crusade was the renewal of Nureddin's attacks. The rest of the county of Edessa, including Tell-bashir on the west, was now conquered (1150); while Raymund of Antioch was defeated and killed (in 1149), and several towns in the east of his principality were captured. Baldwin III. attempted to make head against these troubles, partly by renewing the old alliance with Damascus, partly by drawing closer to Manuel of Constantinople. For the next twenty years, during the reigns of Baldwin and his brother Amalric I., there is indeed a close connexion between the kingdom of Jerusalem and the East Roman empire. Baldwin and Amalric both married into the Comnenian house, while Manuel married Mary of Antioch, the daughter of Raymund. In the north Manuel enjoyed the homage of Antioch, which his father had gained in 1137, and the nominal possession of Tell-bashir, which had been ceded to him by Baldwin III.: in the south he joined with Amalric I. in the attempt to acquire Egypt (1168-1171). In this way he acquired a certain ascendancy over the Latin kings: Baldwin III. rode behind him at Antioch in 1159 without any of the insignia of royalty, and in an inscription at Bethlehem of 1172 Amalric I. had the name of the emperor written above his own.[33] The patronage of Constantinople, to which Jerusalem was thus practically surrendered, contributed to some slight extent in maintaining the kingdom against Nureddin. But there were dissensions within, both between Baldwin and his mother, Melisinda, who sought to protract her regency unduly, and between contending parties in Antioch, where the hand of Constance, Raymund's widow, was a desirable prize[34]; while from without the horns of the crescent were slowly closing in on the kingdom. Nureddin pursued in his policy the tactics which the Mahommedans used against the Franks in battle: he sought to envelop their territories on every side. In 1154 fell Damascus, and the crescent closed perceptibly in the north: the most valuable ally of the kingdom was lost, and the way seemed clear from Aleppo (the peculiar seat of Nureddin's power) into Egypt. On the other hand, in 1153 Baldwin III. had taken Ascalon, which for fifty years had mocked the efforts of successive kings, and by this stroke he might appear to have closed for Nureddin the route to Egypt, and to have opened a path for its conquest by the Franks. For the future, events hinged on the situation of affairs in Egypt, and in Egypt the fate of the kingdom of Jerusalem was finally decided (see EGYPT: *History*, "Mahommedan Period"). There was a race for the possession of the country between Nureddin's lieutenant Shīrgūh or Shīrkūh and Amalric I., the brother and successor of Baldwin III.; and in the race Shīrkūh proved the winner.

Since the days of Godfrey and Baldwin I., Egypt had been a goal of Latin ambition, and the capture of Ascalon must obviously have given form and strength to the projects for its conquest. Plans of attack were sketched: routes were traced: distances were measured; and finally in 1163 there came the impulse from within which turned these plans into action. The Shiite caliphs of Egypt were by this time the playthings of contending viziers, as the Sunnite caliphs of Bagdad had long been the puppets of Turkish sultans or amirs; and in 1164 Amalric I. and Nureddin were fighting in Egypt in support of two rival viziers, Dirgham and Shawar. For Nureddin the fight meant the acquisition of an heretical country for the true faith of the Sunnite, and the final enveloping of the Latin kingdom:[35] for Amalric it meant the escape from Nureddin's net, and a more direct and lucrative contact with Eastern trade. Into the vicissitudes of the fight it is not necessary here to enter; but in the issue Nureddin won, in spite of the support which Manuel gave to Amalric. Nureddin's Kurdish lieutenant, Shīrgūh, succeeded in establishing in power the vizier whom he favoured, and finally in becoming vizier himself (January 1169); and when he died, his nephew Saladin (Sala-ed-din) succeeded to his position (March 1169), and made himself, on the death of the caliph in 1171, sole ruler in Egypt. Thus the Shiite caliphate became extinct: in the mosques of Cairo the name of the caliph of Bagdad was now used; and the long-disunited Mahommedans at last faced the Christians as a solid body. But nevertheless the kingdom of Jerusalem continued almost unmenaced, and practically undiminished, for the next sixteen years. If a religious union had been effected between Egypt and northern Syria, political disunion still remained; and the Franks were safe as long as it lasted. Saladin acted as the peer of Nureddin rather than as his subject; and the jealousy between the two kept both inactive till the death of Nureddin in 1174. Nureddin only left a minor in his place: Amalric, who died in the same year, left a son (Baldwin IV.) who was not only a minor but also a leper; and thus the stage seemed cleared for Saladin. He was confronted, however, by Raymund, count of Tripoli, the one man of ability among the decadent Franks, who acted as guardian of the kingdom; while he was also occupied in trying to win for himself the Syrian possessions of Nureddin. The task engaged his attention for nine years. Damascus he acquired as early as 1174; but Raymund supported the heir of Nureddin in his capital at Aleppo, and it was not until 1183 that Saladin entered the city, and finally brought Egypt and northern Syria under a single rule.

The hour of peril for the Latin kingdom had now at last struck. It had done little to prepare itself for that hour. Repeated appeals had been sent to the West from the beginning of the Egyptian affair (1163) onwards; while in 1184-1185 a great mission, on which the patriarch of Jerusalem and the masters of the Templars and the Hospitallers were all present, came to France and England, and offered the crown of Jerusalem to Philip Augustus and Henry II. in turn, in order to secure their presence in the Holy Land.[36] The only result of these appeals was the rise of a regular system of taxation in France and England, *ad sustentationem Jerosolimitanae terrae*, which starts about 1185 (though there had already been isolated taxes in 1147 and 1166), and which has been described as the beginning of modern taxation. In the East itself, with the exception of the tax of 1183,[37] nothing was done that was good, and two things were done which were evil. Sibylla married her second husband, Guy de Lusignan, in 1180—a marriage destined to be the cause of many dissensions; for Sibylla, the eldest daughter of Amalric I., carried to her husband—a French adventurer—a presumptive title to the crown, which would never be admitted without dispute. In 1186 Guy eventually became king, after the death of Baldwin V. (Sibylla's son by her first marriage); but his coronation was in violation of the promise given to Raymund of Tripoli (that in the event of the death of Baldwin V. without issue the succession should be determined by the pope, the emperor and the kings of France and England), and Guy, with a weak title, was unable to exercise any real control over the kingdom. At this point another French adventurer, who had already made himself somewhat of a name in Antioch, gave the final blow to the kingdom. Raynald of Chatillon, the second husband of Constance of Antioch, after languishing in captivity from 1159 to 1176, had been granted the seignory of Krak, to the east and south of the Dead Sea. From this point of vantage he began depredations on the Red Sea (1182), building a fleet, and seeking to attack Medina and Mecca—a policy which may be interpreted either as mere buccaneering, or as a calculated attempt to deal a blow at Mahommedanism in its very centre. Driven from the Red Sea by Saladin, he turned from buccaneering to brigandage, and infested the great trade-route from Damascus to Egypt, which passed close by his seignory. In 1186 he attacked a caravan in which the sister of Saladin was travelling, thus violating a four years' truce, which, after some two years' skirmishing, Saladin and Raymund of Tripoli had made in the previous year owing to the general prevalence of famine.[38] The coronation of one French adventurer and the conduct of another, whom the first was unable to control, meant the ruin of the kingdom; and Saladin at last delivered in full force his long-deferred attack. The Crusade was now at last answered by the counter-Crusade—the *jihad*; for though for many years past Saladin had, in his attempt to acquire all the inheritance of Nureddin, left Palestine unmenaced and intact, his ultimate aim was always the holy war and the recovery of Jerusalem. The acquisition of Aleppo could only make that supreme object more readily attainable; and so Saladin had spent his time in acquiring Aleppo, but only in order that he might ultimately "attain the goal of his desires, and set the mosque of Asha free, to which Allah once led in the night his servant Mahomet." Thus it was on a kingdom of crusaders who had lost the crusading spirit that a new Crusade swept down; and Saladin's army in 1187 had the spirit and the fire of the Latin crusaders of 1099. The tables were turned; and fighting on their own soil for the recovery of what was to them too a holy place, the Mahommedans easily carried the day. At Tiberias a little squadron of the brethren of the two Orders went down before Saladin's cavalry in May; at Hattin the levy *en masse* of the kingdom, some 20,000 strong, foolishly marching over a sandy plain under the heat of a July sun, was utterly defeated; and after a fortnight's siege Jerusalem capitulated (October 2nd, 1187). In the kingdom itself nothing was left to the Latins by the end of 1189 except the city of Tyre; and to the north of the kingdom they only held Antioch and Tripoli, with the Hospitallers' fortress at Margat. The fingers of the clock had been pushed back; once more things were as they had been at the time of the First Crusade; once more the West must arm itself for the holy war and the recovery of Jerusalem—but now it must face a united Mahommedan world, where in 1096 it had found political and religious dissension, and it must attempt its vastly heavier task without the morning freshness of a new religious impulse, and with something of the weariness of a hundred years of struggle upon its shoulders.

7. *The Forty Years' Crusade for the Recovery of Jerusalem, 1189-1229.*— The forty years from 1189 to 1229 form a period of incessant crusading, occupied by Crusades of every kind. There are the Third, Fifth and Sixth Crusades against the "infidel" Mahommedans encamped in the Holy Land; there is the Albigensian Crusade against the heretic Cathars; there is the Fourth Crusade, directed in the issue against the schismatic Greeks; lastly, there are the Crusades waged by the papacy against revolted Christians—John of England and Frederick II. Our concern lies with the first kind of Crusade, and with the other three only so far as they bear on the first, and as they illustrate the immense widening which the term "Crusade" now underwent—a widening accompanied by its inevitable corollary of shallowness of motive and degradation of impulse.

The Third Crusade, 1189-1192.—Conrad of Montferrat was, as much as any one man, responsible for the Third Crusade. Compelled to leave the court of Constantinople, which he had been serving, he had sailed for the Holy Land and reached Tyre about three weeks after the battle of Hattin. He had saved Tyre; and from it he sent his appeals to the West. Not the least effective of these appeals was a great poster which he had circulated in Europe, and which represented the Holy Sepulchre denied by the horses of the Mahommedans. Meanwhile the papacy, as soon as the news reached Rome, despatched encyclicals throughout Europe; and soon a new Crusade was in full swing. But the Third Crusade, unlike the First, does not spring from the papacy, which was passing through one of its epochs of depression; it springs from the lay power, which, represented by the three strong monarchies of Germany, England and France, was at this time dominant in Europe. In Germany it was the solemn national diet of Mainz (Easter 1188) which "swore the expedition" to the Holy Land; in France and England the agreement of the two kings decided upon a joint Crusade. The very means which Philip Augustus and Henry II. took, in order to further the Crusade, show its lay aspect. A scheme of taxation—the Saladin tithe—was imposed on all who did not take the cross; and this taxation, while on the one hand it drove many to take the cross in order to escape its incidence, on the other hand provided a necessary financial basis for military operations.[39] The lay basis of the Third Crusade made it, in one sense, the greatest of all Crusades, in which all the three great monarchs of western Europe participated; but it also made it a failure, for the kings of France and England, changing *caelum*, *non animum*, carried their political rivalries into the movement, in which it had been agreed that they should be sunk. Spiritually, therefore, the Third Crusade is inferior to the First, however imposing it may be in its material aspects. Yet it must be admitted that the idea of a spiritual regeneration accompanied the crusading movement of 1188. Europe had sinned in the face of God; otherwise Jerusalem would never have fallen; and the idea of a spiritual reform from within, as the necessary corollary and accompaniment of the expedition of Christianity without, breathes in some of the papal letters, just as, during the conciliar movement, the *causa reformationis* was blended with the *causa unionis*.

We may conceive of the Third Crusade under the figure of a number of converging lines, all seeking to reach a common centre. That centre is Acre. The siege of Acre, as arduous and heroic in many of its episodes as the siege of Troy, had been begun in the summer of 1189 by Guy de Lusignan, who, captured by Saladin at the battle of Hattin, and released on parole, had at once broken his word and returned to the attack. The army which was besieging Acre was soon joined by various contingents; for Acre, after all, was the vital point, and its capture would open the way to Jerusalem. Two of these contingents alone concern us here—the German and the Anglo-French. Frederick I. of Germany, using a diplomacy which corresponds to the lay character of the Third Crusade, had sought to prepare his way by embassies to the king of Hungary, the Eastern emperor and the sultan of Iconium. Starting from Regensburg in May 1189, the German army marched quietly through Hungary; but difficulties arose, as they had arisen in 1147, as soon as the frontiers of the Eastern empire were reached. The emperor Isaac Angelus had not only the old grudge of all Eastern emperors against the "upstart" emperor of the West; he had also allied himself with Saladin, in order to acquire for his empire the patronage of the Holy Places and religious supremacy in the Levant. The difficulties between Frederick and Isaac Angelus became acute: in November 1189 Frederick wrote to his son Henry, asking him to induce the pope to preach a Crusade against the schismatic Greeks. But terms were at last arranged, and by the end of March 1190 the Germans had all crossed to the shores of Asia Minor. Taking a route midway between the eastern route of the crusaders of 1097 and the western route of Louis VII. in 1148, Frederick marched by Philadelphia and Iconium, not without dust and heat, until he reached the river Salof, in Armenian territory. Here, with the burden of the day now past, the fine old crusader—he had joined before in the Second Crusade, forty years ago—perished by accident in the river; and of all his fine army only a thousand men won their way through, under his son, Frederick of Swabia, to join the ranks before Acre (October 1190). The Anglo-French detachment achieved a far greater immediate success. War had indeed disturbed the original agreement of Gisors between Philip Augustus and Henry II., but a new agreement was made between Henry's successor, Richard I., and the French king at Nonancourt (December 1189), by which the two monarchs were to meet at Vezelay next year, and then follow the sea route to the Holy Land together. They met, and by different routes they both reached Sicily, where they wintered together (1190-1191). The enforced inactivity of a whole winter was the mother of disputes and bad blood; and when Philip sailed for the Holy Land, at the end of March 1191, the failure of the Crusade was already decided. Richard soon followed; but while Philip sailed straight for Acre, Richard occupied himself by the way in conquering Cyprus—partly out of knight-errantry, and in order to avenge an insult offered to his betrothed wife Berengaria by the despot of the island, partly perhaps out of policy, and in order to provide a basis of supplies and of operations for the armies attempting to recover Palestine. In any case, he is the founder of the Latin kingdom of Cyprus (for he afterwards sold his new acquisition to Guy de Lusignan, who established a dynasty in the island); and thereby he made possible the survival of the institutions and assizes of Jerusalem, which were continued in Cyprus until it was conquered by the Ottoman Turks. From Cyprus Richard sailed to Acre, arriving on the 8th of June, and in little more than a month he was able, in virtue of the large reinforcements he brought, and in spite of dissensions in the Christian camp which he helped to foment, to bring the two years' siege to a successful issue (July 12th, 1191). It was indeed time; the privations of the besiegers during the previous winter had been terrible; and the position of affairs had only been made worse by the dissensions between Guy de Lusignan and Conrad of Montferrat, who had begun to claim the crown in return for his services, and had, on the death of Sibylla, the wife of Guy, reinforced his claim by a marriage with her younger sister, Isabella. In these dissensions it was inevitable that Philip Augustus and Richard I., already discordant, should take contrary sides; and while

Richard naturally sided with Guy de Lusignan, who came from his own county of Poitou, Philip as naturally sided with Conrad. At the end of July it was decided that Guy should remain king for his life, and Conrad should be his successor; but as three days afterwards Philip Augustus began his return to France (pleading ill-health, but in reality eager to gain possession of Flanders), the settlement availed little for the success of the Crusade. Richard stayed in the Holy Land for another year, during which he won a battle at Arsuf and refortified Jaffa. But far more important than any hostilities are the negotiations which, for the whole year, Richard conducted with Saladin. They show the lay aspect of the Third Crusade; they anticipate the Crusade of Frederick II.—for Richard was attempting to secure the same concessions which Frederick secured by the same means which he used. They show again the closer approximation and better understanding with the Mahommedans, which marks this Crusade. Nothing is more striking in these respects than Richard's proposal that Saladin's brother should marry his own sister Johanna and receive Jerusalem and the contiguous towns on the coast. In the event, a peace was made for three years (September 2nd, 1192), by which Lydda and Ramlah were to be equally divided, Ascalon was to be destroyed, and small bodies of crusaders were to be allowed to visit the Holy Sepulchre. Meanwhile Conrad of Montferrat, at the very instant when his superior ability had finally forced Richard to recognize him as king, had been assassinated (April 1192): Guy de Lusignan had bought Cyprus from Richard, and had sailed away to establish himself there;[40] and Henry of Champagne, Richard's nephew, had been called to the throne of Jerusalem, and had given himself a title by marrying Conrad's widow, Isabella. In this condition Richard left the Holy Land, when he began his eventful return, in October 1192. The Crusade had failed—failed because a leaderless army, torn by political dissensions and fighting on a foreign soil, could not succeed against forces united by religious zeal under the banner of a leader like Saladin. Yet it had at any rate saved for the Christians the principality of Antioch, the county of Tripoli, and some of the coast towns of the kingdom;[41] and if it had failed to accomplish its object, it had left behind, none the less, many important results. The difficulties which had arisen between Isaac Angelus and Frederick Barbarossa contain the germs of the Fourth Crusade; the negotiations between Richard and Saladin contain the germs of the Sixth. National rivalries had been accentuated and national differences brought into prominence by the meeting of the nations in a common enterprise; while, on the other hand, Mahommedans and Christians had fraternized as they had never done before during the progress of a Crusade. But what the Third Crusade showed most clearly was that the crusading movement was being lost to the papacy, and becoming part of the demesne of the secular state—organized by the state on its own basis of taxation, and conducted by the state according to its own method of negotiation. This after all is the great change; and even the genius of an Innocent III. "could not make undone what had once been done." On the contrary, the thing once done would go further; and the state would take up the name of Crusade in order to cover, and under such cover to achieve, its own objects and ambitions, as in the future it was destined again and again to do.

The Fourth Crusade, 1202-1204.—The history of the Fourth Crusade is a history of the predominance of the lay motive, of the attempt of the papacy to escape from that predominance, and to establish its old direction of the Crusade, and of the complete failure of its attempt. Until the accession of Innocent III. in 1198 the lay motive was supreme; and its representative was Henry VI.—the greatest politician of his day, and in many ways the greatest emperor since Charlemagne. In 1195 Amalric, the brother of Guy de Lusignan, and his successor in Cyprus, sought the title of king from Henry and did homage; and at the same time Leo of Lesser Armenia, in order to escape from dependence on the Eastern empire, took the same course. Henry thus gained a basis in the Levant; while the death of Saladin in 1193, followed by a civil war between his brother, Malik-al-Adil, and his sons for the possession of his dominions, weakened the position of the Mahommedans. As emperor, Henry was eager to resume the imperial Crusade which had been stopped by his father's death; while both as Frederick's successor and as heir to the Norman kings of Sicily, who had again and again waged war against the Eastern empire, he had an account to settle with the rulers of Constantinople. The project of a Crusade and of an attack on Constantinople wove themselves into a single thread, in a way which very definitely anticipates the Fourth Crusade of 1202-1204. In 1195 Henry took the cross; some time before, he had already sent to Isaac Angelus to demand compensation for the injuries done to Frederick I., along with the cession of all territories ever conquered by the Norman kings of Sicily, and a fleet to co-operate with the new Crusade. In the same year, however, Isaac was dethroned by his brother, Alexius III.; but Henry married Isaac's daughter Irene to his brother, Philip of Swabia, and thus attempted to give the Hohenstaufen a new title and a valid claim against the usurper Alexius. Thus armed he pushed forward the preparations for the Crusade in Germany—a Crusade whose first object would have been an attack on Alexius III.; but in the middle of his preparations he died in Sicily in the autumn of 1197, and the Crusade collapsed. Some results were, however, achieved by a body of German crusaders which had sailed in advance of Henry; by its influence Amalric of Cyprus succeeded Henry of Champagne, who died in 1197, as king of Jerusalem, and a vassal of the emperor thus became ruler in the Holy Land; while the Teutonic order, which had begun as a hospital during the siege of Acre (1190-1191), now received its organization. Some of the coast towns, too, were recovered by the German crusaders, especially Beirut; and in 1198 the new king Amalric II. was able to make a truce with Malik-al-Adil for the next five years.

"The true heir of Henry VI.," Ranke has said, "is Innocent III.," and nowhere is this more true than in respect of the crusading movement. Throughout the course of his crowded and magnificent pontificate, Innocent III. made the Crusade his ultimate object, and attempted to bring it back to its old religious basis and under its old papal direction. By the spring of 1200, owing to Innocent's exertions, a new Crusade was in full progress, especially in France, where Fulk of Neuilly played the part once played by Peter the Hermit. Like the First Crusade, the Fourth Crusade also—in its personnel, but not its direction—was a French enterprise; and its leading members were French feudatories like Theobald of Champagne (who was chosen leader of the Crusade), Baldwin of Flanders (the future emperor of Constantinople), and the count of Blois. The objective, which these three original chiefs of the Fourth Crusade proposed to themselves, was Egypt.[42] Since 1163 the importance of acquiring Egypt had, as we have seen, been definitely understood, and in the summer of 1192 Richard I. had been advised by his counsellors that Cairo and not Jerusalem was the true point of attack; while in 1200 there was the additional reason for preferring an attack on Egypt, that the truce in the Holy Land between Amalric II. and Malik-al-Adil had still three years to run. It is Egypt therefore—to which, it must be remembered, the centre of Mahommedan power had now been virtually shifted, and to which motives of trade impelled the Italian towns (since from it they could easily reach the Red Sea, and the commerce of the Indian Ocean)—it is Egypt which is henceforth the normal goal of the Crusades. This is one of the many facts which differentiate the Crusades of the 13th from those of the preceding century. But, with Syria in the hands of the Mahommedans, the attack on Egypt must necessarily be directed by sea; and thus the Crusade henceforth becomes—what the Third Crusade, here as elsewhere the turning-point in crusading history, had already in part been—a maritime enterprise. Accordingly, early in 1201, envoys from each of the three chiefs of the Fourth Crusade (among whom was Villehardouin, the historian of the Crusade) came to Venice to negotiate for a passage to Egypt. An agreement was made between the doge and the envoys, by which transport and active help were to be given by Venice in return for 85,000 marks and the cession of half of the conquests made by the crusaders. But the Fourth Crusade was not to be plain sailing to Egypt. It became involved in a maelstrom of conflicting political motives, by which it was swept to Constantinople. Here we must distinguish between cause and occasion. There were three great causes which made for an attack on Constantinople by the West. There was first of all the old crusading grudge against the Eastern empire, and its fatal policy of regarding the whole of the Levant as its lost provinces, to be restored as soon as conquered, or at any rate held in fee, by the Western crusaders—a policy which led the Eastern emperors either to give niggardly aid or to pursue obstructive tactics, and caused them to be blamed for the failure of the Crusades in 1101, and 1149, and in 1190. It is significant of the final result of these things that already in 1147 Roger of Sicily, engaged in war with Manuel, had proposed the sea-route for the Second Crusade, perhaps with some intention of diverting it against Constantinople; and in the winter of 1189-1190 Barbarossa, as we have seen, had actually thought and spoken of an attack on Constantinople. In the second place, there was the commercial grudge of Venice, which had only been given large privileges by the Eastern empire to desire still larger, and had, moreover, been annoyed not only by alterations or revocations of those privileges, such as the usurper Alexius III. had but recently attempted, but also by the temporary destruction of their colony in Constantinople in 1171. Lastly, and perhaps most of all, there is the old Norman blood-feud with Constantinople, as old as the old Norse seeking for Micklegarth, and keen and deadly ever since the Norman conquest of the Greek themes in South Italy (1041 onwards). The heirs of the Norman kings were the Hohenstaufen; and we have already seen Henry VI. planning a Crusade which would primarily have been directed against Constantinople. It is this Hohenstaufen policy which becomes the primary occasion of the diversion of the Fourth Crusade. Philip of Swabia, engaged in a struggle with the papacy, found Innocent III. planning a Guelph Crusade, which should be under the direction of the church; and to this Guelph project he opposed the Ghibelline plan of Henry VI., with such success that he transmuted the Fourth Crusade into a political expedition against Constantinople. To such a policy of transmutation he was urged by two things. On the one hand, the death of the count of Champagne (May 1201) had induced the crusaders to elect as their leader Boniface of Montferrat, the brother of Conrad; and Boniface was the cousin of Philip, and interested in Constantinople, where not only Conrad, but another brother as well, had served, and suffered for their service at the hands of their masters. On the other hand Alexius, the son of the dethroned Isaac Angelus, was related to Philip through his marriage with Irene; and Alexius had escaped to the German court to urge the restoration of his father. On Christmas day 1201, Philip, Alexius and Boniface all met at Hagenau[43] and formulated (one may suppose) a plan for the diversion of the Crusade. Events played into their hands. When the crusaders gathered at Venice in the autumn of 1202, it was found impossible to get together the 85,000 marks promised to Venice. The Venetians—already, perhaps, indoctrinated in the Hohenstaufen plan—indicated to the leaders a way of meeting the difficulty: they had only to lend their services to the republic for certain ends which it desired to compass, and the debt was settled. The conquest of Zara, a port on the Adriatic claimed by the Venetians from the king of Hungary, was the only object overtly mentioned; but the idea of the expedition to Constantinople was in the air, and the crusaders knew what was ultimately expected. It took time and effort to bring them round to the diversion: the pope—naturally enough—set his face sternly against the project, the more as the usurper, Alexius III., was in negotiation with him in order to win his support against the Hohenstaufen, and Innocent hoped to find, as Alexius promised, a support and a reinforcement for the Crusade in an alliance with the Greek empire. But they came round none the less, in spite of Innocent's renewed prohibitions. In November 1202 Zara was taken; and at Zara the fatal decision was made. The young Alexius joined the army; and in spite of the opposition of stern crusaders like Simon de Montfort, who sailed away ultimately to Palestine, he succeeded by large promises in inducing the army to follow in his train to Constantinople. By the middle of July 1203 Constantinople was reached, the usurper was in flight, and Isaac Angelus was restored to his throne. But when the time came for Alexius to fulfil his promises, the difficulty which had arisen at Venice in the autumn of 1202 repeated itself. Alexius's resources were insufficient, and he had to beg the crusaders to wait at Constantinople for a year in order that he might have time. They waited; but the closer contact of a prolonged stay only brought into fuller play the essential antipathy of the Greek and the Latin. Continual friction developed at last into the open fire of war; and in March 1204 the crusaders resolved to storm Constantinople, and to divide among themselves the Eastern empire. In April Constantinople was captured; in May Baldwin of Flanders became the first Latin emperor of Constantinople. Venice had her own reward; a Venetian, Thomas Morosini, became patriarch; and the doge of Venice added "a quarter and a half" of the Eastern empire—chiefly the coasts and the islands—to the sphere of his sway. If Venetian cupidity had not originally deflected the Crusade (and it was the view of contemporary writers that Venice had committed her first treason against Christianity by diverting the Crusade from Egypt in order to get commercial concessions from Malik-al-Adil,[44]) yet it had at any rate profited exceedingly from that deflection; and the Hohenstaufen and their protégé Alexius only reaped dust and ashes. For, however Ghibelline might be the original intention, the result was not commensurate with the subtlety of the design, and the power of the pope was rather increased than diminished by the event of the Crusade. The crusaders appealed to Innocent to ratify the subjugation of a schismatic people, and the union of the Eastern and Western Churches; and Innocent, dazzled by the magic of the *fait accompli*, not unwillingly acquiesced. He might soothe himself by reflecting that the basis for the Crusade, which he had hoped to find in Alexius III., was still more securely offered by Baldwin; he could not but feel with pride that he had become "as it were pope and apostolicus of a second world." Yet the result of the Fourth Crusade was on the whole disastrous both for the papacy and for the crusading movement. The pope had been forced to see the helm of the Crusades wrenched from his grasp; and the Albigensian Crusade against the heretics of southern France was soon afterwards to show that the example could be followed, and that the land-hunger of the north French baronage could exploit a Crusade as successfully as ever did Hohenstaufen policy leagued with Venetian cupidity. The Crusade lost its *élan* when it became a move in a political game. If the Third Crusade had been directed by the lay power towards the true spiritual end of all Crusades, the Fourth was directed by the lay power to its own lay ends; and the political and commercial motives, winch were deeply implicit even in the First Crusade, had now become dominantly explicit. In a simpler and more immediate sense, the capture of Constantinople was detrimental to the movement from which it sprang. The precarious empire which had been founded in 1204 drained away all the vigorous adventurers of the West for its support for many years to come, and the Holy Land was starved to feed a land less holy, but equally greedy of men.[45] No basis for the Crusades was ever to be found in the Latin empire of the East; and Innocent, after vainly hoping for the new Crusade which was to emerge from Constantinople, was by 1208 compelled to return to the old idea of a Crusade proceeding simply and immediately from the West to the East.

The Fifth Crusade, 1218-1221.—The glow and the glamour of the Crusades disappear save for the pathetic sunset splendours of St Louis, as Dandolo dies, and gallant Villehardouin drops his pen. But before St Louis sailed for Damietta there intervened the miserable failure of one Crusade, and the secular and diplomatic success of another. The Fifth Crusade is the last which is started in that pontificate of Crusades—the pontificate of Innocent III. It owed its origin to his feverish zeal for the recovery of Jerusalem, rather than to any pressing need in the Holy Land. Here there reigned, during the forty years of the loss of Jerusalem, an almost unbroken peace. Malik-al-Adil, the brother of Saladin, had by 1200 succeeded to his brother's possessions not only in Egypt but also in Syria, and he granted the Christians a series of truces (1198-1203, 1204-1210, 1211-1217). While the Holy Land was thus at peace, crusaders were also being drawn elsewhere by the needs of the Latin empire of Constantinople, or the attractions of the Albigensian Crusade.[46] But Innocent could never consent to forget Jerusalem, as long as his right hand retained its cunning. The pathos of the Children's Crusade of 1212 only nerved him to fresh efforts. A shepherd boy named Stephen had appeared in France, and had induced thousands to follow his guidance: with his boyish army he rode on a wagon southward to Marseilles, promising to lead his followers dry-shod through the seas. In Germany a child from Cologne, named Nicolas, gathered some 20,000 young crusaders by the like promises, and led them into Italy. Stephen's army was kidnapped by slave-dealers and sold into Egypt; while Nicolas's expedition left nothing behind it but an after-echo in the legend of the Pied Piper of Hamelin. But for Innocent these outbursts of the revivalist element, which always accompanied the Crusades, had their moral: "the very children put us to shame," he wrote; "while we sleep they go forth gladly to conquer the Holy Land." In the fourth Lateran council of 1215 Innocent found his opportunity to rekindle the flickering fires. Before this great gathering of all Christian Europe he proclaimed a Crusade for the year 1217, and in common deliberation it was resolved that a truce of God should reign for the next four years, while for the same time all trade with the Levant should cease. Here were two things attempted—neither, indeed, for the first time[47]—which 14th century pamphleteers on the subject of the Crusades unanimously advocate as the necessary conditions of success; there was to be peace in Europe and a commercial war with Egypt. This statesmanlike beginning of a Crusade, preached, as no Crusade had ever been preached before, in a general council of all Europe, presaged well for its success. In Germany (where Frederick II. himself took the cross in this same year) a large body of crusaders gathered together: in 1217 the south-east sent the duke of Austria and the king of Hungary to the Holy Land; while in 1218 an army from the north-west joined at Acre the forces of the previous year. Egypt had already been indicated by Innocent III. in 1215 as the goal of attack, and it was accordingly resolved to begin the Crusade by the siege of Damietta, on the eastern delta of the Nile. The original leader of the Crusade was John of Brienne, king of Jerusalem (who had succeeded Amalric II., marrying Maria, the daughter of Amalric's wife Isabella by her former husband, Conrad of Montferrat); but after the end of 1218 the cardinal legate Pelagius, fortified by papal letters, claimed the command. In spite of dissensions between the cardinal and the king, and in spite of the offers of Malik-al-Kamil (who succeeded Malik-al-Adil at the end of 1218), the crusaders finally carried the siege to a successful conclusion by the end of 1219. The capture of Damietta was a considerable feat of arms, but nothing was done to clinch the advantage which had been won, and the whole of the year 1220 was spent by the crusaders in Damietta, partly in consolidating their immediate position, and partly in waiting for the arrival of Frederick II., who had promised to appear in 1221. In 1221 Hermann of Salza, the master of the Teutonic order, along with the duke of Bavaria, appeared in the camp before Damietta; and as it seemed useless to wait any longer for Frederick II.,[48] the cardinal, in spite of the opposition of King John, gave the signal for the march on Cairo. The army reached a fortress erected by the sultan in 1219 (afterwards, from 1221, the town of Mansura), and encamped there at the end of July. Here the sultan reiterated terms which he had already offered several times before—the cession of most of the kingdom of Jerusalem, the surrender of the cross (captured by Saladin in 1187), and the restoration of all prisoners. King John urged the acceptance of these terms. The legate insisted on a large indemnity in addition: the negotiations failed, and the sultan prepared for war. The crusaders were driven back towards Damietta; and at the end of August 1221 Pelagius had to make a treaty with Malik-al-Kamil, by which he gained a free retreat and the surrender of the Holy Cross at the price of the restoration of Damietta. The treaty was to last for eight years, and could only be broken on the coming of a king or emperor to the East. In pursuance of its terms the crusaders evacuated Egypt, and the Fifth Crusade was at an end. It is difficult to decide whether to blame the legate or the emperor more for its failure. If Frederick had only come in person, a single month of his presence might have meant everything: if Pelagius had only listened to King John, the sultan was ready to concede practically everything which was at issue. Unhappily Frederick preferred to put his Sicilian house in order, and the legate preferred to listen to the Italians, who had their own commercial reasons for wishing to establish a strong position in Egypt, and to the Templars and Hospitallers, who did not feel satisfied by the terms offered by the sultan, because he wished to retain in his hands the two fortresses of Krak and Monreal.

The Sixth Crusade (1228-1229) succeeded as signally as the Fifth Crusade had failed; but the circumstances under which it took place and the means by which it was conducted made its success still more disastrous than the failure of 1221. The last Crusade had, after all, been under papal control: if Richard I. had directed the Third Crusade, and the policy of the Hohenstaufen and the Venetians had directed the Fourth, it was a papal legate who had steered the Fifth to its ultimate fate. The Crusade of Frederick II. in 1228-1229 finds its analogy in the projected Crusade of Henry VI.; it is essentially lay. It is unique in the annals of the Crusades. Alone of all Crusades (though the Fourth Crusade offers some analogy) it was not blessed but cursed by the papacy: alone of all the Crusades it was conducted without a single act of hostility against the Mahommedan. St Louis, the true type of the religious crusader, once said that a layman ought only to argue with a blasphemer against Christian law by running his sword into the bowels of the blasphemer as far as it would go:[49] Frederick II. talked amicably with all unbelievers, if one may trust Arabic accounts, and he achieved by mere negotiation the recovery of Jerusalem, for which men had vainly striven with the sword for the forty years since 1187. It was in 1215 that the leader of this strange Crusade had first taken the vow; it was twelve years afterwards when he finally attempted to carry the vow into effective execution. Again and again he had excused himself to the pope, and been excused by the pope, because the exigencies of his policy in Germany or Sicily tied his hands. After the failure of the Fifth Crusade—for which these delays were in part responsible—Honorius III. had attempted to bind him more intimately to the Holy Land by arranging a marriage with Isabella, the daughter of John of Brienne, and the heiress of the kingdom of Jerusalem. In 1225 Frederick married Isabella, and immediately after the marriage he assumed the title of king in right of his wife, and exacted homage from the vassals of the kingdom.[50] It was thus as king of Jerusalem that Frederick began his Crusade in the autumn of 1227. Scarcely, however, had he sailed from Brindisi when he fell sick of a fever which had been raging for some time among the ranks of his army, while they waited for the crossing. He sailed back to Otranto in order to recover his health, but the new pope, Gregory IX., launched in hot anger the bolt of excommunication, in the belief that Frederick was malingering once more. None the less the emperor sailed on his Crusade in the summer of 1228, affording to astonished Europe the spectacle of an excommunicated crusader, and leaving his territories to be invaded by papal soldiers, whom Gregory IX. professed to regard as crusaders against a non-Christian king, and for whom he accordingly levied a tithe from the churches of Europe. The paradox of Frederick's Crusade is indeed astonishing. Here was a crusader against whom a Crusade was proclaimed in his own territories; and when he arrived in the Holy Land he found little obedience and many insults from all but his own immediate followers. Yet by adroit use of his powers of diplomacy, and by playing upon the dissensions which raged between the descendants of Saladin's brother (Malik-al-Adil), he was able, without striking a blow, to conclude a treaty with the sultan of Egypt which gave him all that Richard I. had vainly attempted to secure by arduous fighting and patient negotiations. By the treaty of the 18th of February 1229, which was to last for ten years, the sultan conceded to Frederick, in addition to the coast towns already in the possession of the Christians, Nazareth, Bethlehem and Jerusalem, with a strip of territory connecting Jerusalem with the port of Acre. As king of Jerusalem Frederick was now able to enter his capital: as one under excommunication, he had to see an interdict immediately fall on the city, and it was with his own hands—for no churchman could perform the office—that he had to take his crown from the altar of the church of the Sepulchre, and crown himself king of his new kingdom. He stayed in the Holy Land little more than a month after his coronation; and leaving in May he soon overcame the papal armies in Italy, and secured absolution from Gregory IX. (August 1229). By his treaty with the sultan he had secured for Christianity the last fifteen years of its possession of Jerusalem (1229-1244): no man since Frederick II. has ever recovered the holy places for the religion which holds them most holy. Yet the church might ask, with some justice, whether the means he had used were excused by the end which he had attained. After all, there was nothing of the holy war about the Sixth Crusade: there was simply huckstering, as in an Eastern bazaar, between a free-thinking, semi-oriental king of Sicily and an Egyptian sultan. It was indeed in the spirit of a king of Sicily, and not in the spirit—though it was in the rôle—of a king of Jerusalem, that Frederick had acted. It was from his Sicilian predecessors, who had made trade treaties with Egypt, that he had learned to make even the Crusade a matter of treaty. The Norman line of Sicilian kings might be extinct; their policy lived after them in their Hohenstaufen successors, and that policy, as it had helped to divert the Fourth Crusade to the old Norman objective of Constantinople, helped still more to give the Sixth Crusade its secular, diplomatic, non-religious aspect.

Forty years of struggle ended in fifteen years' possession of Jerusalem. During those fifteen years the kingdom of Jerusalem was agitated by a struggle between the native barons, championing the principle that sovereignty resided in the collective baronage, and taking their stand on the assizes, and Frederick II., claiming sovereignty for himself, and opposing to the assizes the feudal law of Sicily. It is a struggle between the king and the *haute cour*: it is a struggle between the aristocratic feudalism of the Franks and the monarchical feudalism of the Normans. Already in Cyprus, in the summer of 1228, Frederick II. had insisted on the right of wardship which he enjoyed as overlord of the island,[51] and he had appointed a commission of five barons to exercise his rights. In 1229 this commission was overthrown by John of Ibelin, lord of Beirut, against whom it had taken proceedings. John of Beirut, like many of the Cypriot barons, was also a baron of the kingdom of Jerusalem; and resistance in the one kingdom could only produce difficulties in the other. Difficulties quickly arose when Frederick, in 1231, sent Marshal Richard to Syria as his legate. This in itself was a serious matter; according to the assizes, the barons maintained, the king must either personally reside in the kingdom, or, in the event of his absence, be replaced by a regency. The position became more difficult, when the legate took steps against John of Beirut without any authorization from the high court. A gild was formed at Acre—the gild of St Adrian—which, if nominally religious in its origin, soon came to represent the political opposition to Frederick, as was significantly proved by its reception of the rebellious John of Beirut as a member (1232). The opposition was successful: by 1233 Frederick had lost all hold on Cyprus, and only retained Tyre in his own kingdom of Jerusalem. In 1236 he had to promise to recognize fully the laws of the kingdom: and when, in 1239, he was again excommunicated by Gregory IX., and a new quarrel of papacy and empire began, he soon lost the last vestiges of his power. Till 1243 the party of Frederick had been successful in retaining Tyre, and the baronial demand for a regency had remained without effect; but in that year the opposition, headed by the great family of Ibelin, succeeded, under cover of asserting the rights of Alice of Cyprus to the regency, in securing possession of Tyre, and the kingdom of Jerusalem thus fell back into the power of the baronage. The very next year (1244) Jerusalem was finally and for ever lost. Its loss was the natural corollary of these dissensions. The treaty of Frederick with Malik-al-Kamil (d. 1238) had now expired, and new succours and new measures were needed for the Holy Land. Theobald of Champagne had taken the cross as early as 1230, and 1239 he sailed to Acre in spite of the express prohibition of the pope, who, having quarrelled with Frederick II., was eager to divert any succour from Jerusalem itself, so long as Jerusalem belonged to his enemy. Theobald was followed (1240-1241) by Richard of Cornwall, the brother of Henry III., who, like his predecessor, had to sail in the teeth of papal prohibitions; but neither of the two achieved any permanent result, except the fortification of Ascalon. It was, however, by their own folly that the Franks lost Jerusalem in 1244. They consented to ally themselves with the ruler of Damascus against the sultan of Egypt; but in the battle of Gaza they were deserted by their allies and heavily defeated by Bibars, the Egyptian general and future Mameluke sultan of Egypt. Jerusalem, which had already been plundered and destroyed earlier in the year by Chorasmians (Khwarizmians), was the prize of victory, and Ascalon also fell in 1247.

8. *The Crusades of St Louis.*—As the loss of Jerusalem in 1187 produced

the Third Crusade, so its loss in 1244 produced the Seventh: as the preaching of the Fifth Crusade had taken place in the Lateran council of 1215, so that of the Seventh Crusade began in the council of Lyons of 1245. But the preaching of the Crusade by Innocent IV. at Lyons was a curious thing. On the one hand he repeated the provisions of the Fourth Lateran council on behalf of the Crusade to the Holy Land; on the other hand he preached a Crusade against Frederick II., and promised to all who would join the full benefits of absolution and remission of sins. While the papacy thus bent its energies to the destruction of the Crusades in their genuine sense, and preferred to use for its own political objects what was meant for Jerusalem, a layman took up the derelict cause with all the religious zeal which any pope had ever displayed. Paradoxically enough, it was now the turn for the papacy to exploit the name of Crusade for political ends, as the laity had done before; and it was left to the laity to champion the spiritual meaning of the Crusade even against the papacy.[52] It was at the end of the year in which Jerusalem had fallen that St Louis had taken the cross, and by all the means in his power he attempted to ensure the success of his projected Crusade. He sought to mediate, though with no success, between the pope and the emperor; he descended to a whimsical piety, and took his courtiers by guile in distributing to them, at Christmas, clothing on which a cross had been secretly stitched. He started in 1248 with a gallant company, which contained his three brothers and the sieur de Joinville, his biographer; and after wintering in Cyprus he directed his army in the spring of 1249 against Egypt. The objective was unexpected: it may have been chosen by St Louis, because he knew how seriously the power of the sultan was undermined by the Mamelukes, who were in the very next year to depose the Ayyubite dynasty, which had reigned since 1171, and to substitute one of their number as sultan. Damietta was taken without a blow, and the march for Cairo was begun, as it had been begun by the legate Pelagius in 1221. Again the invading army halted before Mansura (December 1249); again it had to retreat. The retreat became a rout. St Louis was captured, and a treaty was made by which he had to consent to evacuate Damietta and pay a ransom of 800,000 pieces of gold.

Eventually St Louis was released on surrendering Damietta and paying one-half of his ransom, and by the middle of May 1230 he reached Acre, having abandoned the Egyptian expedition. For the next four years he stayed in the Holy Land, seeking to do what he could for the establishing of the kingdom of Jerusalem. He was able to do but little. The struggle of papacy and empire paralysed Europe, and even in France itself there were few ready to answer the calls for help which St Louis sent home from Acre. The one answer was the Shepherds' Crusade, or Crusade of the Pastoureaux—"a religious Jacquerie," as it has been called by Dean Milman. It had some of the features of the Children's Crusade of 1212. That, too, had begun with a shepherd boy: the leader of the Pastoureaux, like the leader of the children, promised to lead his followers dry-shod through the seas; and tradition even said that this leader, "the master of Hungary," as he was called, was the Stephen of the Children's Crusade. But the anti-clerical feeling and action of the Shepherds was new and ominous; and moved by its enormities the government suppressed the new movement ruthlessly. None came to the aid of St Louis; and in 1254, on the death of his mother Blanche, the regent, he had to return to France.

The final collapse of the kingdom of Jerusalem had been really determined by the battle of Gaza in 1244, and by the deposition of the Ayyubite dynasty by the Mamelukes. The Ayyubites had always been, on the whole, chivalrous and tolerant: Saladin and his successors, Malik-al-Adil and Malik-al-Kamil, had none of them shown an implacable enmity to the Christians. The Mamelukes, who are analogous to the janissaries of the Ottoman Turks, were made of sterner and more fanatical stuff; and Bibars, the greatest of these Mamelukes, who had commanded at Gaza in 1244, had been one of the leaders in 1250, and was destined to become sultan in 1260, was the sternest and most fanatical of them all. The Christians were, however, able to maintain a footing in Syria for forty years after St Louis' departure, not by reason of their own strength, but owing to two powers which checked the advance of the Mamelukes. The first of these was Damascus. The kingdom of Jerusalem, as we have seen, had profited by the alliance of Damascus as early as 1130, when the fear of the atabegs of Mosul had first drawn the two together; and when Damascus had been acquired by the rule of Mosul, the hostility between the house of Nureddin in Damascus and Saladin in Egypt had still for a time preserved the kingdom (from 1171 onwards). Saladin had united Egypt and Damascus; but after his death dissensions broke out among the members of his family,[53] which more than once led to wars between Damascus and Cairo. It has already been noticed that such a war between the sons of Malik-al-Adil accounts in large measure for the success of the Sixth Crusade; and it has been seen that the battle of Gaza was an act in the long drama of strife between Egypt and northern Syria. The revolution in Egypt in 1250 separated Damascus from Cairo more trenchantly than they had ever been separated since 1171: while a Mameluke ruled in Cairo, Malik-al-Nāsir of Aleppo was elected as sultan by the emirs of Damascus. But an entirely new and far more important factor in the affairs of the Levant was the extension of the empire of the Mongols during the 13th century. That empire had been founded by Jenghiz Khan in the first quarter of the century; it stretched from Peking on the east to the Euphrates and the Dnieper on the west. Two things gave the Mongols an influence on the history of the Holy Land and the fate of the Crusades. In the first place, the south-western division of the empire, comprising Persia and Armenia, and governed about 1250 by the Khan Hulaku or Hulagu, was inevitably brought into relations, which were naturally hostile, with the Mahommedan powers of Syria and Egypt. In the second place, the Mongols of the 13th century were not as yet, in any great numbers, Mahommedans; the official religion was "Shamanism," but in the Mongol army there were many Christians, the results of early Nestorian missions to the far East. This last fact in particular caused western Europe to dream of an alliance with the great khan "Prester John," who should aid in the reconquest of Jerusalem and the final conversion to Christianity of the whole continent of Asia. The Crusades thus widen out, towards their close, into a general scheme for the christianization of all the known world.[54] About 1220 James of Vitry was already hoping that 4000 knights would, with the assistance of the Mongols, recover Jerusalem; but it is in 1245 that the first definite sign of an alliance with the Mongols appears. In that year Innocent IV. sent a Franciscan friar, Joannes de Piano Carpini, to the Mongols of southern Russia, and despatched a Dominican mission to Persia. Nothing came of either of these missions; but through them Europe first began to know the interior of Asia, for Carpini was conducted by the Mongols as far as Karakorum, the capital of the great khan, on the borders of China. Again in 1252 St Louis (who had already begun to negotiate with the Mongols in the winter of 1248-1249) sent the friar William of Rubruquis to the court of the great khan; but again nothing came of the mission save an increase of geographical knowledge. It was in the year 1260 when it first seemed likely that any results definitely affecting the course of the Crusades would flow from the action of the Mongols. In that year Hulagu, the khan of Persia, invaded Syria and captured Damascus. His general, a Christian named Kitboga, marched southwards to attack the Mamelukes of Egypt, but he was beaten by Bibars (who in the same year became sultan of Egypt), and Damascus fell into the hands of the Mamelukes. Once more, in spite of Mongol intervention, Damascus and Cairo were united, as they had been united in the hands of Saladin; once more they were united in the hands of a devout Mahommedan, who was resolved to extirpate the Christians from Syria.

While these things were taking place around them, the Christians of the kingdom of Jerusalem only hastened their own fall by internal dissensions which repeated the history of the period preceding 1187. In part the war of Guelph and Ghibelline fought itself out in the East; and while one party demanded a regency, as in 1243, another argued for the recognition of Conrad, the son of Frederick II., as king. In part, again, a commercial war raged between Venice and Genoa, which attracted into its orbit all the various feuds and animosities of the Levant (1257). Beaten in the war, the Genoese avenged themselves for their defeat by an alliance with the Palaeologi, which led to the loss of Constantinople by the Latins

(1261), and to the collapse of the Latin empire after sixty years of infirm and precarious existence. On a kingdom thus divided against itself, and deprived of allies, the arm of Bibars soon fell with crushing weight. The sultan, who had risen from a Mongolian slave to become a second Saladin, and who combined the physique and audacity of a Danton with the tenacity and religiosity of a Philip II., dealt blow after blow to the Franks of the East. In 1265 fell Caesarea and Arsuf; in 1268 Antioch was taken, and the principality of Bohemund and Tancred ceased to exist.[55] In the years which followed on the loss of Antioch several attempts were made in the West to meet the progress of the new conqueror. In 1269 James the Conqueror of Aragon, at the bidding of the pope, turned from the long Spanish Crusade to a Crusade in the East in order to atone for his offences against the law matrimonial. An opportune storm, however, gave the king an excuse for returning home, as Frederick II. had done in 1227; and though his followers reached Acre, they hardly dared venture outside its walls, and returned home promptly in the beginning of 1270. More serious were the plans and the attempts of Charles of Anjou and Louis IX., in which the Crusades may be said to have finally ended, save for sundry disjointed epilogues in the 14th and 15th centuries.

Charles of Anjou had succeeded, as a result of the long "crusade" waged by the papacy against the Hohenstaufen from the council of Lyons to the battle of Tagliacozzo (1245-1268), in establishing himself in the kingdom of Sicily. With the kingdom of Frederick II. and Henry VI. he also took over their policy—the "forward" policy in the East which had also been followed by the old Norman kings. On the one hand he aimed at the conquest of Constantinople as Henry VI. had done before; and by the treaty of Viterbo of 1267 he secured from the last Latin emperor of the East, Baldwin II., a right of eventual succession. On the other hand, like Frederick II., he aimed at uniting the kingdom of Jerusalem with that of Sicily; and here, too, he was able to provide himself with a title. On the death of Conradin, Hugh of Cyprus had been recognized in the East as king of Jerusalem (1269); but his pretensions were opposed by Mary of

Antioch, a granddaughter of Amalric II., who was prepared to bequeath her claims to Charles of Anjou, and was therefore naturally supported by him. But the policy of Charles, which thus prepared the way for a Crusade similar to those of 1197 and 1202, was crossed by that of his brother Louis IX. Already in 1267 St Louis had taken the cross a second time, moved by the news of Bibars' conquests; and though the French baronage, including even Joinville himself, refused to follow the lead of their king, Prince Edward of England imitated his example. Louis had been led to think that the bey of Tunis might be converted, and in that hope he resolved to begin this eighth and last of the Crusades by an expedition to Tunis. Charles, as anxious to attack Constantinople as he was reluctant to attack Tunis, with which Sicily had long had commercial relations, was forced to abandon his own plans and to join in those of his brother.[56] St Louis had barely landed in Tunis when he sickened and died, murmuring "Jerusalem, Jerusalem" (August 1270); but Charles, who appeared immediately after his brother's death, was able to conduct the Crusade to a successful conclusion. Negotiating in the spirit of a Frederick II., and acting not as a Crusader but as a king of Sicily, he not only wrested a large indemnity from the bey for himself and the new king of France, but also secured a large annual tribute for his Sicilian exchequer. So ended the Eighth Crusade—much as the Sixth had done—to the profound disgust of many of the crusaders, including Prince Edward of England, who only arrived on the eve of the conclusion of the treaty. Baulked of any opportunity of joining in the main Crusade, Edward, after wintering in Sicily, conducted a Crusade of his own to Acre in the spring of 1271. For over a year he stayed in the Holy Land, making little sallies from Acre, and negotiating with the Mongols, but achieving no permanent results. He returned home at the end of 1272, the last of the western crusaders; and thus all the attempts of St Louis and Charles of Anjou, of James of Aragon and Edward of England left Bibars still in possession of all his conquests.

Two projects of Crusades were started before the final expulsion of the Latins from Syria. In 1274, at the council of Lyons, Gregory X., who had been the companion of Edward in the Holy Land, preached the Crusade to an assembly which contained envoys from the Mongol khan and Michael Palaeologus as well as from many western princes. All the princes of western Europe took the cross; not only so, but Gregory was successful in uniting the Eastern and Western churches for the moment, and in securing for the new Crusade the aid of the Palaeologi, now thoroughly alarmed by the plans of Charles of Anjou. Thus was a papal Crusade begun, backed by an alliance with Constantinople, and thus were the plans of Charles of Anjou temporarily thwarted. But in 1276 Gregory X. died, and all his plans died with him; there was to be no union of the monarchs of the West with the emperor of the East in a common Crusade. Charles was able to resume his plans. In 1277 Mary of Antioch ceded to him her claims, and he was able to establish himself in Acre; in 1278 he took possession of the principality of Achaea. With these bases at his disposal he began to prepare a new Crusade, to be directed primarily (like that of Henry VI. in 1197, and like his own projected Crusade of 1270) against Constantinople. Once more his plans were crossed finally and fatally: the Sicilian Vespers, and the coronation of Peter of Aragon as Sicilian king (1282), gave him troubles at home which occupied him for the rest of his days. This was the last serious attempt at a Crusade on behalf of the dying kingdom of Jerusalem which was made in the West; and its collapse was quickly followed by the final extinction of the kingdom. A precarious peace had reigned in the Holy Land since 1272, when Bibars had granted a truce of ten years; but the fall of the great power of Charles of Anjou set free Kalā'ūn the successor of Bibars' son (who reigned little more than two years), to complete the work of the great sultan. In 1289 Kalā'ūn took Tripoli, and the county of Tripoli was extinguished; in 1290 he died while preparing to besiege Acre, which was captured after a brave defence by his son and successor Khālil in 1291. Thus the kingdom of Jerusalem came to an end. The Franks evacuated Syria altogether, leaving behind them only the ruins of their castles to bear witness, to this very day, of the Crusades they had waged and the kingdom they had founded and lost.

9. *The Ghost of the Crusades.*—The loss of Acre failed to stimulate the powers of Europe to any new effort. France, always the natural home of the Crusades, was too fully occupied, first by war with England and then by a struggle with the papacy, to turn her energies towards the East. But it is often the case that theory develops as practice fails; and as the theory of the Holy Roman Empire was never more vigorous than in the days of its decrepitude, so it was with the Crusades. Particularly in the first quarter of the 14th century, writers were busy in explaining the causes of the failures of past Crusades, and in laying down the lines along which a new Crusade must proceed. Several causes are recognized by these writers as accounting for the failure of the Crusades. Some of them lay the blame on the papacy; and it is true that the papacy had contributed towards the decay of the Crusades when it had allowed its own particular interests to overbear the general welfare of Christianity, and had dignified with the name and the benefits of a Crusade its own political war against the Hohenstaufen. Others again find in the princes of Europe the authors of the ruin of the Crusades; they too had preferred their own national or dynastic interests to the cause of a common Christianity. They had indeed, as has been already noticed, done even more; they had used the name of Crusade, from the days of Henry VI. onwards, as a cover and an excuse for secular ambitions of their own; and in this way they had certainly helped, in very large measure, to discourage the old religious zeal for the Holy War. Other writers, again, blame the commercial cupidity of the Italian towns; of what avail, they asked with no little justice, was the Crusade, when Venice and Genoa destroyed the naval bases necessary for its success by their internecine quarrels in the Levant (as in 1257), or—still worse—entered into commercial treaties with the common enemy against whom the Crusades were directed? On the very eve of the Fifth Crusade, Venice had concluded a commercial treaty with Malik-al-Kamil of Egypt; just before the fall of Acre the Genoese, the king of Aragon and the king of Sicily had all concluded advantageous treaties with the sultan Kalā'ūn. A fourth cause, on which many writers dwelt, particularly at the time when the suppression of the Templars was in question, was the

dissensions between the two orders of Templars and Hospitallers, and the selfish policy of merely pursuing their own interest which was followed by both in common. But one might enumerate *ad infinitum* the causes of the failure of the Crusades. It is simplest, as it is truest, to say that the Crusades did not fail—they simply ceased; and they ceased because they were no longer in joint with the times. The moral character of Europe in 1300 was no longer the moral character of Europe in 1100; and the Crusades, which had been the active and objective embodiment of the other worldly Europe of 1100, were alien to the secular, legal, scholastic Europe of 1300. While Edward I. was seeking to found a united kingdom in Great Britain; while the Habsburgs were entrenching themselves in Austria; above all, while Philippe le Bel and his legists were consolidating the French monarchy on an absolutist basis, there could be little thought of the holy war. These were hard-headed men of affairs—men who would not lightly embark on joyous ventures, or seek for an ideal San Grail; nor were the popes, doomed to the Babylonian captivity for seventy long years at Avignon, able to call down the spark from on high which should consume all earthly ambitions in one great act of sacrifice.

But it is long before the death of any institution is recognized; and it was inevitable that men should busy themselves in trying to rekindle the dead embers into new life. Pierre Dubois, in a pamphlet "*De recuperatione Sanctae Terrae*," addressed to Edward I. in 1307, advocates a general council of Europe to maintain peace and prevent the dissensions which—as, for instance, in 1192—had helped to cause the failure of past Crusades. Along with this advocacy of internationalism goes a plea for the disendowment of the Church, in order to provide an adequate financial basis for the future Crusade. Other proposals, made by men well acquainted with the East, are more definitely practical and less political in their intention. A blockade of Egypt by an international fleet, an alliance with the Mongols, the union of the two great orders—these are the three staple heads of these proposals. Something, indeed, was attempted, if little was actually done, under each of these three heads. The plan of an international fleet to coerce the Mahommedan is even to this day ineffective; but the Hospitallers, who acquired a new basis by the conquest of Rhodes in 1310, used their fleet to enforce a partial and, on the whole, ineffective blockade of the coast of the Levant. The union of the two orders, already suggested at the council of Lyons in 1245, was nominally achieved by the council of Vienne in 1311; but the so-called "union" was in reality the suppression of the Templars, and the confiscation of all their resources by the cupidity of Philippe le Bel. The alliance with the Mongols remained, from the first to the last, something of a chimera; and the last visionary hope vanished when the Mongols finally embraced Mahommedanism, as, by the end of the 14th century, they had almost universally done.

Isolated enterprises somewhat of the character of a Crusade, but hardly serious enough to be dignified by that name, recur during the 14th century. The French kings are all crusaders—in name—until the beginning of the Hundred Years' War; but the only crusader who ever carried war in Palestine and sought to shake the hold of the Mamelukes on the Holy Land was Peter I., king of Cyprus from 1359 to 1369. Peter founded the order of the Sword for the delivery of Jerusalem; and instigated by his chancellor, P. de Mézières (one of the last of the theorists who speculated and wrote on the Crusades), he attempted to revive the old crusading spirit throughout the west of Europe. The mission which he undertook with his chancellor for this purpose (1362-1365) only produced a crop of promises or excuses from sovereigns like Edward III. or the Emperor Charles IV.; and Peter was forced to begin the Crusade with such volunteers as he could collect for himself. In the autumn of 1365 he sacked Alexandria; in 1367 he ravaged the coast of Syria, and inflicted serious damages on the sultan of Egypt. But in 1369 he was assassinated, and the last romantic figure of the Crusades died, leaving only the legacy of his memory to his chancellor de Mézières, who for nearly forty years longer continued to be the preacher of the Crusades to Europe, advocating—what always continued to be the "dream of the old pilgrim"—a new order of knights of the Passion of Christ for the recovery and defence of Jerusalem. De Mézières was the last to advocate seriously, as Peter I. was the last to attempt, a Crusade after the old fashion—an offensive war against Egypt for the recovery of the Holy Sepulchre.[57] From 1350 onwards the Crusade assumes a new aspect; it becomes defensive, and it is directed against the Ottoman Turks, a tribe of Turcomans who had established themselves in the sultanate of Iconium at the end of the 13th century, during the confusion and displacement of peoples which attended the Mongol invasions. As early as 1308 the Ottoman Turks had begun to settle in Europe; by 1350 they had organized their terrible army of janissaries. They threatened at once the débris of the old Latin empire in Greece and the archipelago, and the relics of the Byzantine empire round Constantinople; they menaced the Hospitallers in Rhodes and the Lusignans in Cyprus. It was natural that the popes should endeavour to form a coalition between the various Christian powers which were threatened by the Turks; and Venice, anxious to preserve her possessions in the Aegean, zealously seconded their efforts. In 1344 a Crusade, in which Venice, the Cypriots, and the Hospitallers all joined, ended in the conquest of Smyrna; in 1345 another Crusade, led by Humbert, dauphin of Vienne, ended in failure. The Turks continued their progress; in 1363 they captured Philippopolis, and in 1365 they entered Adrianople; the whole Balkan peninsula was threatened, and even Hungary itself seemed doomed. Already in 1365 Urban VI. sought to unite the king of Hungary and the king of Cyprus in a common Crusade against the Turks; but it was not till 1396 that an attempt was at last made to supplement by a land Crusade the naval Crusades of 1344 and 1345. Master of Servia and of Bulgaria, as well as of Asia Minor, the sultan Bayezid was now threatening Constantinople itself. To arrest his progress, a Crusade, preached by Boniface IX., led by John the Fearless of Burgundy, and joined chiefly by French knights, was directed down the valley of the Danube into the Balkans; but the old faults stigmatized by de Mézières, *divisio* and *propria voluntas*, were the ruin of the crusading army, and at the battle of Nicopolis it was signally defeated. Not the Western Crusades but an Eastern rival, Timur (Tamerlane), king of Transoxiana and conqueror of southern Russia and India, was destined to arrest the progress of Bayezid; and from the battle of Angora (1402) till the days of Murad II. (1422) the Ottoman power was paralysed. Under Murad, however, it rose to its old height. To meet the new danger a new union of the churches of the East and the West was attempted. As in 1074 Gregory VII. had dreamed of such a union, to be followed by a joint attack of East and West on the Seljuks, so in 1439, at the council of Florence, a new union of the two churches was again attempted and temporarily secured, in order that a united Christendom might face the new Turkish danger.[58] The logical result of the union was the Crusade of 1443. An army of cosmopolitan adventurers, led by the Cardinal Caesarini, joined the forces of Wladislaus of Poland and John Hunyadi of Transylvania, and succeeded in forcing on Murad II. a truce of ten years at Szegedin in 1444. But the crusaders broke the truce, to which Caesarini had never consented; and, attempting to better what was already good enough, they were defeated at Varna. Here the last Crusade ended; and nine years afterwards, in 1453, Mahommed II., the successor of Murad, captured Constantinople. It was in vain that the popes sought to gather a new Crusade for its recovery; Pius II., who had vowed to join the crusade in person, only reached Ancona in 1464 to find the crusaders deserting and to die. Yet the ghost of the Crusades still lingered. It became a convention of diplomacy, designed to cover any particularly sharp piece of policy which needed some excuse; and the treaty of Granada, formed between Louis XII. and Ferdinand of Aragon for the partition of Naples in 1500, was excused as a thing necessary in the interests of the Crusades. In a more noble fashion the Crusade survived in the minds of the navigators; "Vasco da Gama, Christopher Columbus, Albuquerque, and many others dreamed, and not insincerely, that they were labouring for the deliverance of the Holy Land, and they bore the Cross on their breasts."[59] "Don Henrique's scheme," it has been said, "represents the final effort of the crusading spirit; and the naval campaigns against the Moslem in the Indian seas, in which it culminated, forty years after Don Henrique's death, may be described as the last Crusade."[60]

10. *Results of the Crusades.*—In one vital respect the result of the Crusades may be written down as failure. They ended, not in the occupation of the East by the Christian West, but in the conquest of the West by the Mahommedan East. The Crusades began with the Seljukian Turk planted at Nicaea; they ended with the Ottoman Turk entrenched by the Danube. Nothing is more striking in history than the recession of Christianity in the East after the 13th century. In the 13th century the whole of Europe was Christian; part of Asia Minor still belonged to Greek Christianity, and there was a Christian kingdom in Palestine. Nor was this all. A wide missionary activity had begun in the 13th century—an activity which was the product of the Crusades and the contact with the Moslem which they brought, but which yet helped to check the Crusades, substituting as it did peaceful and spiritual conquests of souls for the violence and materialism of even a Holy War. The Eastern mission had been begun by St Francis, who had visited and attempted to convert the sultan of Egypt during the Fifth Crusade (1220); within a hundred years the little seed had grown into a great tree. A great field for missionary enterprise opened itself in the Mongol empire, in which, as has already been mentioned, there were many Christians to be found; and by 1350 this field had been so well worked that Christian missions and Christian bishops were established from Persia to Peking, and from the Dnieper to Tibet itself. But a Mahommedan reaction came, thanks in large measure to the zeal of Timur; and central Asia was lost to Christianity. Everywhere in the 15th century, in Europe and in Asia, the crescent was victorious over the cross; and Crusade and mission, whether one regards them as complementary or inimical, perished together.[61]

But the history of the Crusades must be viewed rather as a chapter in the history of civilization in the West itself, than as an extension of Western dominion or religion to the East. It is a chapter very difficult to write, for while on the one hand an ingenious and speculative historian may refer to the influence of the Crusades almost everything which was thought or done between 1100 and 1300, a cautious writer who seeks to find documentary evidence for every assertion may be rather inclined to attribute to that influence little or nothing.[62] The dissolution of feudalism, the development of towns, the growth of scholasticism, all these and much more have been ascribed to the Crusades, when in truth they were concomitants rather than results, or at any rate, if in part the results of the Crusades, were in far larger part the results of other things. At most, therefore, it may be admitted that the Crusades *contributed* to the dissolution of feudalism by putting property on the market and disturbing the validity of titles; that they aided the development of towns by vastly increasing the volume of trade; and that they furthered the growth of scholasticism by bringing the West into contact with the mind of the East. If we seek the peculiar and definite results of the Crusades, we must turn to narrower issues. In the first place, the Crusades represent the attempt of a feudal system, bound under the law of primogeniture to dispose of its younger sons. They are attempts at feudal colonization; and as such they resulted in a number of colonies—the kingdom of Jerusalem, the kingdom of Cyprus, the Latin empire of Constantinople. They resulted too in a number of "chartered companies"—that is to say, the three military orders, which, beginning as charitable societies, developed into military clubs, and developed again from military clubs into chartered companies, possessed of banks, navies and considerable territories. In the second place, as has already been noticed, the Crusades represent the attempt of Western commerce to find new and more easy routes to the wealth of the East; and in this respect they led to various results. On the one hand they led to the establishment of emporia in the East—for instance, Acre, and after the fall of Acre Famagusta, both in their day great centres of Levantine trade. On the other hand, the commodities which poured into Venice and Genoa from the East had to find a route for their diffusion through Europe. The great route was that which led from Venice over the Brenner and up the Rhine to Bruges; and this route became the long red line of municipal development, along which—in Lombardy, Germany and Flanders—the great towns of the middle ages sprang to life. Partly as a result of this trade, ever pushing its way farther east, and partly as a result of the Asiatic missions, which were themselves an accompaniment and effect of the Crusades, a third great result of the Crusades came to light in the 13th century—the discovery of the interior of Asia, and an immense accession to the sphere of geography. When one remembers that missionaries like Piano Carpini, and traders like the Venetian Polos, either penetrated by land from Acre to Peking, or circumnavigated southern Asia from Basra to Canton, one realizes that there was, about 1300, a discovery of Asia as new and tremendous as the discovery of America by Columbus two centuries later. At the same time the old knowledge of nearer Asia was immensely deepened. It has already been noticed how military reconnaissances of the routes to Egypt came to be made; but more important were the guide-books, of which a great number were written to guide the pilgrims from one sacred spot of Bible history to another. There were medieval Baedekers in abundance for the use of the annual flow of tourists, who were carried every Easter by the vessels of the Italian towns or of the Orders to visit the Holy Land and to bathe in Jordan, to gather palms, and to see the miracle of fire at the Sepulchre.

Colonization, trade, geography—these then are three things closely connected with the history of the Crusades. The development of the art of war, and the growth of a systematic taxation, are two debts which medieval Europe also owed to the Crusades. Partly by contact with the Byzantines, partly by conflict with the Mahommedans, the Franks learned new methods both of building and of attacking fortifications. The concentric castle, with its rings of walls, began to displace the old keep and bailey with their single wall, as the crusaders brought back news from the East.[63] The art of the sapper and miner, the use of siege instruments like the mangonel, and the employment of various "fires" as missiles, were all known among the Mahommedans; and in all these respects the Franks learned from their enemies. The common use of armorial bearings, and the practice of the tournament, may be Oriental in their origin; the latter has its affinities with the equestrian exercises of the Jerid, and the former, though of prehistoric antiquity, may have received a new impulse from contact with the Arabs. The military development which sprang from the Crusades is thus largely a matter of borrowing; the financial development is independent and indigenous in the West. As early as 1147 Louis VII. had imposed a tax in the interests of the Crusades; and that tax had been repeated by Louis, and imitated by Henry II. in 1166, while it had been still further extended in the Saladin tithe of 1188. The taxation of 1166 is important as the first to fall on "moveables"; the whole scheme of taxation may be regarded as the beginning of a modern system of taxation. But it was not only to the lay power that the Crusades gave an excuse for taxation; the papacy also profited. Tithes for the Crusades were first imposed on the clergy by Innocent III. at the Lateran council of 1215; and clerical taxation was thus part of the whole statesmanlike project of the Fifth Crusade as it was sketched by the great pope. Henceforth tithes for the Crusades are regular; under Gregory IX. they become a great part of the papal resources in the Crusade against the Hohenstaufen; and in the 16th century they are still a normal part of the government of the Church.

SYRIA
in the 12th. Century, before the conquests of Saladin

In many other ways the Europe over which the Crusades had passed was different from the Europe of the 11th century. In the first place, many political changes had been wrought, largely under its influence. Always in large part French, the Crusades had on the whole contributed to exalt the prestige of France, until it stood at the end of the 13th century the most considerable power in Europe. It was France which had colonized the Levant; it was the French tongue which was used in the Levant; and the results of the ancient and continuous connexion with the East are still to be traced to-day. Of the other great powers of Europe, England and Germany had been little changed by the Crusades, save that Germany had been extended towards the East by the conquests of the Teutonic Order; but the Eastern empire had been profoundly modified, and the papacy had suffered a great change. The Eastern empire had been for a time annihilated by the movement which in 1095 it had helped to evoke; and if it rose from its ashes in 1261 for two centuries of renewed life, it was never more than the shadow of its old self, with little hold on Asia Minor and less on Greece and the Archipelago, which the Latins still continued to occupy until they were finally conquered by the Ottoman Turks. The papacy, on the other hand, had grown as a result of the Crusades. Popes had preached them; popes had financed them; popes had sent their legates to lead them. Through them the popes had deposed the emperors of the West from their headship of the world, partly because through the Crusades the popes were able to direct the common Christianity of Europe in a foreign policy of their own without consultation with the emperor, partly because in the 13th century they were ultimately able to direct the Crusade itself against the empire. Yet while they had magnified, the Crusades had also corrupted the papacy. They became an instrument in its hands which it used to its own undoing. It cried Crusade when there was no Crusade; and the long Crusade against the Hohenstaufen, if it gave the papacy an apparent victory, only served in the long run to lower its prestige in the eyes of Europe. When we turn from the sphere of politics to the history of civilization and culture, we find the effects of the Crusades as deeply impressed, if not so definitely marked. The Crusades had sprung from the policy of a theocratic government counting on the motive of otherworldliness; they had helped in their course to overthrow that motive, and with it the government which it had made possible. In part they had provided a field in which the layman could prove that he too was a priest; in part they had brought the West into a living and continuous contact with a new faith and a new civilization. They had torn men loose from the ancestral custom of home to walk in new ways and see new things and hear new thoughts; and some broadening of view, some lessening in the intensity of the old one-sidedness, was the inevitable result. It is not so much that the West came into contact with a particular civilization in the East, or borrowed from that civilization; it is simply that the West came into contact with something unlike itself, yet in many ways as high as, if not higher than, itself. The spirit of *Nathan der Weise* may not have been exactly the spirit engendered by the Crusades; and yet it is not without reason that Lessing stages the fable which teaches toleration in the Latin kingdom of Jerusalem. In any case the accusations made against the Templars at the time of their suppression prove that there was, at any rate in the ranks of those who knew the East, too little of absolute orthodoxy. While a new spirit which compares and tolerates thus sprang from the Crusades, the large sphere of new knowledge and experience which they gave brought new material at once for scientific thought and poetic imagination. Not only was geography more studied; the Crusades gave a great impulse to the writing of history, and produced, besides innumerable other works, the greatest historical work of the middle ages—the *Historia transmarina* of William of Tyre. Mathematics received an impulse, largely, it is true, from the Arabs of Spain, but also from the East; Leonardo Fibonacci, the first Christian algebraist, had travelled in Syria and Egypt. The study of Oriental languages began in connexion with the Christian missions of the East; Raymond Lull, the indefatigable missionary, induced the council of Vienne to decide on the creation of six schools of Oriental languages in Europe (1311). But the new field of poetic literature afforded by the Crusades is still more striking than this development of science. New poems in abundance dealt with the history of the Crusades, either in a faithful narrative, like that of the *Chanson* of Ambroise, which narrates the Third Crusade, or in a free and poetical spirit, such as breathes in the *Chanson d'Antioche*. Nor was this all. The Crusades afforded new details which might be inserted into old matters, and a new spirit which might be infused into old subjects; and a crusading complexion thus came to be put upon old tales like those of Arthur and Charlemagne. By the side of these greater things it may seem little, and yet, just because it is little, it is all the more significant that the Crusades should have familiarized Europe with new plants, new fruits, new manufactures, new colours, and new fashions in dress. Sugar and maize; lemons, apricots and melons; cotton, muslin and damask; lilac and purple (azure and gules are words derived from the Arabic); the use of powder and of glass mirrors, and also of the rosary itself—all these things came to Europe from the East and as a result of the Crusades. To this day there are many Arabic words in the vocabulary of the languages of western Europe which are a standing witness of the Crusades—words relating to trade and seafaring, like tariff and corvette, or words for musical instruments, like lute or the Elizabethan word "naker."

GENEALOGY OF THE KINGS OF JERUSALEM

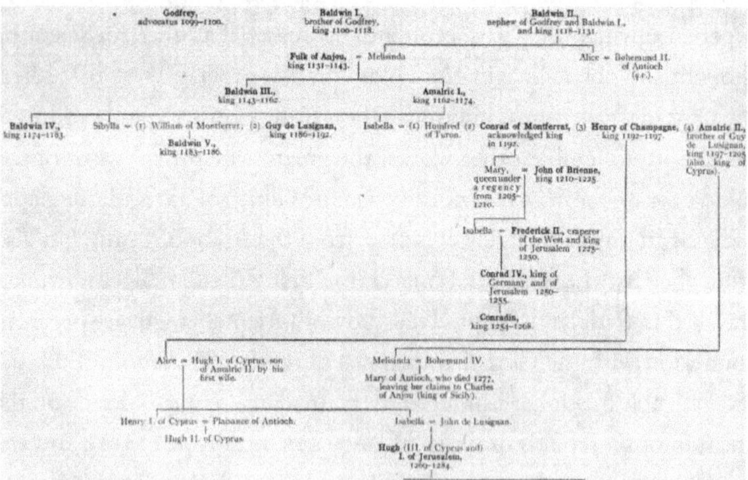

When all is said, the Crusades remain a wonderful and perpetually astonishing act in the great drama of human life. They touched the summits of daring and devotion, if they also sank into the deep abysms of shame. Motives of self-interest may have lurked in them—otherworldly motives of buying salvation for a little price, or worldly motives of achieving riches and acquiring lands. Yet it would be treason to the majesty of man's incessant struggle towards an ideal good, if one were to deny that in and through the Crusades men strove for righteousness' sake to extend the kingdom of God upon earth. Therefore the tears and the blood that were shed were not unavailing; the heroism and the chivalry were not wasted. Humanity is the richer for the memory of those millions of men, who followed the pillar of cloud and fire in the sure and certain hope of an eternal reward. The ages were not dark in which Christianity could gather itself together in a common cause, and carry the flag of its faith to the grave of its Redeemer; nor can we but give thanks for their memory, even if for us religion is of the spirit, and Jerusalem in the heart of every man who believes in Christ.

LITERATURE.—In dealing with the literature of the Crusades, it is perhaps better, though ideally less scientific, to begin with chronicles and narratives rather than with documents. One of the results of the Crusades, as has just been suggested above, was a great increase in the writing of history. Crusaders themselves kept diaries or *itineraria*; while home-keeping ecclesiastics in the West—monks like Robert of Reims, abbots like Guibert of Nogent, archbishops like Balderich of Dol—found a fertile subject for their pens in the history of the Crusades. The history of a series of actions like the Crusades must primarily be based on these accounts, and more particularly on the former: narratives must precede documents where one is dealing, not with the continuous life of an organized kingdom, but with a number of enterprises—especially when those enterprises have been, as in this case, excellently narrated by contemporary writers.

I. *Chronicles and Narratives of the Crusades*—(1) Collections. The authorities for the Crusades have been collected in Bongars, *Gesta Dei per Francos* (Hanover, 1611) (incomplete); Michaud, *Bibliothèque des croisades* (Paris, 1829) (containing translations of select passages in the authorities); the *Recueil des historiens des croisades*, published by the Académie des Inscriptions (Paris, 1841 onwards) (the best general collection, containing many of the Latin, Greek, Arabic and Armenian authorities, and also the text of the assizes; but sometimes poorly edited and still incomplete); and the publications of the Société de l'Orient Latin (founded in 1875), especially the *Archives*, of which two volumes were published in 1881 and 1884, and the volumes of the *Revue*, published yearly from 1893 to 1902, and containing not only new texts, but articles and reviews of books which are of great service. (2) Particular authorities. The Crusades—a movement which engaged all Europe and brought the East into contact with the West—must necessarily be studied not only in the Latin authorities of Europe and of Palestine, but also in Byzantine, Armenian and Arabic writers. There are thus some four or five different points of view to be considered.

The *First Crusade*, far more than any other, became the theme of a multitude of writings, whose different degrees of value it is all-important to distinguish. Until about 1840 the authority followed for its history was naturally the great work of William of Tyre. For the First Crusade William had followed Albert of Aix; and he had consequently depicted Peter the Hermit as the prime mover in the Crusade. But about 1840 Ranke suggested, and von Sybel in his *Geschichte des ersten Kreuzzüges* proved, that Albert of Aix was *not* a good authority, and that consequently William of Tyre must be set aside for the history of the First Crusade, and other and more contemporary authorities used. In writing his account of the First Crusade, von Sybel accordingly based himself on the three contemporary Western authorities—the *Gesta Francorum*, Raymond of Agiles, and Fulcher. His view of the value of Albert of Aix, and his account of the First Crusade, have been generally followed (Kugler alone having attempted, to some extent, to rehabilitate Albert of Aix); and thus von Sybel's work may be said to mark a revolution in the history of the First Crusade, when its legendary features were stripped away, and its real progress was first properly discovered.

Taking the Western authorities for the First Crusade separately, one may divide them, in the light of von Sybel's work, into four kinds—the accounts of eye-witnesses; later compilations based on these accounts; semi-legendary and legendary narratives; and lastly, in a class by itself, the "History" of William of Tyre, who is rather a scientific historian than a chronicler.

(*a*) The three chief eye-witnesses are the anonymous author of the *Gesta Francorum*, Raymund of Agiles, and Fulcher. The anonymous author of the *Gesta* (see Hagenmeyer's edition, Heidelberg, 1890) was a Norman of South Italy, who followed Bohemund, and accordingly depicts the progress of the First Crusade from a Norman point of view. He was a layman, marching and fighting in the ranks; and thus he is additionally valuable as representing the opinion of the ordinary crusader. Finally he was an eye-witness throughout, and absolutely contemporary, in the sense that he wrote his account of each great event practically at the time of the event. He is the primary authority for the First Crusade. Raymund of Agiles, a Provençal clerk and a follower of Raymund of Toulouse, writes his *Historia Francorum qui ceperunt Jerusalem* from the Provençal point of view. He gives an ecclesiastic's account of the First Crusade, and is specially full on the spiritualistic phenomena which accompanied and followed the finding of the Holy Lance. His book might almost be called the "Visions of Peter Bartholomew and others," and it is written in the plain matter-of-fact manner of Defoe's narratives. He too was an eye-witness throughout, and thoroughly honest; and his account ranks second to the *Gesta*. Fulcher of Chartres originally followed Robert of Normandy, but in October 1097 he joined Baldwin of Lorraine in his expedition to Edessa, and afterwards followed his fortunes. His *Historia Hierosolymitana*, which extends to 1127, and embraces not only the history of the First Crusade, but also that of the foundation of the kingdom of Jerusalem, is written on the whole from a Lotharingian point of view, and is thus a natural complement to the accounts of the Anonymus and Raymund. His account of the First Crusade itself is poor (he was absent at Edessa during its course), but otherwise he is an excellent authority. A kindly old pedant, Fulcher interlards his history with much discourse on geography, zoology and sacred history. Besides these three chief eye-witnesses we may also mention the *Annales Genuenses* by the Genoese consul Caffarus,[64] and the *Annales Pisani* of Bernardus Marago, useful as giving the mercantile and Italian side of the Crusade; the *Hierosolymita* of Ekkehard, the German abbot of Aura, who first came to Jerusalem about 1101 (partly based on the *Gesta*, but also of independent value: see Hagenmeyer's edition, Tübingen, 1877); and Raoul of Caen's *Gesta Tancredi*, composed on the basis of information supplied by Tancred himself. The last two works, if not actually the works of eye-witnesses, are at any rate first-hand, and belong to the category of primary writers rather than to that of later compilations. Finally, to contemporary writers we may add contemporary letters, especially those written by Stephen of Blois and Anselm of Ribemont, and the three letters sent to the West by the crusading princes during the First Crusade (see Hagenmeyer, *Epistulae et Chartae*, &c., Innsbruck, 1901).[65]

(*b*) The later compilations are chiefly based on the *Gesta*, whose uncouth style many writers set themselves to mend. In the first place, there is the *Historia de Hierosolymitano itinere* of Tudebod, which according to Besly, writing in 1641, is the original from which the *Gesta* was a mere plagiarism—an absolute inversion of the truth, as von Sybel first proved two centuries later. Secondly, besides the plagiarist Tudebod, there are the artistic *rédacteurs* of the *Gesta*, who confess their indebtedness, but plead the bad style of their original—Guibert of Nogent, Balderich of Dol, Robert of Reims (all *c.* 1120-1130), and Fulco, the author of a Virgilian poem on the Crusades, continued by Gilo (*ob. c.* 1142). Of these, the monk Robert was more popular in the middle ages than either the pompous abbot Guibert or the quiet garden-loving archbishop of Dol.

(*c*) The growth of a legend, or perhaps better, a saga of the First Crusade began, according to von Sybel, even during the Crusade itself. The basis of this growth is partly the story-telling instinct innate in all men, which loves to heighten an effect, sharpen a point or increase a contrast—the instinct which breathes in Icelandic sagas like that of *Burnt Njal*; partly the instinct of idolization, if it may be so called, which leads to the perversion into impossible greatness of an approved character, and has created, in this instance, the legendary figures of Peter the Hermit and Godfrey of Bouillon (*qq.v.*); partly the religious impulse, which counted nothing wonderful in a holy war, and imported miraculous elements even into the sober pages of the *Gesta*. These instincts and impulses would be at work already among the soldiers during the Crusade, producing a saga all the more readily, as there were poets in the camp; for we know that a certain Richard, who joined the First Crusade, sang its exploits in verse, while still more famous is the princely troubadour, William of Aquitaine, who joined the Crusade of 1100. If we are to follow von Sybel rather than Kugler, this saga of the First Crusade found one of its earliest expressions (*c.* 1120) in the prose work of Albert of Aix (*Historia Hierosolymitana*)—genuine saga in its inconsistencies, its errors of chronology and topography, its poetical colour, and its living descriptions of battles. Kugler, however, regards Albert as a copyist, somewhat in the manner of Tudebod, of an unknown writer of value, who belonged to the Lotharingian ranks during the Crusade, and settled in the kingdom of Jerusalem afterwards (see Kugler, *Albert von Aachen*, Stuttgart, 1885).[66] In the *Chanson des chétifs* and the *Chanson d'Antioche* the legend of the Crusades more certainly finds its expression. The former, composed at Antioch about 1130, contained an idolization of the Hermit: the latter is a poem written about 1180 by Graindor of Douai, who used as his basis the verses of the crusader Richard (see the edition of P. Paris, 1848). It shows the growth of the legend that Graindor regards the vision of the Hermit as responsible for the Crusade, and makes the Crusade led by him precede, and indeed occasion by its failure, the meeting at Clermont (which is dated in May instead of November). Into the legendary overgrowth of the First Crusade we cannot here enter any further[67]; but it is perhaps worth while to mention that the French legend of the Third Crusade equally perverted the truth, making Richard I. return home in disgrace, while Philip Augustus stays, captures Damascus and mortally wounds Saladin (cf. G. Paris, *L'Estoire de la guerre sainte*, Paris, 1897; Introduction).

(*d*) William of Tyre is the scientific historian and rationalizer, weaving into a harmonious account, which was followed by historians for centuries, the sober accounts of eye-witnesses and the picturesque details of the saga—with somewhat of a bias towards the latter in regard to the First Crusade. He was a native of Palestine, born about 1130, and educated in the West. On his return he was happy in winning the good opinion of Amalric I.; he was made first canon and then archdeacon of Tyre, and tutor of the future Baldwin IV. (1170); while on Baldwin's accession he became chancellor of the kingdom and archbishop of Tyre (1174-1175). He was a man often employed on missions and negotiations, and as chancellor he had in his care the archives of the kingdom. His temper was naturally that of a trimmer; and he had thus many qualifications for the writing of well-informed and unbiassed history. He knew Greek and Arabic; and he was well acquainted with the affairs of Constantinople, to which he went at least twice on political business, and with the history of the Mahommedan powers, on which he had written a work (now lost) at the command of Amalric. It was Amalric also who set him to write the history of the Crusades which we still possess (in twenty-two books, with a fragment of a twenty-third)—the *Historia rerum in partibus transmarinis gestarum*. He wrote the book at different times between 1170 and 1183, when it abruptly ends, and its author as abruptly disappears from sight. The book falls into two parts, the first (books i.-xv.) derivative, the second (books xvi.-xxiii.) original. In the second part he had his own knowledge of events and the information of his contemporaries as his source: in the first he used the same authorities which we still possess—the *Gesta*, Fulcher, and Albert of Aix—in somewhat of an eclectic spirit, choosing now here, now there, according as he could best weave a pleasant narrative, but not according to any real critical principle. His book thus begins to be a real authority only from the date of the Second Crusade onwards; but the perfection of his form (for he is one of the greatest stylists of the middle ages) and the prestige of his position conspired to make his book the one authority for the whole history of the first century of the Crusades. Nor was he (apart from his reception of legendary elements into his narrative) unworthy of the honour in which he was held; for he is really a great historian, in the form of his matter and in his conception of his subject—diligent, impartial, well-informed and interesting, if somewhat rhetorical in style and vague in chronology. [During the middle ages his work was current in a French translation, known as the *Chronique d'outre-mer*, or the *Livre* or *Roman d'Éracles* (so called from the reference at the beginning to the emperor Heraclius). This translation also contained a continuation by various hands down to 1277; while besides the continuation embedded in the *Livre d'Éracles*, there are separate continuations, of the nature of independent works, by Ernoul and Bernard the Treasurer. These latter cover the period from 1183 to 1228; and of the two Ernoul's account seems primary, while that of Bernard is in large part a mere copy of Ernoul. But the whole subject of the continuators of William of Tyre is dubious.]

To the Western authorities for the First Crusade must be added the Eastern—Byzantine, Arabic and Armenian. Of these the Byzantine authority, the *Alexiad* of Anna Comnena, is most important, partly from the position of the authoress, partly from the many points of contact between the Byzantine empire and the crusaders. Anna's narrative both furnishes a useful corrective of the prejudiced Western accounts of Alexius, and serves to bring Bohemund forward into his proper prominence. The Armenian view of the First Crusade and of Baldwin's principality of Edessa is presented in the *Armenian Chronicle* of Matthew of Edessa. There is little in Arabic bearing on the First Crusade: the Arabic authorities only begin to be of value with the rise of the atabegs of Mosul (*c.* 1127). But Kemal-ud-din's *History of Aleppo* (composed in the 13th century) contains some details on the history of the First Crusade; and the *Vie d'Ousāma* (the autobiography of a sheik at Caesarea in northern Syria, edited and paraphrased by Derenbourg in the *Publications de l'École des langues orientales vivantes*) presents the point of view of an Arab whose life covered the first century of the Crusades (1095-1188).

For the *Second Crusade* the primary authority in the West is the work of Odo de Deuil, *De profectione Ludovici VII regis Francorum in Orientem.* Odo was a monk attached by Suger to Louis VII. during the Second Crusade; and he wrote home to Suger during the Crusade seven short letters, afterwards pieced together in a single work. The *Gesta Friderici Primi* of Otto of Freising (who joined in the Second Crusade) gives some details from the German point of view (i. c. 44 sqq.). The former is supplemented by the letters of Louis VII. to Suger; the latter by the letters of Conrad III. to Wibald, abbot of Stablo and Corvey. The Byzantine point of view is presented in the Ἐπιτομή of Cinnamus, the private secretary of Manuel, who continued the *Alexiad* of Anna Comnena in a work describing the reigns of John and Manuel. It is from the Second Crusade that William of Tyre, representing the attitude of the Franks of Jerusalem, begins to be a primary authority; while on the Mahommedan side a considerable authority emerges in Ibn Athīr. His history of the Atabegs was written about 1200, and it presents in a light favourable to Zengi and Nureddin, but unfavourable to Saladin (who thrust Nureddin's descendants aside), the history of the great Mahommedan power which finally crushed the kingdom of Jerusalem.[68]

Side by side with Beha-ud-dīn's life of Saladin, Ibn Athīr's work is the most considerable historical record written by the Arabs. Generally speaking the Arabic writings are late in point of date, and cold and jejune in style; while it must also be remembered that they are set religious works written to defend Islam. On the other hand they are generally written by men of affairs—governors, secretaries or ambassadors; and a fatalistic temper leads their authors to a certain impartial recording of everything, good or evil, which seems of moment.

The *Third Crusade* was narrated in the West from very different points of view by Anglo-Norman, French and German authorities. The primary Anglo-Norman authority is the *Carmen Ambrosii*, or, as it is called by M. Gaston Paris, *L'Estoire de la guerre sainte*. This is an octosyllabic poem in French verse, written by Ambroise, a Norman *trouvère* who followed Richard I. to the Holy Land. The poem first came to be known by scholars about 1873, and has been edited by M. Gaston Paris (Paris, 1897). The *Itinerarium Peregrinorum*, a work in ornate Latin prose, is (except for the first book) a translation of the *Carmen* masquerading under the guise of an independent work. There seems no doubt that it is a piece of plagiary, and that its writer, Richard, "canon of the Holy Trinity" in London, stands to the *Carmen* as Tudebod to the *Gesta*, or Albert of Aix to his supposed original. The Third Crusade is also described from the English point of view by all contemporary writers of history in England, *e.g.* Ralph of Coggeshall, who used information gained from crusaders, and William of Newburgh, who had access to a work by Richard I.'s chaplain Anselm, which is now lost.[69] The French side is presented in Rigord's *Gesta Philippi Augusti* and in the *Gesta* (an abridgment and continuation of Rigord) and the *Philippeis* of William the Breton. The two French writers represent Richard as a faithless vassal: in the German writers—Tagino, dean of Passau, who wrote a *Descriptio* of Barbarossa's Crusade (1189-1190); and Ansbert, an Austrian clerk, who wrote *De expeditione Friderici Imperatoris* (1187-1196)—Richard appears rather as a monster of pride and arrogance. From the

Arabic point of view the life of Richard's rival, Saladin, is described by Beha-ud-din, a high official under Saladin, who writes a panegyric on his master, somewhat confused in chronology and partial in its sympathies, but nevertheless of great value. The various continuations of William of Tyre above mentioned represent the opinion of the native Franks (which is hostile to Richard I.); while in Nicetas, who wrote a history of the Eastern empire from 1118 to 1206, we have a Byzantine authority who, as Professor Bury remarks, "differs from Anna and Cinnamus in his tone towards the crusaders, to whom he is surprisingly fair."

For the *Fourth Crusade* the primary authority is Villehardouin's *La Conquête de Constantinople*, an official apology for the diversion of the Crusade written by one of its leaders, and concealing the arcana under an appearance of frank naïveté. His work is usefully supplemented by the narrative (*La Prise de Constantinople*) of Robert de Clary, a knight from Picardy, who presents the non-official view of the Crusade, as it appeared to an ordinary soldier. The χρονικὸν τῶν ἐν Ῥωμανίᾳ (composed in Greek verse some time after 1300, apparently by an author of mixed Frankish and Greek parentage, and translated into French at an early date under the title "The Book of the Conquest of Constantinople and the Empire of Rumania") narrates in a prologue the events of the Fourth (as indeed also of the First) Crusade. The *Chronicle of the Morea* (as this work is generally called) is written from the Frankish point of view, in spite of its Greek verse; and the Byzantine point of view must be sought in Nicetas.[70]

The history of the later Crusades, from the Fifth to the Eighth, enters into the continuations of William of Tyre above mentioned; while the *Historia orientalis* of Jacques de Vitry, who had taken part in the Fifth Crusade, and died in 1240, embraces the history of events till 1218 (the third book being a later addition). The *Secreta fidelium Crucis* of Marino Sanudo, a history of the Crusades written by a Venetian noble between 1306 and 1321, is also of value, particularly for the Crusade of Frederick II. The minor authorities for the Fifth

Crusade have been collected by Röhricht, in the publications of the Société de l'Orient Latin for 1879 and 1882; the ten valuable letters of Oliver, bishop of Paderborn, and the *Historia Damiettina*, based on these letters, have also been edited by Röhricht in the *Westdeutsche Zeitschrift für Geschichte und Kunst* (1891). The Sixth Crusade, that of Frederick II., is described in the chronicle of Richard of San Germano, a notary of the emperor, and in other Western authorities, *e.g.* Roger of Wendover. For the Crusades of St Louis the chief authorities are Joinville's life of his master (whom he accompanied to Egypt on the Seventh Crusade), and de Nangis' *Gesta Ludovici regis*. Several works were written on the capture of Acre in 1291, especially the *Excidium urbis Acconensis*, a treatise which emerges to throw light, after many years of darkness, on the last hours of the kingdom. The Oriental point of view for the 13th century appears in Jelaleddin's history of the Ayyubite sultans of Egypt, written towards the end of the 13th century; in Maqrizi's history of Egypt, written in the middle of the 15th century; and in the compendium of the history of the human race by Abulfeda (†1332); while the omniscient Abulfaragius (whom Rey calls the Eastern St Thomas) wrote, in the latter half of the 13th century, a chronicle of universal history in Syriac, which he also issued, in an Arabic recension, as a *Compendious History of the Dynasties*.

II. The documents bearing on the history of the Crusades and the Latin kingdom of Jerusalem are various. Under the head of charters come the *Regesta regni Hierosolymitani*, published by Röhricht, Innsbruck, 1893 (with an Additamentum in 1904); the *Cartulaire générale des Hospitaliers*, by Delaville Leroulx (Paris, 1894 onwards); and the *Cartulaire de l'église du St Sépulcre*, by de Rozière (Paris, 1849). Under the head of laws come the assizes of the Kingdom, edited by Beugnot in the *Recueil des historiens des croisades*; and the assizes of Antioch, printed at Venice in 1876. G. Schlumberger has written on the coins and seals of the Latin East in various publications; while Rey has written an *Étude sur les monuments de l'architec-*

ture militaire (Paris, 1871). The genealogy of the Levant is given in *Le Livre des lignages d'outre-mer* (published along with the assizes).

Bibliographies.—The best modern account of the original authorities for the Crusades is that of A. Molinier, *Les Sources de l'histoire de France*, vols. ii. and iii. W. Wattenbach's *Deutschlands Geschichtsquellen* gives an account of Albert of Aix (vol. ii., ed. 1894, pp. 170-180) and of Ekkehard of Aura (*ibid.* pp. 189-198). Von Sybel's *Geschichte des ersten Kreuzzüges* contains a full study of the authorities for the First Crusade; while the prefaces to Hagenmeyer's editions of the *Gesta* and of Ekkehard are also valuable. Gaston Dodu, in the work mentioned below, begins by a brief account of the original authorities, which is chiefly of value so far as it deals with William of Tyre and the history of the assizes; and H. Prutz has also a short account of some of the historians of the Crusades (*Kulturgeschichte*, pp. 453-469). Finally reference may be made to the works of Kugler and Klimke above mentioned, and to J. F. Michaud's *Bibliographie des croisades* (Paris, 1822).

Modern Writers.—The various works of R. Röhricht present the soundest, if not the brightest, account of the Crusades. There is a *Geschichte des ersten Kreuzzugs* (Innsbruck, 1901), a *Geschichte des Königreichs Jerusalem* (*ibid.* 1898) and a *Geschichte der Kreuzzüge in Umris* (*ibid.* 1898). For the First Crusade von Sybel's work and Chalandon's *Alexis Ier Comnène* may also be mentioned; for the Fourth A. Luchaire's volume on *Innocent III: La Question d'Orient*; while for the whole of the Crusades Norden's *Papstum und Byzanz* is of value. B. Kugler's *Geschichte der Kreuzzüge* (in Oncken's series) still remains a suggestive and valuable work; and L. Bréhier's *L'Église et l'orient au moyen âge* (Paris, 1907) contains not only an up-to-date account of the Crusades, but also a full and useful bibliography, which should be consulted for fuller information. On points of chronology, and on the relations between the crusaders and their Mahommedan neighbours, W. B. Stevenson's *The Crusaders in the East* (Cambridge, 1907) is very valuable. On the constitutional and social history of the Latin kingdom of Jerusalem Dodu's *Histoire des institutions du royaume latin de Jérusalem* is very useful; E. G. Rey's *Les Colonies franques en Syrie* contains many interesting details; and Prutz's *Kulturgeschichte der Kreuzzüge* contains both an account of the Latin East and an attempt to sketch the effects of the Crusades on the progress of civilization. The works of Gmelin and J. Delaville-Leroulx on the Templars and Hospitallers respectively are worth consulting; while for Eastern affairs the English reader may be referred to G. Lestrange's *Palestine under the Moslem*, and to Stanley Lane-Poole's *Life of Saladin* and his *Mahommedan Dynasties* (the latter a valuable work of reference).

(E. Br.)

---

1 Fulcher of Chartres, 1, i. For what follows, with regard to the Church's conversion of *guerra* into the Holy War, cf. especially the passage—"Procedant contra infideles ad pugnam jam incipi dignam ... qui abusive *privatum certamen* contra fideles consuescebant distendere quondam."

2 Tradition credits a pope still earlier than Gregory VII. with the idea of a crusade. Silvester II. is said to have preached a general expedition for the recovery of Jerusalem; and the same preaching is attributed to Sergius IV. in 1011. But the supposed letter of Silvester is a later forgery; and in 1000 the way of the Christian to Jerusalem was still free and open.

3 The comte de Riant impugned the authenticity of Alexius' letter to the count of Flanders. It is very probable that the versions of this letter which we possess, and which are to be found only in later writings like Guibert de Nogent, are apocryphal; Alexius can hardly have held out the bait of the beauty of Greek women, or have written that he preferred to fall under the yoke of the Latins rather than that of the Turks. But it is also probable that these apocryphal versions are based on a genuine original.

4 Ekkehard, *Chronica*, p. 213.

5 The *Chanson de Roland*, which cannot be posterior to the First Crusade—for the poem never alludes to it—already contains the idea of the Holy War against Islam. The idea of the crusade had thus already ripened in French poetry, before Urban preached his sermon.

6 Book i. c. iii. (in Muratori, *S.R.I.*, v. 550).

7 Ekkehard, *Chronica*, 214.

8 Later legend ascribed the origin of the First Crusade to the preaching of Peter

the Hermit. The legend has been followed by modern historians; but in point of fact Peter is a figure of secondary importance.(See PETER THE HERMIT.)

9 Godfrey's army numbered some 30,000 infantry and 10,000 cavalry (Röhricht, *Erst. Kreuzz.* 61): Urban II. reckons Bohemund's knights as 7000 in number (*ibid.* 71, n. 7).

10 The Genoese had been invited by Urban II. in September 1096 "to go with their gallies to Eastern parts in order to set free the path to the Lord's Sepulchre."

11 Thus already on the First Crusade the path of negotiation is attempted simultaneously with the Holy War. On the Third Crusade, and above all on the Sixth, this path was still more seriously attempted. It is interesting, too, to notice the part which the laity already plays in directing the course of the Crusade. From the first the Crusade, however clerical in its conception, was largely secular in its conduct; and thus, somewhat paradoxically, a religious enterprise aided the growth of the secular motive, and contributed to the escape of the laity from that tendency towards a papal theocracy, which was evident in the pontificate of Gregory VII.

12 Before he left, Raymund had played in Jerusalem the same part of dog in the manger which he had also played at Antioch, and had given Godfrey considerable trouble. See the articles, GODFREY OF BOUILLON and RAYMUND OF TOULOUSE.

13 For an account of the kings of Jerusalem see the articles on the five BALDWINS, on the two AMALRICS, on FULK and JOHN OF BRIENNE and on the LUSIGNAN (family).

14 The genuineness of the letter (on which, by the way, depends the story of Godfrey's agreement with Dagobert) has been impeached by Prutz and Kugler, and doubted by Röhricht. It is accepted by von Sybel and Hagenmeyer.

15 Yet the north always continued to be more populous than the south; and the Latins maintained themselves in Antioch and Tripoli a century after the loss of Jerusalem. The land was richer in the north: it was protected by its connexion with Cyprus and Armenia: it was more remote from Egypt—the basis of Mahommedan power from the reign of Saladin onwards.

16 Pisa naturally connected itself with Antioch, because Antioch was hostile to Constantinople, and Pisa cherished the same hostility, since Alexius I. had in 1080 given preferential treatment to Venice, the enemy of Pisa.

17 This is the year in which the kingdom may be regarded as definitely founded. The period of conquest practically ends at this date, though isolated gains were afterwards made. The year 1110 is additionally important by reason of the accession of Maudud al Mosul, which marks the beginning of a Moslem reaction.

18 Ilghazi died in 1122. His successor was Balak, who ruled from 1122 to 1124, and succeeded in capturing in 1123 Baldwin II. of Jerusalem. The union of Mardin and Aleppo under the sway of these two amirs, connecting as it did Mesopotamia with Syria, marks an important stage in the revival of Mahommedan power (Stevenson, *Crusades in the East*, p. 109).

19 Maudud (the brother of the sultan Mahommed) may be regarded as the first to begin the *jihad*, or counter-crusade, and his attack expedition of 1113, which carried him so far into the heart of Palestine, may be considered as the first act of the *jihad* (Stevenson, op. cit. pp. 87, 96).
20 Aleppo had passed from the rule of Timurtash (son of Ilghazi and successor of Balak) into the possession of Aksunkur, 1125.
21 Stevenson, however, believes that Zengi was *not* animated by the idea of recovering Jerusalem. He thinks that his principal aim was simply the formation of a compact Mahommedan state, which was, indeed, in the issue destined to be the instrument of the *jihad*, but was not so intended by Zengi (op. cit. pp. 123-124).
22 There are certain connexions and analogies between the kingdom of Sicily and that of Jerusalem during the twelfth century. In either case there is an importation of Western feudalism into a country originally possessed of Byzantine institutions, but affected by an Arabic occupation. The subject deserves investigation.
23 The holders of fiefs (*sodeers*) both held fiefs of land and received pay; the paid force of *soudoyers* only received pay. An instance of the latter is furnished by John of Margat, a vassal of the seignory of Arsuf. He has 200 bezants along with a quantity of wheat, barley, lentils and oil; and in return he must march with four horses (Rey, *Les Colonies franques en Syrie*, p. 24).
24 For the history of the orders see the articles on the TEMPLARS; ST JOHN OF JERUSALEM, KNIGHTS OF; KNIGHTS, and the TEUTONIC ORDER. The Templars were founded about the year 1118 by a Burgundian knight, Hugh de Paganis; the Hospitallers sprang from a foundation in Jerusalem erected by merchants of Amalfi before the First Crusade, and were reorganized under Gerard le Puy, master until 1120. The Teutonic knights date from the Third Crusade.
25 As was noticed above, there were apparently separate assizes for the three principalities, in addition to the assizes of the kingdom. The assizes of Antioch have been discovered and published. The assizes of the kingdom itself are twofold—the assizes of the high court and the assizes of the court of burgesses. (1) The assizes of the high court are preserved for us in works by legists—John of Ibelin, Philip of Novara and Geoffrey of Tort—composed in the 13th century. We possess, in other words, *law-books* (like Bracton's treatise *De legibus*), but not *laws*—and law-books made after the loss of the kingdom to which the laws belonged. There are two vexed questions with regard to these law-books. (*a*) The first concerns the origin and character of the laws which the law-books profess to expound. According to the story of the legists who wrote these books—*e.g.* John of Ibelin—the laws of the kingdom were laid down by Godfrey, who is thus regarded as the great νομοθέτης of the kingdom. These laws (progressively modified, it is admitted) were kept in Jerusalem, under the name of "Letters of the Sepulchre," until 1187. In that year they were lost; and the legists tell us that they are attempting to reconstruct *par oir dire* the gist of the lost archetype. The story of the legists is now

generally rejected. Godfrey never legislated: the customs of the kingdom gradually grew, and were gradually defined, especially under kings like Baldwin III. and Amalric I. If there was thus only a customary and unwritten law (and William of Tyre definitely speaks of a *jus consuetudinarium* under Baldwin III., *quo regnum regebatur*), then the "Letters of the Sepulchre" are a myth—or rather, if they ever existed, they existed not as a code of written law, but, perhaps, as a register of fiefs, like the Sicilian *Defetarii*. Thus the story of the legists shrinks down to the regular myth of the primitive legislator, used to give an air of respectability to law-books, which really record an unwritten custom. The fact is that until the 13th century the Franks lived *consuetudinibus antiquis et jure non scripto*. They preferred an unwritten law, as Prutz suggests, partly because it suited the barristers (who often belonged to the baronage, for the Frankish nobles were "great pleaders in court and out of court"), and partly because the high court was left unbound so long as there was no written code. In the 13th century it became necessary for the legists to codify, as it were, the unwritten law, because the upheavals of the times necessitated the fixing of some rules in writing, and especially because it was necessary to oppose a definite custom of the kingdom to Frederick II., who sought, as king of Jerusalem, to take advantage of the want of a written law, to substitute his own conceptions of law in the teeth of the high court. (*b*) The second difficulty concerns the text of the law-books themselves. The text of Ibelin became a *textus receptus*—but it also became overlaid by glosses, for it was used as authoritative in the kingdom of Cyprus after the loss of the kingdom of Jerusalem, and it needed expounding. Recensions and revisions were twice made, in 1368 and 1531; but how far the true Ibelin was recovered, and what additions or alterations were made at these two dates, we cannot tell. We can only say that we have the text of Ibelin which was used in Cyprus in the later middle ages. At the same time, if our text is thus late, it must be remembered that its content gives us the earliest and purest exposition of French feudalism, and describes for us the organization of a kingdom, where all rights and duties were connected with the fief, and the monarch was only a suzerain of feudatories. (2) The assizes of the court of burgesses became the basis of a treatise at an earlier date than the assizes of the high court. The date of the redaction (which was probably made by some learned burgess) may well have been the reign of Baldwin III., as Kugler suggests: he was the first native king, and a king learned in the law; but Beugnot would refer the assizes to the years immediately preceding Saladin's capture of Jerusalem. These assizes do not, of course, appear in Ibelin, who was only concerned with the feudal law of the high court. They were used, like the assizes of the high court, in Cyprus; and, like the other assizes, they were made the subject of investigation in 1531, with the object of discovering a good text. The law which is expounded in these assizes is a mixture of Frankish law with the Graeco-Roman law of the Eastern empire which prevailed among the native population of Syria.

In regard to both assizes, it is most important to bear in mind that we possess not laws, but law-books or custumals—records made by lawyers for their fellows of what they conceived to be the law, and supported by legal arguments and citations of cases. But, as Prutz remarks, Philip of Novara *lehrt nicht die Wissenschaft des Rechts, sondern die des Unrechts*: he does not explain the law so much as the ways of getting round it.

26 For instance, the abbey of Mount Sion had large possessions, not only in the Holy Land (at Ascalon, Jaffa, Acre, Tyre, Caesarea and Tarsus), but also in Sicily, Calabria, Lombardy, Spain and France (at Orleans, Bourges and Poitiers).

27 One must remember that these reinforcements would often consist of desperate characters. It was one of the misfortunes of Palestine that it served as a Botany Bay, to which the criminals of the West were transported for penance. The natives, already prone to the immorality which must infect a mixed population living under a hot sun, the immorality which still infects a place like Aden, were not improved by the addition of convicts.

28 The manorial system in the Latin kingdom of Jerusalem was a continuation of the village system as it had existed under the Arabs. In each village (*casale)* the *rustici* were grouped in families (*foci*): the tenants paid from ¼ to ⅓ of the crop, besides a poll-tax and labour-dues. The villages were mostly inhabited by Syrians: it was rarely that Franks settled down as tillers of the soil. Prutz regards the manorial system as oppressive. Absentee landlords, he thinks, rack-rented the soil (p. 167), while the "inhuman severity" of their treatment of villeins led to a progressive decay of agriculture, destroyed the economic basis of the Latin kingdom, and led the natives to welcome the invasion of Saladin (pp. 327-331). The French writers Rey and Dodu are more kind to the Franks; and the testimony of contemporary Arabic writers, who seem favourably impressed by the treatment of their subjects by the Franks, bears out their view, while the tone of the assizes is admittedly favourable to the Syrians. One must not forget that there was a brisk native manufacture of carpets, pottery, ironwork, gold-work and soap; or that the Syrians of the towns had a definite legal position.

29 After 1143 one may therefore speak of the period of the Epigoni—the native Franks, ready to view the Moslems as joint occupants of Syria, and to imitate the dress and habits of their neighbours.

30 Doubt has been cast on the view that a troubled conscience drove Louis to take the cross; and his action has been ascribed to simple religious zeal (cf. Lavisse, *Histoire de France*, iii. 12).

31 We speak of First, Second and Third Crusades, but, more exactly, the Crusades were one continuous process. Scarcely a year passed in which new bands did not come to the Holy Land. We have already noticed the great if disastrous Crusade of 1100-1101, and the Venetian Crusade of 1123-1124; and we may also refer to the Crusade of Henry the Lion in 1172, and to that of Edward I. in 1271-1272—all

famous Crusades, which are not reckoned in the usual numbering. Crusades appear to have been dignified by numbers when they followed some crushing disaster—the loss of Edessa in 1144, or the fall of Jerusalem in 1187—and were led by kings and emperors; or when, like the Fourth and Fifth Crusades, they achieved some conspicuous success or failure. But it is important to bear in mind the continuity of the Crusades—the constant flow of new forces eastward and back again westward; for this alone explains why the Crusades formed a great epoch in civilization, familiarizing, as they did, the West with the East.

32 This body of crusaders ultimately reached the Holy Land, where it joined Conrad (who had lost his own original forces), and helped in the fruitless siege of Damascus. The services which it rendered to Portugal were repeated by later crusaders. Crusaders from the Low Countries, England and the Scandinavian north took the coast route round western Europe; and it was natural that, landing for provisions and water, they should be asked, and should consent, to lend their aid to the natives against the Moors. Such aid is recorded to have been given on the Third and the Fifth Crusades.

33 Manuel was an ambitious sovereign, apparently aiming at a world-monarchy, such as was afterwards attempted from the other side by Henry VI. As Henry VI. had designs on Constantinople and the Eastern empire, so Manuel cherished the ambition of acquiring Italy and the Western empire, and he negotiated with Alexander III. to that end in 1167 and 1169: cf. the life of Alexander III. in Muratori, *S. R. I.* iii. 460.

34 The prize was won by Raynald of Chatillon (*q.v.*).

35 Nureddin, unlike his father, was definitely animated by a religious motive: he fought first and foremost against the Latins (and not, like his father, against Moslem states), and he did so as a matter of religious duty.

36 Henry II., as an Angevin, was the natural heir of the kingdom of Jerusalem on the extinction of the line descended from Fulk of Anjou. This explains the part played by Richard I. in deciding the question of the succession during the Third Crusade.

37 The taxation levied in the West was also attempted in the East, and in 1183 a universal tax was levied in the kingdom of Jerusalem, at the rate of 1% on movables and 2% on rents and revenues. Cf. Dr A. Cartellieri, *Philipp II. August*, ii. pp. 3-18 and p. 85.

38 Stevenson argues (op. cit. p. 240) that this truce was already practically dissolved before Raynald struck, and that Raynald's "action may reasonably be viewed as the practical outcome of the feeling of a party."

39 The "economic" motive for taking the cross was strengthened by the papal regulations in favour of debtors who joined the Crusade. Thousands must have joined the Third Crusade in order to escape paying either their taxes or the interest

on their debts; and the atmosphere of the gold-digger's camp (or of the cave of Adullam) must have begun more than ever to characterize the crusading armies.
40 The Crusades in their course established a number of new states or kingdoms. The First Crusade established the kingdom of Jerusalem (1100); the Third, the kingdom of Cyprus (1195); the Fourth, the Latin empire of Constantinople (1204); while the long Crusade of the Teutonic knights on the coast of the Baltic led to the rise of a new state east of the Vistula. The kingdom of Lesser Armenia, established in 1195, may also be regarded as a result of the Crusades. The history of the kingdom of Jerusalem is part of the history of the Crusades: the history of the other kingdoms or states touches the history of the Crusades less vitally. But the history of Cyprus is particularly important—and for two reasons. In the first place, Cyprus was a natural and excellent basis of operations; it sent provisions to the crusaders in 1191, and again at the siege of Damietta in 1219, while its advantages as a strategic basis were proved by the exploits of Peter of Cyprus in the 14th century. In the second place, as the Latin kingdom of Jerusalem fell, its institutions and assizes were transplanted bodily to Cyprus, where they survived until the island was conquered by the Ottoman Turks. But the monarchy was stronger in Cyprus than in Jerusalem: the fiefs were distributed by the monarch, and were smaller in extent; while the feudatories had neither the collective powers of the haute cour of Jerusalem, nor the individual privileges (such as jurisdiction over the bourgeoisie), which had been enjoyed by the feudatories of the old kingdom. Till 1489 the kingdom of Cyprus survived as an independent monarchy, and its capital, Famagusta, was an important centre of trade after the loss of the coast-towns in the kingdom of Jerusalem. In 1489 it was acquired by Venice, which claimed the island on the death of the last king, having adopted his widow (a Venetian lady named Catarina Cornaro) as a daughter of the republic. On the history of Cyprus, see Stubbs, *Lectures on Medieval and Modern History*, 156-208. The history of the kingdom of Armenia is closely connected with that of Cyprus. The Armenians in the south-east of Asia Minor borrowed feudal institutions from the Franks and the feudal vocabulary itself. The kingdom was involved in a struggle with Antioch in the early part of the 13th century. Later, it allied itself with the Mongols and fought against the Mamelukes, to whom, however, it finally succumbed in 1375.
41 The kingdom of Jerusalem is thus from 1192 to its final fall a strip of coast, to which it is the object of kings and crusaders to annex Jerusalem and a line of communication connecting it with the coast. This was practically the aim of Richard I.'s negotiations; and this was what Frederick II. for a time secured.
42 M. Luchaire, in the volume of his biography of Innocent III. called *La Question d'Orient*, shows how, in spite of the pope, the Fourth Crusade was in its very beginnings a lay enterprise. The crusading barons of France chose their own leader, and determined their own route, without consulting Innocent.

43 As a matter of fact, there is some doubt whether Alexius arrived in Germany before the spring of 1202. But there seems to be little doubt of Philip's complicity in the diversion of the Fourth Crusade to Constantinople (cf. M. Luchaire, *La Question d'Orient*, pp. 84-86).

44 It is true that in 1208 Venice received commercial concessions from the court of Cairo. But this *ex post facto* argument is the sole proof of this view; and it is quite insufficient to prove the accusation. Venice is *not* the primary agent in the deflection of the Fourth Crusade.

45 Already under Innocent III. the benefits of the Crusade were promised to those who went to the assistance of the Latin empire of the East.

46 In 1208 Innocent excommunicated Raymund VI. of Toulouse on account of the murder of a papal legate who was attempting to suppress Manichaeism, and offered all Catholics the right to occupy and guard his territories. Thus was begun the First Crusade against heresy. Raymund at once submitted to the pope, but the Crusade continued none the less, because, as Luchaire says, "the baronage of the north and centre of France had finished their preparations," and were resolved to annex the rich lands of the south. In this way land-hunger exploited the Albigensian, as political and commercial motives had helped to exploit the Fourth Crusade; and in the former, as in the latter, Innocent had reluctantly to consent to the results of the secular motives which had infected a spiritual enterprise. The Albigensian Crusades, however, belong to French history; and it can only be noted here that their ultimate result was the absorption of the fertile lands, and the extinction of the peculiar civilization, of southern France by the northern monarchy. (See the article ALBIGENSES.)

47 A canon of the third Lateran council (1179) forbade traffic with the Saracens in munitions of war; and this canon had been renewed by Innocent in the beginning of his pontificate.

48 He had promised the pope, at his coronation in 1220, to begin his Crusade in August 1221. But he declared himself exhausted by the expenses of his coronation; and Honorius III. consented to defer his Crusade until March 1222. The letter of the pope informing Pelagius of this delay is dated the 20th of June: it would probably reach his hands *after* his departure from Damietta; and thus the Cardinal gave the signal for the march, when, as he thought, the emperor's coming was imminent.

49 Joinville, ch. x.

50 John of Brienne had only ruled in right of his wife Mary. On her death (1212) John might be regarded as only ruling "by the courtesy of the kingdom" until her daughter Isabella was married, when the husband would succeed. That, at any rate, was the view Frederick II. took.

51 Amalric I. of Cyprus had done homage to Henry VI., from whom he had received the title of king (1195).

52 It may be argued that the Crusade against a revolted Christian like Frederick II. was not misplaced, and that the pope had a true sense of religious values when he attacked Frederick. The answer is partly that men like St Louis *did* think that the Crusade was misplaced, and partly that Frederick was really attacked *not* as a revolted Christian, but as the would-be unifier of Italy, the enemy of the states of the church.

53 The following table of the Ayyubite rulers serves to illustrate the text:—

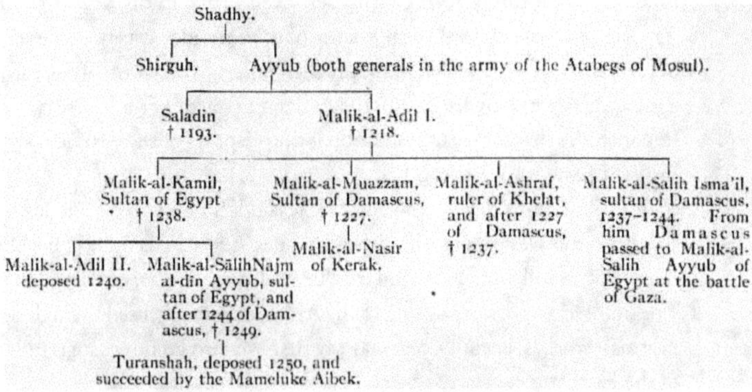

54 Though Europe indulged in dreams of Mongol aid, the eventual results of the extension of the Mongol Empire were prejudicial to the Latin East. The sultans of Egypt were stirred to fresh activity by the attacks of the Mongols; and as Syria became the battleground of the two, the Latin principalities of Syria were fated to fall as the prize of victory to one or other of the combatants.

55 Of the four Latin principalities of the East, Edessa was the first to fall, being extinguished between 1144 and 1150. Antioch fell in 1268; Tripoli in 1289; and the kingdom itself may be said to end with the capture of Acre, 1291.

56 Michael Palaeologus had actually appealed to Louis IX. against Charles of Anjou, who in 1270 had actively begun preparations for the attack on Constantinople.

57 The dream of a Crusade to Jerusalem survived de Mézières; a society which read "romaunts" of the Crusades, could not but dream the dream. Henry V., whose father had fought with the Teutonic knights on the Baltic, dreamed of a voyage to Jerusalem.

58 The union of 1274, conceded by the Palaeologi at the council of Lyons in order to defeat the plans of Charles of Anjou, had only been temporary.

59 Bréhier, *L'Église el l'Orient*, p. 347.

60 *Cambridge Modern History*, i. 11. It is perhaps worth remarking that something of the old crusading spirit seems still to linger in the movement of Russia towards Constantinople.

61 While from this point of view the Crusades appear as a failure, it must not be forgotten that elsewhere than in the East Crusades did attain some success. A Crusade won for Christianity the coast of the eastern Baltic (see TEUTONIC ORDER); and the centuries of the Spanish Crusade ended in the conquest of the whole of Spain for Christianity.
62 Authors like Heeren (*Versuch einer Entwickelung der Folgen der Kreuzzüge*) and Michaud (in the last volume of his *Histoire des croisades*) fall into the error of assigning all things to the Crusades. Even Prutz, in his *Kulturgeschichte der Kreuzzüge*, over-estimates the influence of the Crusades as a chapter in the history of civilization. He depreciates unduly the Western civilization of the early middle ages, and exalts the civilization of the Arabs; and starting from these two premises, he concludes that modern civilization is the offspring of the Crusades, which first brought East and West together.
63 It is difficult to decide how far Arabic models influenced ecclesiastical architecture in the West as a result of the Crusades. Greater freedom of moulding and the use of trefoil and cinquefoil may be, but need not be, explained in this way. The pointed arch owes nothing to the Arabs; it is already used in England in early Norman work. Generally, one may say that Western architecture is independent of the East.
64 His somewhat legendary treatise, *De liberatione civitatum Orientis*, was only composed about 1155.
65 There is also an *Inventaire critique* of these letters by the comte de Riant (Paris, 1880).
66 Von Sybel's view must be modified by that of Kugler, to which a scholar like Hagenmeyer has to some extent given his adhesion (cf. his edition of the *Gesta*, pp. 62-68). Hagenmeyer inclines to believe in an original author, distinct from Albert the copyist; and he thinks that this original author (whether or no he was present during the Crusade) used the *Gesta* and also Fulcher, though he had probably also "*eigene Notizen und Aufzeichnungen*."
67 See Pigonneau, *Le Cycle de la croisade*, &c. (Paris, 1877); and Hagenmeyer, *Peter der Eremite* (Leipzig, 1879).
68 On the bibliography of the Second Crusade see Kugler, *Studien zur Geschichte des zweiten Kreuzzüges* (Stuttgart, 1866).
69 Of these writers see Archer's *Crusade of Richard I.*, Appendix (in Nutt's series of Histories from Contemporary Writers).
70 The bibliography of the Fourth Crusade is discussed in Klimke, *Die Quellen zur Geschichte des vierten Kreuzzüges* (Breslau, 1875).

CRUSENSTOLPE, MAGNUS JAKOB

(1795-1865), Swedish historian, early became famous both as a political and a historical writer. His first important work was a *History of the Early Years of the Life of King Gustavus IV. Adolphus*, which was followed by a series of monographs and by some politico-historical novels, of which *The House of Holstein-Gottorp in Sweden* is considered the best. He obtained a great influence over King Charles XIV. (Bernadotte), who during the years 1830-1833 gave him his fullest confidence, and sanctioned the official character of Crusenstolpe's newspaper *Fäderneslandet*. In the last-mentioned year, however, the historian suddenly became the king's bitterest enemy, and used his acrid pen on all occasions in attacking him. In 1838 he was condemned, for one of these angry utterances, to be imprisoned three years in the castle of Waxholm. He continued his literary labours until his death in 1865. Few Swedish writers have wielded so pure and so incisive a style as Crusenstolpe, but his historical work is vitiated by political and personal bias.

CRUSIUS, CHRISTIAN AUGUST

(1715-1775), German philosopher and theologian, was born on the 10th of January 1715 at Lenau near Merseburg in Saxony. He was educated at Leipzig, and became professor of theology there in 1750, and principal of the university in 1773. He died on the 18th of October 1775. Crusius first came into notice as an opponent of the philosophy of Leibnitz and Wolff from the standpoint of religious orthodoxy. He attacked it mainly on the score of the moral evils that must flow from any system of determinism, and exerted himself in particular to vindicate the freedom of the will. The most important works of this period of his life are *Entwurf der nothwendigen Vernunftwahrheiten* (1745), and *Weg zur Gewissheit und Zuverlässigkeit der menschlichen Erkenntniss* (1747). Though diffusely written, and neither brilliant nor profound, Crusius' philosophical books had a great but short-lived popularity. His criticism of Wolff, which is generally based on sound sense, had much influence upon Kant at the time when his system was forming; and his ethical doctrines are mentioned with respect in the *Kritik of Practical Reason*. Crusius's later life was devoted to theology. In this capacity his sincere piety and amiable character gained him great influence, and he led the party in the university which became known as the "Crusianer" as opposed to the "Ernestianer," the followers of J. A. Ernesti. The two professors adopted opposite methods of exegesis. Ernesti wished to subject the Scripture to the same laws of exposition as are applied to other ancient books; Crusius held firmly to orthodox ecclesiastical tradition. Crusius's chief theological works are *Hypomnemata ad theologiam propheticam* (1764-1778), and *Kurzer Entwurf der Moraltheologie* (1772-1773). He sets his face against innovation in such matters as the accepted authorship of canonical writings, verbal inspiration, and the treatment of persons and events in the Old Testament as types of the New. His views, unscholarly and uncritical as they seem to us now, have had influence on later evangelical students of the Old Testament, such as E. W. Hengstenberg and F. Delitzsch.

There is a full notice of Crusius in Ersch and Gruber's *Allgemeine Encyclopädie*. Consult also J. E. Erdmann's *History of Philosophy*; A. Marquardt, *Kant und Crusius*; and art. in Herzog-Hauck, *Realencyklopädie* (1898).

(H. St.)

CRUSTACEA

A very large division of the animal kingdom, comprising the familiar crabs, lobsters, crayfish, shrimps and prawns, the sandhoppers and wood-lice, the strangely modified barnacles and the minute water-fleas. Besides these the group also includes a multitude of related forms which, from their aquatic habits and generally inconspicuous size, and from the fact that they are commonly neither edible nor noxious, are little known except to naturalists and are undistinguished by any popular names. Collectively, they are ranked as one of the classes forming the sub-phylum ARTHROPODA, and their distinguishing characters are discussed under that heading. It will be sufficient here to define them as Arthropoda for the most part of aquatic habits, having typically two pairs of antenniform appendages in front of the mouth and at least three pairs of post-oral limbs acting as jaws.

As a matter of fact, however, the range of structural variation within the group is so wide, and the modifications due to parasitism and other causes are so profound, that it is almost impossible to frame a definition which shall be applicable to all the members of the class. In certain parasites, for instance, the adults have lost every trace not only of Crustacean but even of Arthropodous structure, and the only clue to their zoological position is that afforded by the study of their development. In point of size also the Crustacea vary within very wide limits. Certain water-fleas (Cladocera) fall short of one-hundredth of an inch in total length; the giant Japanese crab (*Macrocheira*) can span over 10 ft. between its outstretched claws.

The habits of the Crustacea are no less diversified than their structure. Most of them inhabit the sea, but representatives of all the chief groups are found in fresh water (though the Cirripedia have hardly gained a footing there), and this is the chief home of the primitive Phyllopoda. A terrestrial habitat is less common, but the widely-distributed land Isopoda or woodlice and the land-crabs of tropical regions have solved the problem of adaptation to a subaërial life.

Swimming is perhaps the commonest mode of locomotion, but numerous forms have taken to creeping or walking, and the robber-crab (*Birgus latro*) of the Indo-Pacific islands even climbs palm-trees. None has the power of flight, though certain pelagic Copepoda are said to leap from the surface of the sea like flying-fish. Apart from the numerous parasitic forms, the only Crustacea which have adopted a strictly sedentary habit of life are the Cirripedia, and here, as elsewhere, profound modifications of structure have resulted, leading ultimately to a partial assumption of the radial type of symmetry which is so often associated with a sedentary life.

Many, perhaps the majority, of the Crustacea are omnivorous or carrion-feeders, but many are actively predatory in their habits, and are provided with more or less complex and efficient instruments for capturing their prey, and there are also many plant-eaters. Besides the sedentary Cirripedia, numbers of the smaller forms, especially among the Entomostraca, subsist on floating particles of organic matter swept within reach of the jaws by the movements of the other limbs.

Symbiotic association with other animals, in varying degrees of interdependence, is frequent. Sometimes the one partner affords the other merely a convenient means of transport, as in the case of the barnacles which grow on, or of the gulf-weed crab which clings to, the carapace of marine turtles. From this we may pass through various grades of "commensalism," like that of the hermit-crab with its protective anemones, to the cases of actual parasitism. The parasitic habit is most common among the Copepoda and Isopoda, where it leads to complex modifications of structure and life-history. Perhaps the most complete degeneration is found in the Rhizocephala, which are parasitic on other Crustacea. In these the adult consists of a simple saccular body containing the reproductive organs and attached by root-like filaments which ramify throughout the body of the host and serve for the absorption of nourishment (fig. 1).

Many of the larger species of Crustacea are used as food by man, the most valuable being the lobster, which is caught in large quantities on both sides of the North Atlantic. Perhaps the most important of all Crustacea, however, with respect to the part which they play in the economy of nature, are the minute pelagic Copepoda, of which incalculable myriads form an important constituent of the "plankton" in all the seas of the globe. It is on the plankton that a great part of the higher animal life of the sea ultimately depends for food. The Copepoda live upon the diatoms and other important microscopic vegetable life at the surface of the sea, and in their turn serve as food for fishes and other larger forms and thus, indirectly, for man himself.

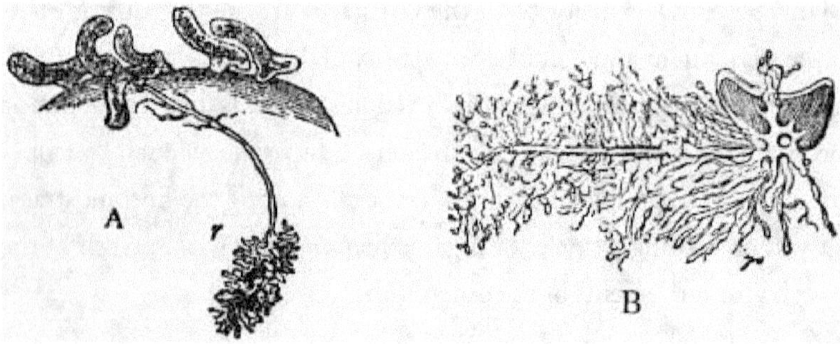

FIG. I.

A, Group of Peltogaster socialis on the abdomen of a small hermit-crab; in one of them the fasciculately ramified roots, r, in the liver of the crab are shown (Fritz Müller).
B, Young of Sacculina purpurea with its roots. (Fritz Müller.)

Historical Sketch.—In common with most branches of natural history, the science of Carcinology may be traced back to its beginnings in the writings of Aristotle. It received additions of varying importance at the hands of medieval and later naturalists, and first began to assume systematic form under the influence of Linnaeus. The application of the morphological method to the Crustacea may perhaps be dated from the work of J. C. Fabricius towards the end of the 18th century.

In the first quarter of the 19th century important advances in classification were made by P. A. Latreille, W. E. Leach and others, and J. Vaughan Thompson demonstrated the existence of metamorphosis in the development of the higher Crustacea. A new epoch may be said to begin with H. Milne-Edwards' classical *Histoire naturelle des crustacés* (1834-1840). It is noteworthy that even at this late date the Cirripedia (Thyrostraca) were still excluded from the Crustacea, though Darwin's Monograph (1851-1854) was soon to make them known with a wealth of anatomical and systematic detail such as was available, at that time, for few other groups of Crustacea. About the same period three authors call for special mention, W. de Haan, J. D. Dana and H. Kröyer. The new impulse given to biological research by the publication of the *Origin of Species* bore fruit in Fritz Müller's *Für Darwin*, in which an attempt was made to reconstruct the phylogenetic history of the class. The same line of work was followed in the long series of important memoirs from the pen of K. F. W. Claus, and noteworthy contributions were made, among many others, by A. Dohrn, Ray Lankester and Huxley. In more recent years the long and constantly increasing list of writers on Crustacea contains no name more honoured than that of the veteran G. O. Sars of Christiania.

Morphology.

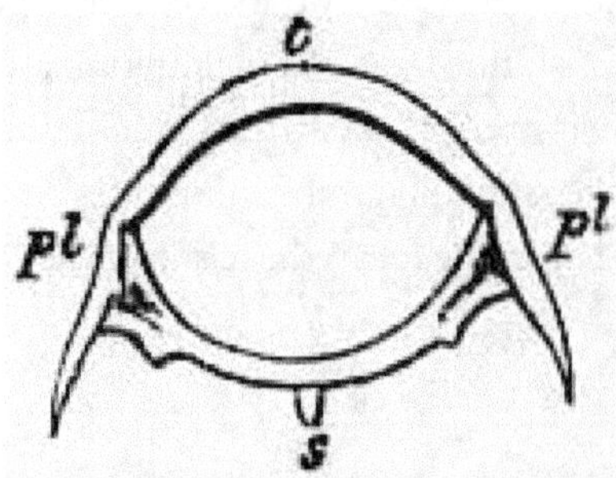

FIG. 2.—Abdominal Somite of a Lobster, separated and viewed from in front. *t*, tergum; *s*, sternum; *pl*, pleuron.

External Structure: Body.—As in all Arthropoda the body consists of a series of segments or somites which may be free or more or less coalesced together. In its simplest form the exoskeleton of a typical somite is a ring of chitin defined from the rings in front and behind by areas of thinner integument forming moveable joints, and having a pair of appendages articulated to its ventral surface on either side of the middle line. Frequently, however, this exoskeletal somite may be differentiated into various regions. A dorsal and a ventral plate are often distinguished, known respectively as the tergum and the sternum, and the tergum may overhang the insertion of the limb on each side as a free plate called the pleuron. The name epimeron is sometimes applied to what is here called the pleuron, but the word has been used in widely different senses and it seems better to abandon it. The typical form of a somite is well seen, for example, in the segments which make up the abdomen or "tail" of a lobster or crayfish (fig. 2). The posterior terminal segment of the body, on which the opening of the anus is situated, never bears appendages. The nature of this segment, which is known as the "anal segment" or telson (fig. 3, *T*), has been much discussed, some authorities holding that it is a true somite, homologous with those which precede it. Others have regarded it as representing the fusion of a number of somites, and others again as a "median appendage" or as a pair of appendages fused. Its morphological nature, however, is clearly shown by its development. In the larval development of the more primitive Crustacea, the number of somites, at first small, increases by the successive appearance of new somites between the last-formed somite and the terminal region which bears the anus. The "growing point" of the trunk is, in fact, situated in front of this region, and, when the full number of somites has been reached, the unsegmented part remaining forms the telson of the adult.

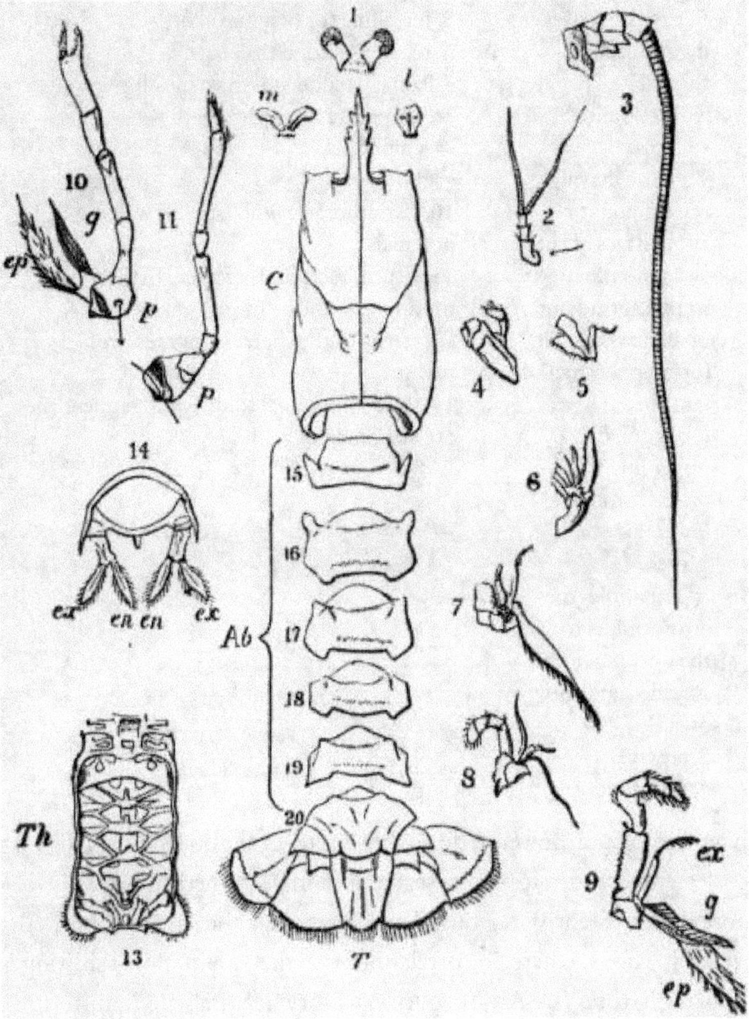

FIG. 3.—The Separated Somites and Appendages of the Common Lobster (*Homarus gammarus*).

C, carapace covering the cephalothorax.
Ab, abdominal somites.
T, telson, having the uropods or appendages of the last abdominal somite spread out on either side of it, forming the "tail-fan."
l, labrum, or upper lip.
m, metastoma, or lower lip.
1, eyes.
2, antennule (the arrow points to the opening of the so-called auditory organ).
3, antenna.
4, mandible.
5, maxillula (or first maxilla).
6, maxilla (second maxilla).
7-9, first, second and third maxillipeds.
ex, exopodite.
ep, epipodite.
g, gill.
10, sixth thoracic limb (second walking-leg) of female.
11, last thoracic limb of male. In 10 and 11 the arrows indicate the genital apertures.
13, sterna of the thoracic somites, from within.
14, third abdominal somite, with appendages or "swimmerets."

In no Crustacean, however, do all the somites of the body remain distinct. Coalescence, or suppression of segmentation ("lipomerism"), may involve more or less extensive regions. This is especially the case in the anterior part of the body, where, in correlation with the "adaptational shifting of the oral aperture" (see ARTHROPODA), a varying number of somites unite to form the "cephalon" or head. Apart from the possible existence of an ocular somite corresponding to the eyes (the morphological nature of which is discussed below), the smallest number of head-somites so united in any Crustacean is five. Even where a large number of the somites have fused, there is generally a marked change in the character of the appendages after the fifth pair, and since the integumental fold which forms the carapace seems to originate from this point, it is usual to take the fifth somite as the morphological limit of the cephalon throughout the class. It is quite probable, however, that in the primitive ancestors of existing Crustacea a still smaller number of somites formed the head. The three pairs of appendages present in the "nauplius" larva show certain peculiarities of structure and development which seem to place them in a different category from the other limbs, and there is some ground for regarding the three corresponding somites as constituting a "primary cephalon." For practical purposes, however, it is convenient to include the two following somites also as cephalic.

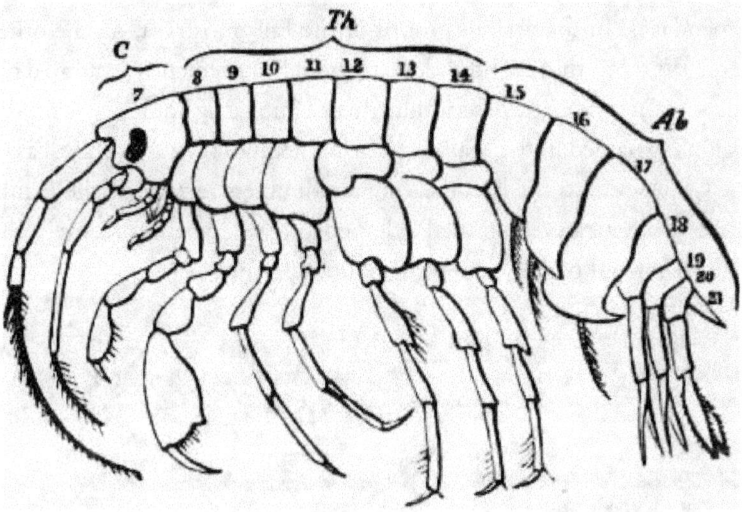

FIG. 4.—Diagram of an Amphipod. (After Spence Bate and Westwood.)

C, cephalon.
Th, thorax. (Only seven of the eight thoracic somites are visible, the first being fused with the cephalon.)
Ab, abdomen.
The numbers appended to the somites do not correspond to the enumeration adopted in the text. 21 is the telson.

A remarkable feature found only in the Stomatopoda is the reappearance of segmentation in the anterior part of the cephalic region. Whether the movably articulated segments which bear the eye-stalks and the antennules in this aberrant group correspond to the primitive head somites or not, their distinctness is certainly a secondarily acquired character, for it is not found in the larvae, nor in any of the more primitive groups of Malacostraca.

The body proper is usually divisible into two regions to which the names *thorax* and *abdomen* are applied. Throughout the whole of the Malacostraca the thorax consists of eight and the abdomen of six somites (fig. 4), and the two regions are sharply distinguished by the character of their appendages. In the various groups of the Entomostraca, on the other hand, the terms thorax and abdomen, though conveniently employed for purposes of systematic description, do not imply any homology with the regions so named in the Malacostraca. Sometimes they are applied, as in the Copepoda, to the limb-bearing and limbless regions of the trunk, while in other cases, as in the Phyllopoda, they denote, respectively, the regions in front of and behind the genital apertures.

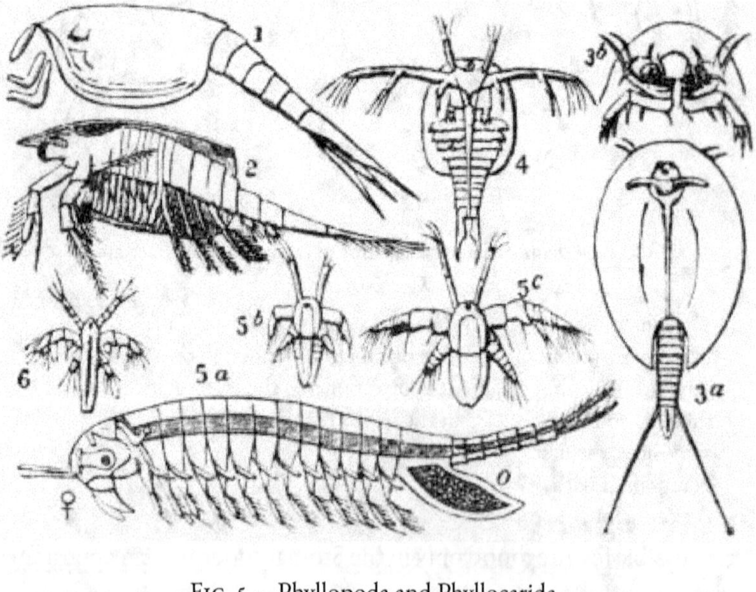

FIG. 5.—Phyllopoda and Phyllocarida.

1, *Ceratiocaris papilio*, U. Silurian, Lanark.
2, *Nebalia bipes*(one side of carapace removed).
3, *Lepidurus Angassi*: a, dorsal aspect; b, ventral aspect of head showing the labrum and mouth-parts.
4, larva of *Apus cancriformis*.
5, *Branchipus stagnalis*: *a*, adult female; *b*, first larval stage (Nauplius); *c*, second larval stage.
6, Nauplius of *Artemia salina*.

A character which recurs in the most diverse groups of the Crustacea, and which is probably to be regarded as a primitive attribute of the class, is the possession of a carapace or shell, arising as a dorsal fold of the integument from the posterior margin of the head-region. In its most primitive form, as seen in the *Apodidae* (fig. 5, 3) and in *Nebalia* (fig. 5, 2), this shell-fold remains free from the trunk, which it envelops more or less completely. It may assume the form of a bivalve shell entirely enclosing the body and limbs, as in many Phyllopoda (fig. 6) and in the Ostracoda. In the Cirripedia it forms a fleshy "mantle" strengthened by shelly plates or valves which may assume a very complex structure. In many cases, however, the shell-fold coalesces with some of the succeeding somites. In the Decapoda (fig. 3), this coalescence affects only the dorsal region of the thoracic somites, and the lateral portions of the carapace overhang on each side, enclosing a pair of chambers within which lie the gills. The arrangement is similar in Schizopoda and Stomatopoda (fig. 7), except that the coalescence does not usually involve the posterior thoracic somites, several of which remain free, though they may be overlapped by the carapace.

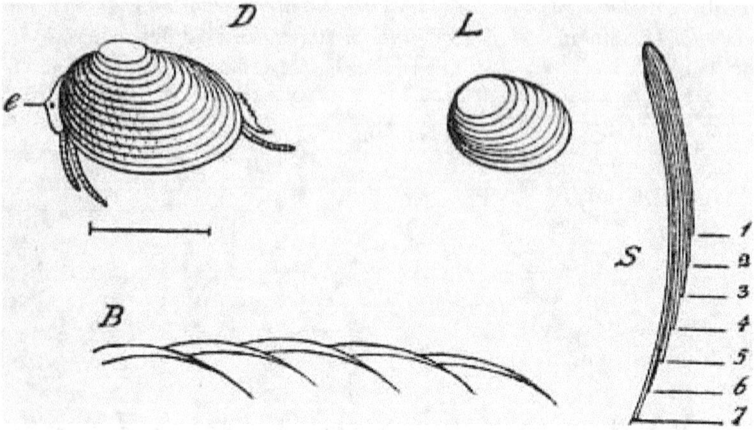

From Morse's Zoology.

Fig. 6.—Estheria, sp.; D from Dubuque, Iowa; (e) the eye. L from Lynn, Massachusetts (nat. size). S presents a highly magnified section of one of the valves to show the successive moults. B an enlarged portion of the edge of the shell along the back, showing the overlap of each growth.

In the Isopoda and Amphipoda, where, as a rule, all the thoracic somites except the first are distinct (fig. 4), there seems at first sight to be no shell-fold. A comparison with the related Tanaidacea (fig. 8) and Cumacea (or Sympoda), however, leads to the conclusion that the coalescence of the first thoracic somite with the cephalon really involves a vestigial shell-fold, and, indeed, traces of this are said to be observed in the embryonic development of some Isopoda. It seems likely that a similar explanation is to be applied to the coalescence of one or two trunk-somites with the head in the Copepoda, and, if this be so, the only Crustacea remaining in which no trace of a shell-fold is found in the adult are the Anostracous Phyllopoda such as Branchipus (fig. 5, 5).

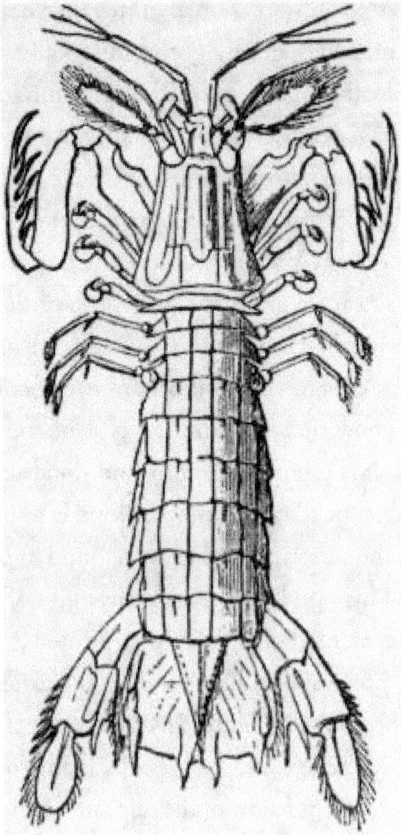

FIG. 7.—*Squilla mantis* (Stomatopoda), showing the last four thoracic (leg-bearing) somites free from the carapace.

General Morphology of Appendages.—Amid the great variety of forms assumed by the appendages of the Crustacea, it is possible to trace, more or less plainly, the modifications of a fundamental type consisting of a peduncle, the protopodite, bearing two branches, the endopodite and exopodite. This simple biramous form is shown in the swimming-feet of the Copepoda and Branchiura, the "cirri" of the Cirripedia, and the abdominal appendages of the Malacostraca (fig. 3, 14). It is also found in the earliest and most primitive form of larva, known as the *Nauplius*. As a rule the protopodite is composed of two segments, though one may be reduced or suppressed and occasionally three may be present. In many cases, one of the branches, generally the endopodite, is more strongly developed than the other. Thus, in the thoracic limbs of the Malacostraca, the endopodite generally forms a walking-leg while the exopodite becomes a swimming-branch or may disappear altogether. Very often the basal segment of the protopodite bears, on the outer side, a lamellar appendage (more rarely, two), the epipodite, which may function as a gill. In the appendages near the mouth one or both of the protopodal segments may bear inwardly-turned processes, assisting in mastication and known as gnathobases. The frequent occurrence of epipodites and gnathobases tends to show that the primitive type of appendage was more complex than the simple biramous limb, and some authorities have regarded the leaf-like appendages of the Phyllopoda as nearer the original form from which the various modifications found in other groups have been derived. In a Phyllopod such as *Apus* the limbs of the trunk consist of a flattened, unsegmented or obscurely segmented axis or corm having a series of lobes or processes known as endites and exites on its inner and outer margins respectively. In all the Phyllopoda the number of endites is six, and the proximal one is more or less distinctly specialized as a gnathobase, working against its fellow of the opposite side in seizing food and transferring it to the mouth. The Phyllopoda are the only Crustacea in which distinct and functional gnathobasic processes are found on appendages far removed from the mouth. The two distal endites are regarded as corresponding to the endopodite and exopodite of the higher Crustacea, the axis or corm of the Phyllopod limb representing the protopodite. The number of exites is less constant, but, in *Apus*, two are present, the proximal branchial in function and the distal forming a stiffer plate which probably aids in swimming. It is not altogether easy to recognize the homologies of the endites and exites even within the order Phyllopoda, and the identification of the two distal endites as corresponding to the endopodite and exopodite of higher Crustacea is not free from difficulty. It is highly probable, however, that the biramous limb is a simplification of a more complex primitive type, to which the Phyllopod limb is a more or less close approximation.

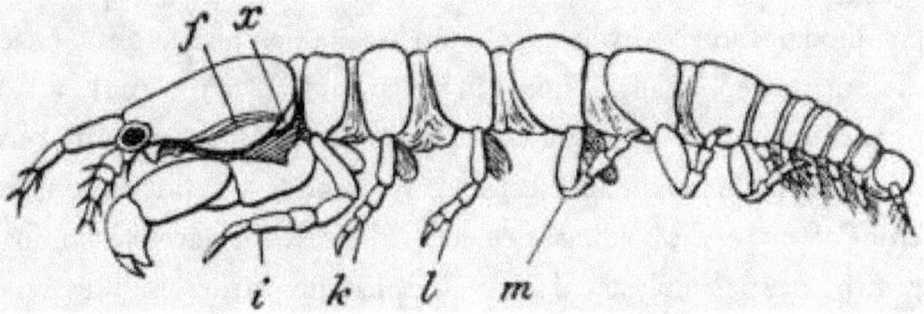

Fig. 8.—Tanais dubius (?) Kr. ☒, showing the orifice of entrance (x) into the cavity overarched by the carapace in which an appendage of the maxilliped (f) plays. On four feet (i, k, l, m) are the rudiments of the lamellae which subsequently form the brood-cavity. (Fritz Müller.)

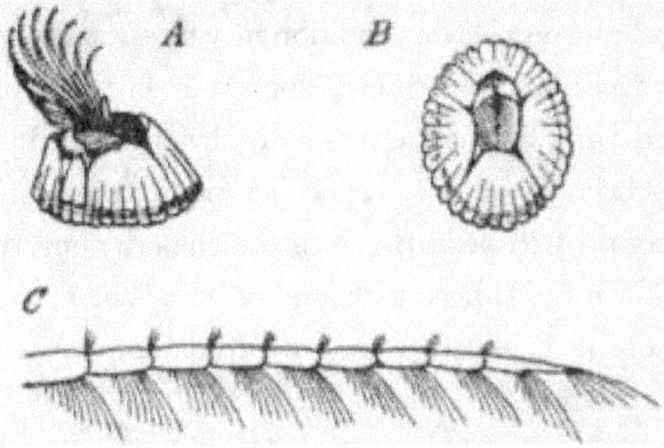

FIG. 9.—*A*, *Balanus* (young), side view with cirri protruded. *B*, Upper surface of same; valves closed. *C*, Highly magnified view of one of the cirri. (Morse.)

The modifications which this original type undergoes are usually more or less plainly correlated with the functions which the appendages have to discharge. Thus, when acting as swimming organs, the appendages, or their rami, are more or less flattened, or oar-like, and often have the margins fringed with long plumose hairs. When used for walking, one of the rami, usually the inner, is stout and cylindrical, terminating in a claw, and having the segments united by definite hinge-joints. The jaws have the gnathobasic endites developed at the expense of the rest of the limb, the endopodite and exopodite persisting only as sensory "palps" or disappearing altogether. When specialized as bearers of sensory (olfactory or tactile) organs, the rami are generally elongated, many-jointed and flagelliform. This modification is usually only found in the antennules and antennae, but it may exceptionally be found in the appendages of the trunk, as, for instance, in the thoracic legs of some Decapods (*e.g. Mastigocheirus*). Very often one or other of the appendages may be modified for prehension, the seizing of prey or the holding of a mate. In this case, the claw-like terminal segment may be simply flexed against the preceding in the same way as the blade of a penknife shuts up against the handle. The penultimate segment is often broadened, so that the terminal claw shuts against a transverse edge (fig. 4), or, finally, the penultimate segment may be produced into a thumb-like process opposed to the movable terminal segment or finger, forming a perfect chela or forceps, as, for instance, in the large claws of a crab or lobster. This chelate condition may be assumed by almost any of the appendages, and sometimes it appears in different appendages in closely related forms, so that no very great phylogenetic importance can in most cases be attached to it. A peculiar modification is found in the trunk-limbs of the Cirripedia (fig. 9), in which both rami are multiarticulate and filiform and fringed with long bristles. When protruded from the opening of the shell these "cirri" are spread out to form a casting-net for the capture of minute floating prey.

Gills or branchiae may be developed by parts of an appendage becoming thin-walled and vascular and either expanded into a thin lamella or ramified. Some of the special modifications of branchiae are referred to below.

Special Morphology of Appendages.—In many Crustacea the eyes are borne on stalks which are movably articulated with the head and which may be divided into two or three segments. The view is commonly held that these eye-stalks are really limbs, homologous with the other appendages. In spite of much discussion, however, it cannot be said that this point has been finally settled. The evidence of embryology is decidedly against the view that the eye-stalks are limbs. They are absent in the earliest and most primitive larval forms (nauplius), and appear only late in the course of development, after many of the trunk-limbs are fully formed. In the development of the Phyllopod *Branchipus*, the eyes are at first sessile, and the lateral lobes of the head on which they are set grow out and become movably articulated, forming the peduncles. The most important evidence in favour of their appendicular nature is afforded by the phenomena of regeneration. When the eye-stalk is removed from a living lobster or prawn, it is found that under certain conditions a many-jointed appendage like the flagellum of an antennule or antenna may grow in its place. It is open to question, however, how far the evidence from such "heteromorphic regeneration" can be regarded as conclusive on the points of homology. The fact that in certain rare cases among insects a leg may apparently be replaced by a wing tends to show that under exceptional conditions similar forms may be assumed by non-homologous parts.

The antennules (or first antennae) are almost universally regarded as true appendages, though they differ from all the other appendages in the fact that they are always innervated from the "brain" (or preoral ganglia), and that they are uniramous in the nauplius larva and in all the Entomostracan orders. As regards their innervation an apparent exception is found in the case of *Apus*, where the nerves to the antennules arise, behind the brain, from the oesophageal commissures, but this is, no doubt, a secondary condition, and the nerve-fibres have been traced forwards to centres within the brain. In the Malacostraca, the antennules are often biramous, but there is considerable doubt as to whether the two branches represent the endopodite and exopodite of the other limbs, and three branches are found in the Stomatopoda and in some Caridea. In the great majority of Crustacea the antennules are purely sensory in function and carry numerous "olfactory" hairs. They may, however, be natatory as in many Ostracoda and Copepoda, or prehensile, as in some Copepoda. The most peculiar modification, perhaps, is that found in the Cirripedia (Thyrostraca), in the larvae of which the antennules develop into organs of attachment, bearing the openings of the cement-glands, and becoming, in the adult, involved in the attachment of the animal to its support.

The antennae (second antennae) are of special interest on account of the clear evidence that, although preoral in position in all adult Crustacea, they were originally postoral appendages. In the nauplius larva they lie rather at the sides than in front of the mouth, and their basal portion carries a hook-like masticatory process which assists the similar processes of the mandibles in seizing food. In the primitive Phyllopoda, and less distinctly in some other orders, the nerves supplying the antennae arise, not from the brain, but from the circum-oesophageal commissures, and even in those cases where the nerves and the ganglia in which they are rooted have been moved forwards to the brain, the transverse commissure of the ganglia can still be traced, running behind the oesophagus.

The functions of the antennae are more varied than is the case with the antennules. In many Entomostraca (Phyllopoda, Cladocera, Ostracoda, Copepoda) they are important, and sometimes the only, organs of locomotion. In some male Phyllopoda they form complex "claspers" for holding the female. They are frequently organs of attachment in parasitic Copepoda, and they may be completely pediform in the Ostracoda. In the Malacostraca they are chiefly sensory, the endopodite forming a long flagellum, while the exopodite may form a lamellar "scale," probably useful as a balancer in swimming, or may disappear altogether. A very curious function sometimes discharged by the antennules or antennae of Decapods is that of forming a respiratory siphon in sand-burrowing species.

The mandibles, like the antennae, have, in the nauplius, the form of biramous swimming limbs, with a masticatory process originating from the proximal part of the protopodite. This form is retained, with little alteration in some adult Copepoda, where the biramous "palp" still aids in locomotion. A somewhat similar structure is found also in some Ostracoda. In most cases, however, the palp loses its exopodite and it often disappears altogether, while the coxal segment forms the body of the mandible, with a masticatory edge variously armed with teeth and spines. In a few Ostracoda, by a rare exception, the masticatory process is reduced or suppressed, and the palp alone remains, forming a pediform appendage used in locomotion as well as in the prehension of food. In parasitic blood-sucking forms the mandibles often have the shape of piercing stylets, and are enclosed in a tubular proboscis formed by the union of the upper lip (labrum) with the lower lip (hypostome or paragnatha).

The maxillulae and maxillae (or, as they are often termed, first and second maxillae) are nearly always flattened leaf-like appendages, having gnathobasic lobes or endites borne by the segments of the protopodite. The endopodite, when present, is unsegmented or composed of few segments and forms the "palp," and outwardly-directed lobes representing the exopodite and epipodites may also be present. These limbs undergo great modification in the different groups. The maxillulae are sometimes closely connected with the "paragnatha" or lobes of the lower lip, when these are present, and it has been suggested that the paragnatha are really the basal endites which have become partly separated from the rest of the appendage.

The limbs of the post-cephalic series show little differentiation among themselves in many Entomostraca. In the Phyllopoda they are for the most part all alike, though one or two of the anterior pairs may be specialized as sensory (*Apus*) or grasping (*Estheriidae*) organs. In the Cirripedia (Thyrostraca) the six pairs of biramous cirriform limbs differ only slightly from each other, and in many Copepoda this is also the case. In other Entomostraca considerable differentiation may take place, but the series is never divided into definite "tagmata" or groups of similarly modified appendages. It is highly characteristic of the Malacostraca, however, that the trunk-limbs are divided into two sharply defined tagmata corresponding to the thoracic and abdominal regions respectively, the limit between the two being marked by the position of the male genital openings. The thoracic limbs have the endopodites converted, as a rule, into more or less efficient walking-legs, and the exopodites are often lost, while the abdominal limbs more generally preserve the biramous form and are, in the more primitive types, natatory. These tagmata may again be subdivided into groups preserving a more or less marked individuality. For example, in the Amphipoda (fig. 4) the abdominal appendages are constantly divided into an anterior group of three natatory "swimmerets" and a posterior group of three limbs used chiefly in jumping or in burrowing. In nearly all Malacostraca the last pair of abdominal appendages (uropods) differ from the others, and in the more primitive groups they form, with the telson, a lamellar "tail-fan" (fig. 3, *T*), used in springing backwards through the water. In the thoracic series it is usual for one or more of the anterior pairs to be pressed into the service of the mouth, forming "foot-jaws" or maxillipeds. In the Decapoda three pairs are thus modified, and in the Tanaidacea, Isopoda and Amphipoda only one. In the Schizopoda and Cumacea the line of division is less sharp, and the varying number of so-called maxillipeds recognized by different authors gives rise to some confusion of terminology in systematic literature.

Gills.—In many of the smaller Entomostraca (Copepoda and most Ostracoda) no special gills are present, and respiration is carried on by the general surface of the body and limbs. When present, the branchiae are generally differentiations of parts of the appendages, most often the epipodites, as in the Phyllopoda. In the Cirripedia, however, they are vascular processes from the inner surface of the mantle or shell-fold, and in some Ostracoda they are outgrowths from the sides of the body. In the primitive Malacostraca the gills were probably, as in the Phyllopoda and in *Nebalia*, the modified epipodites of the thoracic limbs, and this is the condition found in some Schizopoda. In the Cumacea and Tanaidacea only the first thoracic limb has a branchial epipodite. In the Amphipoda, the gills though arising from the inner side of the bases of the thoracic legs are probably also epipodial in nature. In the Isopoda the respiratory function has been taken over by the abdominal appendages, both rami or only the inner becoming thin or flattened. In the Decapoda the branchial system is more complex. The gills are inserted at the base of the thoracic limbs, and lie within a pair of branchial chambers covered by the carapace. Three series are distinguished, *podobranchiae*, attached to the proximal segments of the appendages, *pleurobranchiae*, springing from the body-wall, and an intermediate series, *arthrobranchiae*, inserted on the articular membrane of the joint between the limb and the body. The podobranchiae are clearly epipodites, or, more correctly, parts of the epipodites, and it is probable that the arthro- and pleurobranchiae are also epipodial in origin and have migrated from the proximal segment of the limbs on to the adjacent body-wall.

Adaptations for aërial respiration are found in some of the land-crabs, where the lining membrane of the gill-chamber is beset with vascular papillae and acts as a lung. In some of the terrestrial Isopoda or woodlice (Oniscoidea) the abdominal appendages have ramified tubular invaginations of the integument, filled with air and resembling the tracheae of insects.

Internal Structure: Alimentary System.—In almost all Crustacea the food-canal runs straight through the body, except at its anterior end, where it curves downwards to the ventrally-placed mouth. In a few cases its course is slightly sinuous or twisted, but the only cases in which it is actually coiled upon itself are found in the Cladocera of the family *Lynceidae* (*Alonidae*) and in a single recently-discovered genus of Cumacea (Sympoda). As in all Arthropoda, it is composed of three divisions, a fore-gut or stomodaeum, ectodermal in origin and lined by an inturning of the chitinous cuticle, a mid-gut formed by endoderm and without a cuticular lining, and a hind-gut or proctodaeum, which, like the fore-gut, is ectodermal and is lined by cuticle. The relative proportions of these three divisions vary considerably, and the extreme abbreviation of the mid-gut found in the common crayfish (*Astacus*) is by no means typical of the class. Even in the closely-related lobster (*Homarus*) the mid-gut may be 2 or 3 in. long.

In a few Entomostraca (some Phyllopoda and Ostracoda) the chitinous lining of the fore-gut develops spines and hairs which help to triturate and strain the food, and among the Ostracods there is occasionally (*Bairdia*) a more elaborate armature of toothed plates moved by muscles. It is among the Malacostraca, however, and especially in the Decapoda, that the "gastric mill" reaches its greatest perfection. In most Decapods the "stomach" or dilated portion of the fore-gut is divided into two chambers, a large anterior "cardiac" and a smaller posterior "pyloric." In the narrow opening between these, three teeth (fig. 10) are set, one dorsally and one on each side. These teeth are connected with a framework of movably articulated ossicles developed as thickened and calcified portions of the lining cuticle of the stomach and moved by special muscles in such a way as to bring the three teeth together in the middle line. The walls of the pyloric chamber bear a series of pads and ridges beset with hairs and so disposed as to form a straining apparatus.

The mid-gut is essentially the digestive and absorptive region of the alimentary canal, and its surface is, in most cases, increased by pouch-like or tubular outgrowths which not only serve as glands for the secretion of the digestive juices, but may also become filled by the more fluid portion of the partially digested food and facilitate its absorption. These outgrowths vary much in their arrangement in the different groups. Most commonly there is a pair of lateral caeca, which may be more or less ramified and may form a massive "hepato-pancreas" or "liver."

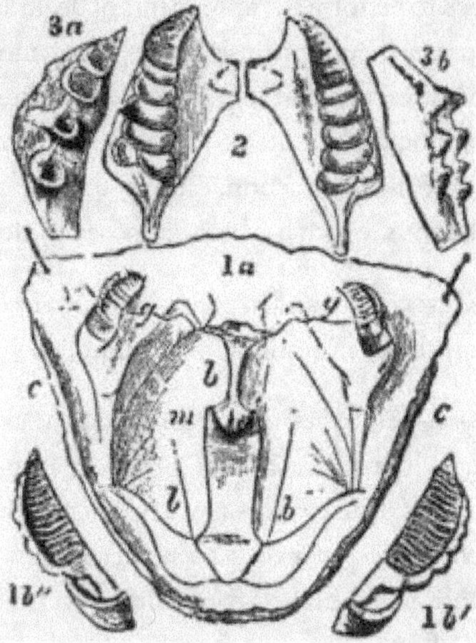

FIG. 10.—Gastric Teeth of Crab and Lobster. FIG. 10.—Gastric Teeth of Crab and Lobster.

1*a*, Stomach of common crab, *Cancer pagurus*, laid open, showing *b*, *b*, *b*, some of the calcareous plates inserted in its muscular coat; *g*, *g*, the lateral teeth, which when in use are brought in contact with the sides of the median tooth *m*; *c*, *c*, the muscular coat.

1*b'* and 1*b''*, The gastric teeth enlarged to show their grinding surfaces. 2, Gastric teeth of common lobster, *Homarus vulgaris*.

3*a* and 3*b*, Two crustacean teeth (of *Dithyrocaris*) from the Carboniferous series of Renfrewshire (these, however, may be the toothed edges of the mandibles).

The whole length of the alimentary canal is provided, as a rule, with muscular fibres, both circular and longitudinal, running in its walls, and, in addition, there may be muscle-bands running between the gut and the body-wall. In the region of the oesophagus these muscles are more strongly developed to perform the movements of deglutition, and, where a gastric mill is present, both intrinsic and extrinsic muscles co-operate in producing the movements of its various parts. The hind-gut is also provided with sphincter and dilator muscles, and these may produce rhythmic expansion and contraction, causing an inflow and outflow of water through the anus, which has been supposed to aid in respiration.

In the parasitic Rhizocephala and in a few Copepoda (*Monstrillidae*) the alimentary canal is absent or vestigial throughout life.

Circulatory System.—As in the other Arthropoda, the circulatory system in Crustacea is largely lacunar, the blood flowing in spaces or channels without definite walls. These spaces make up the apparent body-cavity, the true body-cavity or coelom having been, for the most part, obliterated by the great expansion of the blood-containing spaces. The heart is of the usual Arthropodous type, lying in a more or less well-defined pericardial blood-sinus, with which it communicates by valvular openings or ostia. In the details of the system, however, great differences exist within the limits of the class. There is every reason to believe that, in the primitive Arthropoda, the heart was tubular in form, extending the whole length of the body, and having a pair of ostia in each somite. This arrangement is retained in some of the Phyllopoda, but even in that group a progressive abbreviation of the heart, with a diminution in the number of the ostia, can be traced, leading to the condition found in the closely related Cladocera, where the heart is a subglobular sac, with only a single pair of ostia. In the Malacostraca, an elongated heart with numerous segmentally arranged ostia is found only in the aberrant group of Stomatopoda and in the transitional Phyllocarida. In the other Malacostraca the heart is generally abbreviated, and even where, as in the Amphipoda, it is elongated and tubular, the ostia are restricted in number, three pairs only being usually present. In many Entomostraca the heart is absent, and it is impossible to speak of a "circulation" in the proper sense of the term, the blood being merely driven hither and thither by the movements of the body and limbs and of the alimentary canal.

A very remarkable condition of the blood-system, unique, as far as is yet known among the Arthropoda, is found in a few genera of parasitic Copepoda (*Lernanthropus*, *Mytilicola*). In these there is a closed system of vessels, not communicating with the body-cavity, and containing a coloured fluid. There is no heart. The morphological nature of this system is unknown.

Excretory System.—The most important excretory or renal organs of the Crustacea are two pairs of glands lying at the base of the antennae and of the second maxillae respectively. The two are probably never functional together in the same animal, though one may replace the other in the course of development. Thus, in the Phyllopoda, the antennal gland develops early and is functional during a great part of the larval life, but it ultimately atrophies, and in the adult (as in most Entomostraca) the maxillary gland is the functional excretory organ. In the Decapoda, where the antennal gland alone is well-developed in the adult, the maxillary gland sometimes precedes it in the larva. The structure of both glands is essentially the same. There is a more or less convoluted tube with glandular walls connected internally with a closed "end-sac" and opening to the exterior by means of a thin-walled duct. Development shows that the glandular tube is mesoblastic in origin and is of the nature of a coelomoduct, while the end-sac is to be regarded as a vestigial portion of the coelom. In the Branchiopoda the maxillary gland is lodged in the thickness of the shell-fold (when this is present), and, from this circumstance, it often receives the somewhat misleading name of "shell-gland." In the Decapoda the antennal gland is largely developed and is known as the "green gland." The external duct of this gland is often dilated into a bladder, and may sometimes send out diverticula, forming a complex system of sinuses ramifying through the body. The green gland and the structures associated with it in Decapods were at one time regarded as constituting an auditory apparatus.

In addition to these two pairs of glands, which are in all probability the survivors of a series of segmentally arranged coelomoducts present in the primitive Arthropoda, other excretory organs have been described in various Crustacea. Although the excretory function of these has been demonstrated by physiological methods, however, their morphological relations are not clear. In some cases they consist of masses of mesodermal cells, within which the excretory products appear to be stored up instead of being expelled from the body.

Nervous System.—The central nervous system is constructed on the same general plan as in the other Arthropoda, consisting of a supra-oesophageal ganglionic mass or brain, united by circum-oesophageal connectives with a double ventral chain of segmentally arranged ganglia. In the primitive Phyllopoda the ventral chain retains the ladder-like arrangement found in some Annelids and lower worms, the two halves being widely separated and the pairs of ganglia connected together across the middle line by double transverse commissures. In the higher groups the two halves of the chain are more or less closely approximated and coalesced, and, in addition, a concentration of the ganglia in a longitudinal direction takes place, leading ultimately, in many cases, to the formation of an unsegmented ganglionic mass representing the whole of the ventral chain. This is seen, for example, in the Brachyura among the Decapoda. The brain, or supra-oesophageal ganglion, shows various degrees of complexity. In the Phyllopoda it consists mainly of two pairs of ganglionic centres, giving origin respectively to the optic and antennular nerves. The centres for the antennal nerves form ganglionic swellings on the oesophageal connectives. In the higher forms, as already mentioned, the antennal ganglia have become shifted forwards and coalesced with the brain. In the higher Decapoda, numerous additional centres are developed in the brain and its structure becomes extremely complex.

Eyes.—The eyes of Crustacea are of two kinds, the unpaired, median or "nauplius" eye, and the paired compound eyes. The former is generally present in the earliest larval stages (nauplius), and in some Entomostraca (*e.g.* Copepoda) it forms the sole organ of vision in the adult. In the Malacostraca it is absent in the adult, or persists only in a vestigial condition, as in some Decapoda and Schizopoda. It is typically tripartite, consisting of three cup-shaped masses of pigment, the cavity of each cup being filled with columnar retinal cells. At their inner ends (towards the pigment) these cells contain rod-like structures, while their outer ends are connected with the nerve-fibres. In some cases three separate nerves arise from the front of the brain, one going to each of the three divisions of the eye. In the Copepoda the median eye may undergo considerable elaboration, and refracting lenses and other accessory structures may be developed in connexion with it.

The compound eyes are very similar in the details of their structure (see ARTHROPODA) to those of insects (Hexapoda). They consist of a varying number of ommatidia or visual elements, covered by a transparent region of the external cuticle forming the cornea. In most cases this cornea is divided into lenticular facets corresponding to the underlying ommatidia.

As has been already stated, the compound eyes are often set on movable peduncles. It is probable that this is the primitive condition from which the sessile eyes of other forms have been derived. In the Malacostraca the sessile eyed groups are certainly less primitive than some of those with stalked eyes, and among the Entomostraca also there is some evidence pointing in the same direction.

Although typically paired, the compound eyes may occasionally coalesce in the middle line into a single organ. This is the case in the Cladocera, the Cumacea and a few Amphipoda.

Mention should also be made of the partial or complete atrophy of the eyes in many Crustacea which live in darkness, either in the deep sea or in subterranean habitats. In these cases the peduncles may persist and may even be modified into spinous organs of defence.

Other Sense-Organs.—As in Arthropoda, the hairs or setae on the surface of the body are important organs of sense and are variously modified for special sensory functions. Many, perhaps all, of them are tactile. They are movably articulated at the base where they are inserted in pits formed by a thinning away of the cuticle, and each is supplied by a nerve-fibril. When feathered or provided with secondary barbs the setae will respond to movements or vibrations in the surrounding water, and have been supposed to have an auditory function. In certain divisions of the Malacostraca more specialized organs are found which have been regarded as auditory. In the majority of the Decapoda there is a saccular invagination of the integument in the basal segment of the antennular peduncle having on its inner surface "auditory" setae of the type just described. The sac is open to the exterior in most of the Macrura, but completely closed in the Brachyura. In the former case it contains numerous grains of sand which are introduced by the animal itself after each moult and which are supposed to act as otoliths. Where the sac is completely closed it generally contains no solid particles, but in a few Macrura a single otolith secreted by the walls of the sac is present. In the *Mysidae* among the Schizopoda a pair of similar otocysts are found in the endopodites of the last pair of appendages (uropods). These contain each a single concretionary otolith.

Recent observations, however, make it very doubtful whether aquatic Crustacea can hear at all, in the proper sense of the term, and it has been shown that one function, at least, of the so-called otocysts is connected with the equilibration of the body. They are more properly termed statocysts.

Another modification of sensory setae is supposed to be associated with the sense of smell. In nearly all Crustacea the antennules and often also the antennae bear groups of hair-like filaments in which the chitinous cuticle is extremely delicate and which do not taper to a point but end bluntly. These are known as olfactory filaments or aesthetascs. They are very often more strongly developed in the male sex, and are supposed to guide the males in pursuit of the females.

Glands.—In addition to the digestive and excretory glands already mentioned, various glandular structures occur in the different groups of

Crustacea. The most important of these belong to the category of dermal glands, and may be scattered over the surface of the body and limbs, or grouped at certain points for the discharge of special functions. Such glands occurring on the upper and lower lips or on the walls of the oesophagus have been regarded as salivary. In some Amphipoda the secretion of glands on the body and limbs is used in the construction of tubular cases in which the animals live. In some freshwater Copepoda the secretion of the dermal glands forms a gelatinous envelope, by means of which the animals are able to survive desiccation. In certain Copepoda and Ostracoda glands of the same type produce a phosphorescent substance, and others, in certain Amphipoda and Branchiura, are believed to have a poisonous function. Possibly related to the same group of structures are the greatly-developed cement-glands of the Cirripedia, which serve to attach the animals to their support.

Phosphorescent Organs.—Many Crustacea belonging to very different groups (Ostracoda, Copepoda, Schizopoda, Decapoda) possess the power of emitting light. In the Ostracoda and Copepoda the phosphorescence, as already mentioned, is due to glands which produce a luminous secretion, and this is the case also in certain members of the Schizopoda and Decapoda. In other cases in the last two groups, however, the light-producing organs found on the body and limbs have a complex and remarkable structure, and were formerly described as accessory eyes. Each consists of a globular capsule pierced at one or two points for the entrance of nerves which end in a central cup-shaped "striated body." This body appears to be the source of light, and has behind it a reflector formed of concentric lamellae, while, in front, in some cases, there is a refracting lens. The whole organ can be rotated by special muscles. Organs of this type are best known in the *Euphausiidae* among the Schizopoda, but a modified form is found in some of the lower Decapods.

Reproductive System.—In the great majority of Crustacea the sexes are separate. Apart from certain doubtful and possibly abnormal instances among Phyllopoda and Amphipoda, the only exceptions are the sessile

Cirripedia and some parasitic Isopoda (*Cymothoidae*), where hermaphroditism is the rule. Parthenogenesis is prevalent in the Branchiopoda and Ostracoda, often in more or less definite seasonal alternation with sexual reproduction. Where the sexes are distinct, a more or less marked dimorphism often exists. The male is very often provided with clasping organs for seizing the female. These may be formed by the modification of almost any of the appendages, often the antennules or antennae or some of the thoracic limbs, or even the mandibular palps (some Ostracoda). In addition, some of the appendages in the neighbourhood of the genital apertures may be modified for the purpose of transferring the genital products to the female, as, for instance, the first and second abdominal limbs in the Decapoda. In the higher Decapoda the male is generally larger than the female and has stronger chelae. On the other hand, in other groups the male is often smaller than the female. In the parasitic Copepoda and Isopoda the disparity in size is carried to an extreme degree, and the minute male is attached, like a parasite, to the enormously larger female.

The Cirripedia present some examples of sexual relationships which are only paralleled, in the animal kingdom, among the parasitic Myzostomida. While the great majority are simple hermaphrodites, capable of cross and self fertilization, it was discovered by Darwin that, in certain species, minute degraded males exist, attached within the mantle-cavity of the ordinary individuals. Since these dwarf males pair, not with females, but with hermaphrodites, Darwin termed them "complemental" males. In other species the large individuals have become purely female by atrophy of the male organs, and are entirely dependent on the dwarf males for fertilization. In spite of the opinion of some distinguished zoologists to the contrary, it seems most probable that the separation of the sexes is in this case a secondary condition, derived from hermaphroditism through the intermediate stage represented by the species having complemental males.

The gonads, as in other Arthropoda, are hollow saccular organs, the cavity communicating with the efferent ducts. They are primitively paired, but often coalesce with each other more or less completely. The ducts are present only as a single pair, except in one genus of parasitic Isopoda (*Hemioniscus*), where two pairs of oviducts are found. Various accessory structures may be connected with the efferent ducts in both sexes. The oviducts may have diverticula serving as receptacles for the spermatozoa (in cases where internal impregnation takes place), and may be provided with glands secreting envelopes or shells around the eggs. The male ducts often have glandular walls, secreting capsules or spermatophores within which the spermatozoa are packed for transference to the female. The terminal part of the male ducts may be protrusible and act as an intromittent organ, or this function may be discharged by some of the appendages, as, for instance, in the Brachyura.

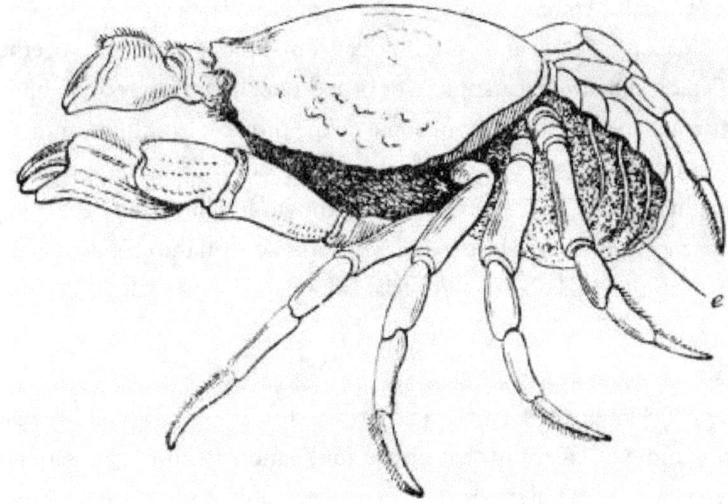

FIG. 11.—Side view of Crab, the abdomen extended and carrying a mass of eggs beneath it; *e*, eggs. (After Morse.)

The position of the genital apertures varies very greatly in the different groups of the class. They are farthest forward in the case of the female organs of the Cirripedia, where the openings are on the first thoracic (fourth postoral) somite. The most posterior position is occupied by the genital apertures of certain Phyllopoda (*Polyartemia*), which lie behind the nineteenth trunk-somite. It is characteristic of the Malacostraca that the position of the genital apertures is constantly different in the two sexes, the female openings being on the sixth, and those of the male on the eighth thoracic somite.

Very few Crustacea are viviparous in the sense that the eggs are retained within the body until hatching takes place (some Phyllopoda), but, on the other hand, the great majority carry the eggs in some way or other after their extrusion. In some Phyllopoda (*Apus*) egg-sacs are formed by modification of certain of the thoracic feet. The eggs are retained between the valves of the shell in some Phyllopoda and in the Cladocera and Ostracoda, and they lie in the mantle cavity in the Cirripedia. In the Copepoda they are agglutinated together into masses attached to the body of the female. Among the Malacostraca some Schizopoda, the Cumacea, Tanaidacea, Isopoda and Amphipoda (sometimes grouped all together as Peracarida) have a marsupium or brood-pouch formed by overlapping plates attached to the bases of some of the thoracic legs. In most of the Decapoda the eggs are carried by the female, attached to the abdominal appendages (fig. 11). A few cases are known in which the developing embryos are nourished by a special secretion while in the brood-chamber of the mother (Cladocera, terrestrial Isopoda).

Embryology.

The majority of the Crustacea are hatched from the egg in a form differing more or less from that of the adult, and pass through a series of free-swimming larval stages. There are many cases, however, in which the metamorphosis is suppressed, and the newly-hatched young resemble the parent in general structure. The relative size of the eggs and the amount of nutritive yolk which they contain are generally much greater in those forms which have a direct development.

The details of the early embryonic stages vary considerably within the limits of the class. They are of interest, however, rather from the point of view of general embryology than from that of the special student of the Crustacea, and cannot be fully dealt with here.

Segmentation is usually of the superficial or centrolecithal type. The hypoblast is formed either by a definite invagination or by the immigration of isolated cells, known as vitellophags, which wander through the yolk and later become associated into a definite mesenteron, or by some combination of these two methods. The blastopore generally occupies a position corresponding to the posterior end of the body. The mesoblast of the cephalic (naupliar) region probably arises in connexion with the lips of the blastopore and consists of loosely-connected cells or mesenchyme. In the region of the trunk, in many cases, paired mesoblastic bands are formed, growing in length by the division of teloblastic cells at the posterior end, and becoming segmented into somites. The existence of true coelom-sacs is somewhat doubtful. The rudiments of the first three pairs of appendages commonly appear simultaneously, and, even in forms with embryonic development, they show differences in their mode of appearance from the succeeding somites. Further, a definite cuticular membrane is frequently formed and shed at this stage, which corresponds to the nauplius-stage of larval development.

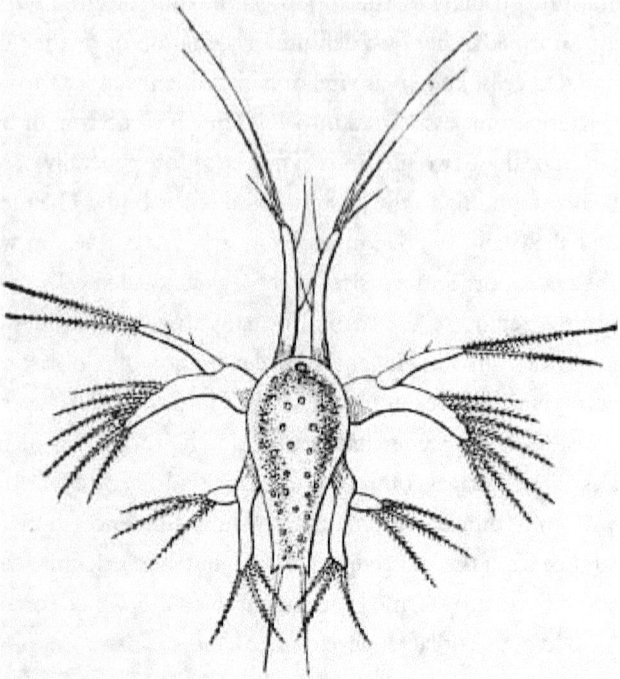

FIG. 12.—Nauplius of a Prawn (*Penaeus*). (Fritz Müller).

The larval metamorphoses of the Crustacea have attracted much attention, and have been the subject of much discussion in view of their bearing on the phylogenetic history of the group. In those Crustacea in which the series of larval stages is most complete, the starting-point is the form already mentioned under the name of *nauplius*. The typical nauplius (fig. 12) has an oval unsegmented body and three pairs of limbs corresponding to the antennules, antennae and mandibles of the adult. The antennules are uniramous, the others biramous, and all three pairs are used in swimming. The antennae have a spiniform or hooked masticatory process at the base, and share with the mandibles, which have a similar process, the function of seizing and masticating the food. The mouth is overhung by a large labrum or upper lip, and the integument of the dorsal surface of the body forms a more or less definite dorsal shield. The paired eyes are, as yet, wanting, but the unpaired eye is large and conspicuous. A pair of frontal papillae or filaments, probably sensory, are commonly present.

A nauplius larva differing only in details from the typical form just described is found in the majority of the Phyllopoda, Copepoda and Cirripedia, and in a more modified form, in some Ostracoda. Among the Malacostraca the nauplius is less commonly found, but it occurs in the *Euphausiidae* among the Schizopoda and in a few of the more primitive Decapoda (*Penaeidea*) (fig. 12). In most of the Crustacea which hatch at a later stage there is, as already mentioned, more or less clear evidence of an embryonic nauplius stage. It seems certain, therefore, that the possession of a nauplius larva must be regarded as a very primitive character of the Crustacean stock.

As development proceeds, the body of the nauplius elongates, and indications of segmentation begin to appear in its posterior part. At successive moults the somites increase in number, new somites being added behind those already differentiated, from a formative zone in front of the telsonic region. Very commonly the posterior end of the body becomes forked, two processes growing out at the sides of the anus and often persisting in the adult as the "caudal furca." The appendages posterior to the mandibles appear as buds on the ventral surface of the somites, and in the most primitive cases they become differentiated, like the somites which bear them, in regular order from before backwards. The limb-buds early become bilobed and grow out into typical biramous appendages which gradually assume the characters found in the adult. With the elongation of the body, the dorsal shield begins to project posteriorly as a shell-fold, which may increase in size to envelop more or less of the body or may disappear altogether. The rudiments of the paired eyes appear under the integument at the sides of the head, but only become pedunculated at a comparatively late stage.

The course of development here outlined, in which the nauplius gradually passes into the adult form by the successive addition of somites and appendages in regular order, agrees so well with the process observed in the development of the typical Annelida that we must regard it as being the most primitive method. It is most closely followed by the Phyllopods such as *Apus* or *Branchipus*, and by some Copepoda.

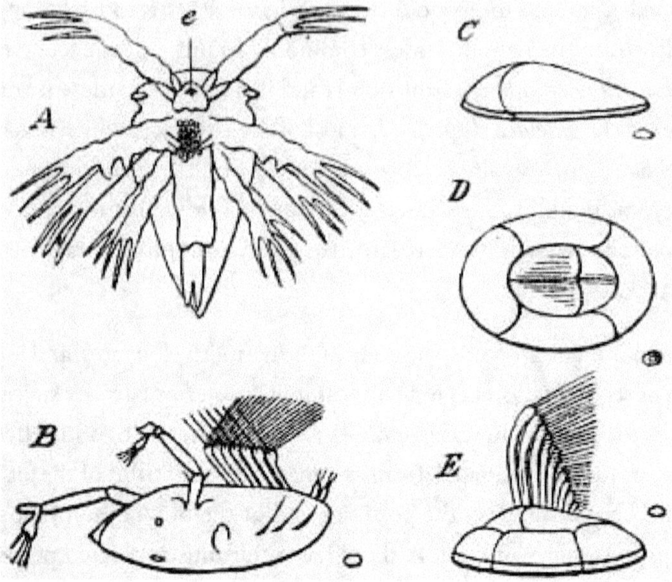

FIG. 13.—Early Stages of *Balanus*. (After Spence Bate.)

A, Nauplius. e, Eye.
B, Cypris-larva with a bivalve shell and just before becoming attached (represented feet upwards for comparison with E, where it is attached).
The dots indicate the actual size.

C, After becoming attached, side views.
D, Later stage, viewed from above.
E, Side view, later stage and with cirri extended.

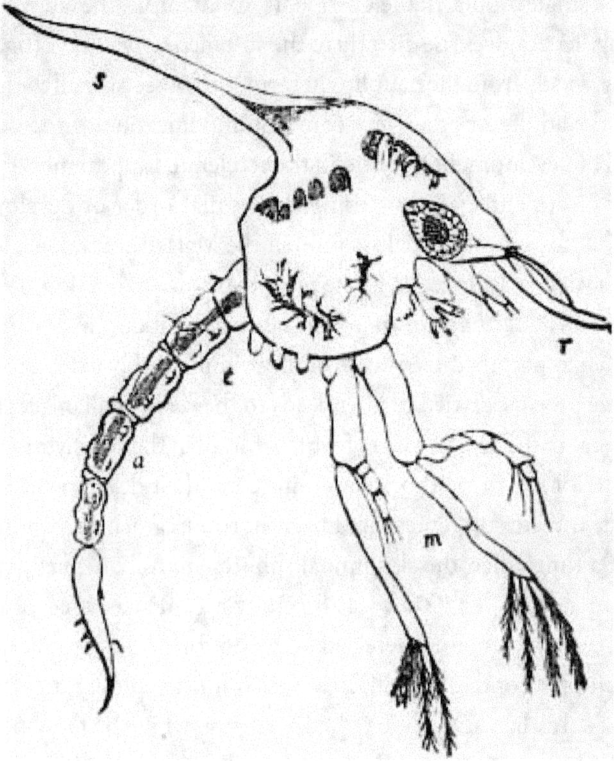

FIG. 14.—Zoea of Common Shore-Crab in its second stage. (Spence Bate.)
r, Rostral spine.
s, Dorsal spine.
m, Maxillipeds.
t, Buds of
thoracic feet.
a, Abdomen.

In most Crustacea, however, this primitive scheme is more or less modified. The earlier stages may be suppressed or passed through within the egg (or within the maternal brood-chamber), so that the larva, on hatching, has reached a stage more advanced than the nauplius. Further, the gradual appearance and differentiation of the successive somites and appendages may be accelerated, so that comparatively great advances take place at a single moult. In the Cirripedia, for example, the latest nauplius stage (fig. 13, A) gives rise directly to the so-called *Cypris*-larva (fig. 13, B), differing widely from the nauplius in form, and possessing all the appendages of the adult. Another very common modification of the primitive method of development is found in the accelerated appearance of certain somites or appendages, disturbing the regular order of development. This modification is especially found in the Malacostraca. Even in those which have most fully retained the primitive order of development, as in the *Penaeidea* and *Euphausiidae*, the last pair of abdominal appendages make their appearance in advance of those immediately in front of them. The same process, carried further, leads to the very peculiar larva known as the *Zoea*, in the typical form of which, found in the Brachyura (fig. 14), the posterior five or six thoracic somites have their development greatly retarded, and are still represented by a short unsegmented region of the body at a time when the abdominal somites are fully formed and even carry appendages. The *Zoea* was formerly regarded as a recapitulation of an ancestral form, but there can be no doubt that its peculiarities are the result of secondary modification. It is most typically developed in the most specialized Decapoda, the Brachyura, while the more primitive groups of Malacostraca, the *Euphausiidae*, *Penaeidea* and Stomatopoda, retain the primitive order of appearance of the somites, and, for the most part, of the limbs. At the same time, the tendency to a retardation in the development of the posterior thoracic somites is very general in Malacostracan larvae, and may perhaps be correlated with the fact that in the primitive Phyllocarida the whole thoracic region is very short and the limbs closely crowded together.

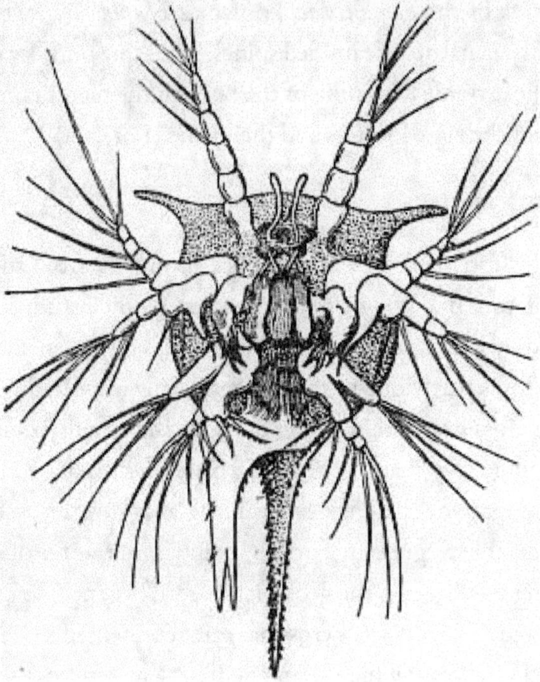

FIG. 15.—Nauplius of *Tetraclita porosa* after the first moult. (Fritz Müller.)

Besides the nauplius and the zoea there are many other types of Crustacean larvae, distinguished by special names, though, as their occurrence is restricted within the limits of the smaller systematic groups, they are of less general interest. We need only mention the *Mysis*-stage (better termed Schizopod-stage) found in many Macrura (as, for example, the lobster), which differs from the adult in having large natatory exopodites on the thoracic legs.

Most of the larval forms swim freely at the surface of the sea, and many show special adaptations to this habit of life. As in many other "pelagic" organisms, spines and processes from the surface of the body are often developed, which are probably less important as defensive organs than as aids to flotation. This is well seen in the nauplius of many Cirripedia (fig. 15) and in nearly all zoeae. Perhaps the most striking example is the zoea-like larva of the *Sergestidae*, known as *Elaphocaris*, which has an extraordinary armature of ramified spines. The same purpose is probably served by the extreme flattening of the body in the membranous *Phyllosoma*-larva of the rock-lobsters and their allies (Loricata).

Past History.

Although fossil remains of Crustacea are abundant, from the most ancient fossiliferous rocks down to the most recent, their study has hitherto contributed little to a precise knowledge of the phylogenetic history of the class. This is partly due to the fact that many important forms must have escaped fossilization altogether owing to their small size and delicate structure, while very many of those actually preserved are known only from the carapace or shell, the limbs being absent or represented only by indecipherable fragments. Further, many important groups were already differentiated when the geological record began. The Phyllopoda, Ostracoda and Cirripedia (Thyrostraca) are represented in Cambrian or Silurian rocks by forms which seem to have resembled closely those now existing, so that palaeontology can have little light to throw on the mode of origin of these groups. With the Malacostraca the case is little better. There is considerable reason for believing that the *Ceratiocaridae*, which are found from the Cambrian onwards, were allied to the existing *Nebalia*, and may possibly include the forerunners of the true Malacostraca, but nothing is definitely known of their appendages. In Palaeozoic formations, from the Upper Devonian onwards, numbers of shrimp-like forms are found which have been referred to the Schizopoda and the Decapoda, but here again the scanty information which may be gleaned as to the structure of the limbs rarely permits of definite conclusions as to their affinities. The recent discovery in the Tasmanian "schizopod" *Anaspides*, of what is believed to be a living representative of the Carboniferous and Permian *Syncarida*, has, however, afforded a clue to the affinities of some of these problematical forms.

True Decapods are first met with in Mesozoic rocks, the first to appear being the *Penaeidea*, a primitive group comprising the *Penaeidae* and *Sergestidae*, which occur in the Jurassic and perhaps in the Trias. Some of the earliest are referred to the existing genus *Penaeus*. The Stenopidea, another primitive group, differing from the Penaeidea in the character of the gills, appear in the Trias and Jurassic. The Caridea or true prawns and shrimps appear later, in the Upper Jurassic, some of them presenting primitive characteristics in the retention of swimming exopodites on the walking-legs. The Eryonidea (fig. 16, 3), a group related to the Loricata but of a more generalized type, are specially interesting since the few existing deep-sea forms appear to be only surviving remnants of what was, in the Mesozoic period, a dominant group. The Mesozoic *Glyphaeidae* have been supposed to stand in the direct line of descent of the modern rock-lobsters and their allies (Loricata). Some of the Loricata have persisted with little change from the Cretaceous period to the present day.

The Anomura are hardly known as fossils. The Brachyura, on the other hand, are well represented (fig 16, 1, 2). The earliest forms, from the Lower Oolite and later, belonging chiefly to the extinct family *Prosoponidae*, have been shown to have close relations with the most generalized of existing Brachyura, the deep-sea *Homolodromiidae*, and to link the Brachyura to the Homarine (lobster-like) Macrura.

A few Isopoda are known from Secondary rocks, but their systematic position is doubtful and they throw no light on the evolution of the group. The Amphipoda are not definitely known to occur till Tertiary times. Stomatopoda of a very modern-looking type, and even their larvae, occur in Jurassic rocks.

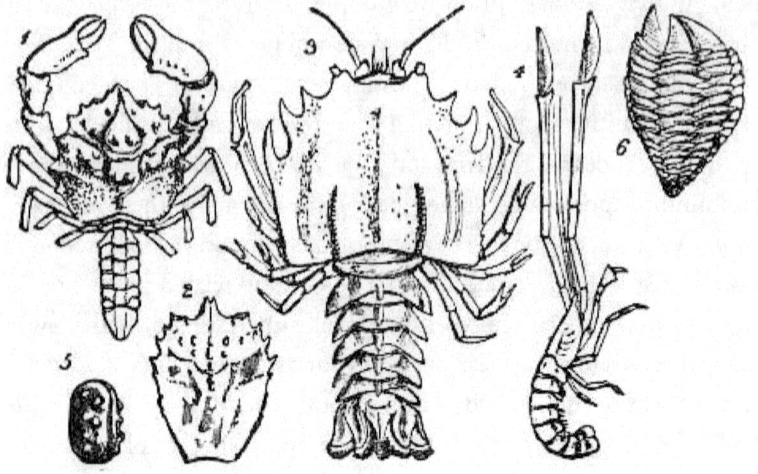

FIG. 16.

1, *Dromilites Lamarckii*, Desm.; London Clay, Sheppey.
2, *Palaeocorystes Stokesii*, Gault; Folkestone.
3, *Eryon arctiformis*, Schl.; Lithographic stone, Solenhofen.
4, *Mecocheirus longimanus*, Schl.; Lithographic stone, Solenhofen.
5, *Cypridea tuberculata*, Sby.; (Ostracoda); Weald, Sussex.
6, *Loricula pulchella*, Sby (Cirripedia); L. Chalk, Sussex.

In the dearth of trustworthy evidence as to the actual forerunners of existing Crustacea, we are compelled to rely wholly on the data afforded by comparative anatomy and embryology in attempting to reconstruct the probable phylogeny of the class. It is unnecessary to insist on the purely speculative character of the conclusions to be reached in this way, so long as they cannot be checked by the results of palaeontology, but, when this is recognized, such speculation is not only legitimate but necessary as a basis on which to build a natural classification.

The first attempts to reconstruct the genealogical history of the Crustacea started from the assumption that the "theory of recapitulation" could be applied to their larval history. The various larval forms, especially the nauplius and zoea, were supposed to reproduce, more or less closely, the actual structure of ancestral types. So far as the zoea was concerned, this assumption was soon shown to be erroneous, and the secondary nature of this type of larva is now generally admitted. As regards the nauplius, however, the constancy of its general character in the most widely diverse groups of Crustacea strongly suggests that it is a very ancient type, and the view has been advocated that the Crustacea must have arisen from an unsegmented nauplius-like ancestor.

The objections to this view, however, are considerable. The resemblances between the Crustacea and the Annelid worms, in such characters as the structure of the nervous system and the mode of growth of the somites, can hardly be ignored. Several structures which must be attributed, to the common stock of the Crustacea, such as the paired eyes and the shell-fold, are not present in the nauplius. The opinion now most generally held is that the primitive Crustacean type is most nearly approached by certain Phyllopods such as *Apus*. The large number and the uniformity of the trunk somites and their appendages, and the structure of the nervous system and of the heart in *Apus*, are Annelidan characters which can hardly be without significance. It is probable also, as already mentioned, that the leaf-like appendages of the Phyllopoda are of a primitive type, and attempts have been made to refer their structure to that of the Annelid parapodium. In many respects, however, the Phyllopoda, and especially *Apus*, have diverged considerably from the primitive Crustacean type. All the cephalic appendages are much reduced, the mandibles have no palps, and the maxillulae are vestigial. In these respects some of the Copepoda have retained characters which we must regard as much more primitive. In those Copepods in which the palps of the mandibles as well as the antennae are biramous and natatory, the first three pairs of appendages retain throughout life, with little modification, the shape and function which they have in the nauplius stage, and must, in all likelihood, be regarded as approximating to those of the primitive Crustacea. In other respects, however, such as the absence of paired eyes and of a shell-fold, as well as in the characters of the post-oral limbs, the Copepoda are undoubtedly specialized.

In order to reconstruct the hypothetical ancestral Crustacean, therefore, it is necessary to combine the characters of several of the existing groups. It may be supposed to have approximated, in general form, to *Apus*, with an elongated body composed of numerous similar somites and terminating in a caudal furca; with the post-oral appendages all similar and all bearing gnathobasic processes; and with a carapace originating as a shell-fold from the maxillary somite. The eyes were probably stalked, the antennae and mandibles biramous and natatory, and both armed with masticatory processes. It is likely that the trunk-limbs were also biramous, with additional endites and exites. Whether any of the obscure fossils generally referred to the Phyllopoda or Phyllocarida may have approximated to this hypothetical form it is impossible to say. It is to be noted, however, that the Trilobita, which, according to the classification here adopted, are dealt with under Arachnida, are not very far removed, except in such characters as the absence of a shell-fold and of eye-stalks, from the primitive Crustacean here sketched.

On this view, the nauplius, while no longer regarded as reproducing an ancestral type, does not altogether lose its phylogenetic significance. It is an ancestral *larval* form, corresponding perhaps to the stages immediately succeeding the trochophore in the development of Annelids, but with some of the later-acquired Crustacean characters superposed upon it. While little importance is to be given to such characters as the unsegmented body, the small number of limbs and the absence of a shell-fold and of paired eyes, it has, on the other hand, preserved archaic features in the form of the limbs and the masticatory function of the antenna.

The probable course of evolution of the different groups of Crustacea from this hypothetical ancestral form can only be touched on here. The Phyllopoda must have branched off very early and from them to the Cladocera the way is clear. The Ostracoda might have been derived from the same stock were it not that they retain the mandibular palp which all the Phyllopods have lost. The Copepoda must have separated themselves very early, though perhaps some of their characters may be persistently larval rather than phylogenetically primitive. The Cirripedia are so specialized both as larvae and as adults that it is hard to say in what direction their origin is to be sought.

For the Malacostraca, it is generally admitted that the Leptostraca (*Nebalia*, &c.) provide a connecting-link with the base of the Phyllopod stem. Nearest to them come the Schizopoda, a primitive group from which two lines of descent can be traced, the one leading from the Mysidacea (*Mysidae + Lophogastridae*) to the Cumacea and the sessile-eyed groups Isopoda and Amphipoda, the other from the Euphausiacea (*Euphausiidae*) to the Decapoda.

Classification.

The modern classification of Crustacea may be said to have been founded by P. A. Latreille, who, in the beginning of the 19th century, divided the class into Entomostraca and Malacostraca. The latter division, characterized by the possession of 19 somites and pairs of appendages (apart from the eyes), by the division of the appendages into two tagmata corresponding to cephalothorax and abdomen, and by the constancy in position of the generative apertures, differing in the two sexes, is unquestionably a natural group. The Entomostraca, however, are certainly a heterogeneous assemblage, defined only by negative characters, and the name is retained only for the sake of convenience, just as it is often useful to speak of a still more heterogeneous and unnatural assemblage of animals as Invertebrata. The barnacles and their allies, forming the group Cirripedia or Thyrostraca, sometimes treated as a separate sub-class, are distinguished by being sessile in the adult state, the larval antennules serving as organs of attachment, and the antennae being lost. An account of them will be found in the article THYROSTRACA. The remaining groups are dealt with under the headings ENTOMOSTRACA and MALACOSTRACA, the annectent group Leptostraca being included in the former.

It may be useful to give here a synopsis of the classification adopted in this encyclopaedia, noting that, for convenience of treatment, it has been thought necessary to adopt a grouping not always expressive of the most recent views o. T. Ca.)

CRUSTUMERIUM

An ancient town of Latium, on the edge of the Sabine territory, near the headwaters of the Allia, not far from the Tiber. It appears several times in the early history of Rome, but was conquered in 500 B.C. according to Livy ii. 19, the *tribus Crustumina* [or *Clustumina*] being formed in 471 B.C. Pliny mentions it among the lost cities of Latium, but the name clung to the district, the fertility of which remained famous. No remains of it exist, and its exact site is uncertain.

See T. Ashby in *Papers of the British School at Rome*, iii. 50.

CRUVEILHIER, JEAN

(1791-1874), French anatomist, was born at Limoges in 1791, and was educated at the university of Paris, where in 1825 he became professor of anatomy. In 1836 he became the first occupant of the recently founded chair of pathological anatomy. He died at Jussac in 1874. His chief works are *Anatomie descriptive* (1834-1836); *Anatomie pathologique du corps humain* (1829-1842), with many coloured plates; *Traité d'anatomie pathologique générale* (1849-1864); *Anatomie du système nerveux de l'homme* (1845); *Traité d'anatomie descriptive* (1851).

CRUZ E SILVA, ANTONIO DINIZ DA

(1731-1799), Portuguese heroic-comic poet, was the son of a Lisbon carpenter who emigrated to Brazil shortly before the poet's birth, leaving his wife to support and educate her young family by the earnings of her needle. Diniz studied Latin and philosophy with the Oratorians, and in 1747 matriculated at Coimbra University, where he wrote his first versus about 1750. In 1753 he took his degree in law, and returning to the capital, devoted much of the next six years to literary work. In 1756 he became one of the founders and drew up the statues of the *Arcadia Lusitana*, a literary society whose aims were the instruction of its members, the cultivation of the art of poetry, and the restoration of good taste. The fault was not his if these ends were not attained, for, taking contemporary French authors as his models, he contributed much, both in prose and verse, to its proceedings, until he left in February 1760 to take up the position of *juiz de fora* at Castello de Vide. On returning to Lisbon for a short visit, he found the *Arcadia* a prey to the internal dissensions that caused its dissolution in 1774, but succeeded in composing them and in 1764 he went to Elvas to act as auditor of one of the regiments stationed there. During a ten years' residence, his wide reading and witty conversation gained him the friendship of the governor of that fortress and the admiration of a circle comprising all that was cultivated in Elvas. As in most cathedral and garrison towns, the clerical and military elements dominated society, and here were mutually antagonistic, because of the enmity between their respective leaders, the bishop and the governor. Moreover, Elvas, being a remote provincial centre, abounded in curious and grotesque types. Diniz, who was a keen observer, noted these, and, treasuring them in his memory, reproduced them, with their vanities, intrigues and ignorance, in his masterpiece, *Hyssope*. In 1768 a quarrel arose between the bishop, a proud, pretentious prelate, and the dean, as to the right of the former to receive holy water from the latter at a private side door of the cathedral, instead of at the principal entrance. The matter being one of principle, neither party would yield what he considered his rights, and it led to a lawsuit, and divided the town into two sections, which eagerly debated the arguments on both sides and enjoyed the ridiculous incidents which accompanied the dispute. Ultimately the dean died, and was succeeded by his nephew, who appealed to the crown with success and the bishop lost his pretension. The *Hyssope* arose out of and deals with this affair. It was dictated in seventeen days, in the years 1770-1772, and, in its final redaction, consists of eight cantos of blank verse. The pressure of absolutism left open only one form of expression, satire, and in this poem Diniz produced an original work which ridicules the clergy and the prevailing Gallomania, and contains episodes full of humour. It has been compared with Boileau's *Lutrin*, because both are founded on a petty ecclesiastical quarrel, but here the resemblance ends, and the poem of Diniz is the superior in everything except matrification.

Returning to Lisbon in 1774, Diniz endeavoured once more to resuscitate the *Arcadia*, but his long absence had withdrawn its chief support, its most talented members Garção (*q.v.*) and Quita were no more, and he only assisted at its demise. In April 1776 he was appointed *disembargador* of the court of Relação in Rio de Janeiro and given the habit of Aviz. He lived in Brazil, devouting his leisure to a study of its natural history and mineralogy, until 1789, when he went back to Lisbon to take up the post of *disembargador* of the Relação of Oporto; in July 1790 he was promoted, and became *disembargador* of the Casa da Supplicação. In this year he was sent again to Brazil to assist in trying the leaders of the Republican conspiracy in Minas, in which Gonzaga (*q.v.*) and the other men of letters were involved, and in December 1792 he became chancellor of the Relação in Rio. Six years later he was named councillor of the *Conselho Ultramarino*, but did not live to return home, dying in Rio on the 5th of October 1799.

Diniz possessed a poetic temperament, but his love of imitating the classics, whose spirit he failed to understand, fettered his muse, and he seems never to have perceived that mythological comparisons and pastoral allegories were poor substitutes for the expression of natural feeling. The conventionalism of his art prejudiced its sincerity, and, inwardly cherishing the belief that poetry was unworthy of the dignity of a judge, he never gave his real talents a chance to display themselves. His Anacreontic odes, dithyrambs and idylls earned the admiration of contemporaries, but his Pindaric odes lack fire, his sonnets are weak, and his idylls have neither the truth nor the simplicity of Quita's work. As a rule Diniz's versification is weak and his verses lack harmony, though the diction is beyond cavil.

His poems were published in 6 vols. (Lisbon, 1807-1817). The best edition of *Hyssope*, to which Diniz owes his lasting fame, is that of J. R. Coelho (Lisbon, 1879), with an exhaustive introductory study on his life and writings. A French prose version of the poem by Boissonade has gone through two editions (Paris, 1828 and 1867), and English translations of selections have been printed in the *Foreign Quarterly Review*, and in the *Manchester Quarterly* (April 1896).

See also Dr Theophilo Braga, *A. Arcadia Lusitana* (Oporto, 1899).

(E. Pr.)

CRYOLITE

A mineral discovered in Greenland by the Danes in 1794, and found to be a compound of fluorine, sodium and aluminium. From its general appearance, and from the fact that it melts readily, even in a candle-flame, it was regarded by the Eskimos as a peculiar kind of ice; from this fact it acquired the name of cryolite (from Gr. κρύος, frost, and λίθος, stone). Cryolite occurs in colourless or snow-white cleavable masses, often tinted brown or red with iron oxide, and occasionally passing into a black variety. It is usually translucent, becoming nearly transparent on immersion in water. The mineral cleaves in three rectangular directions, and the crystals occasionally found in the crevices have a cubic habit, but it has been proved, after much discussion, that they belong to the anorthic system. The hardness is 2.5, and the specific gravity 3. Cryolite has the formula Na_3AlF_6, or $3NaF{\cdot}AlF_3$, corresponding to fluorine 54.4, sodium 32.8, and aluminium 12.8%. It colours a flame yellow, through the presence of sodium, and when heated with sulphuric acid it evolves hydrofluoric acid.

Cryolite occurs almost exclusively at Ivigtut (sometimes written Evigtok) on the Arksut Fjord in S.W. Greenland. There it forms a large deposit, in a granitic vein running through gneiss, and is accompanied by quartz, siderite, galena, blende, chalcopyrite, &c. It is also associated with a group of kindred minerals, some of which are evidently products of alteration of the cryolite, known as pachnolite, thomsenolite, ralstonite, gearksutite, arksutite, &c. Cryolite likewise occurs, though only to a limited extent, at Miyask, in the Ilmen Mountains; at Pike's Peak, Colorado, and in the Yellowstone Park.

Cryolite is a mineral of much economic importance. It has been extensively used as a source of metallic aluminium, and as a flux in smelting the metal. It is largely employed in the manufacture of certain sodium salts, as suggested by Julius Thomsen, of Copenhagen, in 1849; and it has been used for the production of certain kinds of porcelain and glass, remarkable for its toughness, and for enamelled ware.

Although cryolite is known as "ice-stone" (*Eisstein*), it is not to be confused with "ice-spar" (*Eisspath*), which is a vitreous kind of felspar termed "glassy felspar" or rhyacolite.

(F. W. R.*)

CRYPT

(Lat. *crypta*, from the Gr. κρύπτειν, to hide), a vault or subterranean chamber, especially under churches. In classical phraseology "crypta" was employed for any vaulted building, either partially or entirely below the level of the ground. It is used for a sewer (*crypta Suburae*, Juvenal, *Sat.* v. 106); for the "carceres," or vaulted stalls for the horses and chariots in a circus (Sidon. Apoll. *Carm.* xxiii. 319); for the close porticoes or arcades, more fully known as "cryptoporticus," attached by the Romans to their suburban villas for the sake of coolness, and to the theatres as places of exercise or rehearsal for the performers (Plin. *Epist.* ii. 15, v. 6, vii. 21; Sueton. *Calig.* 58; Sidon. Apoll, lib. ii. epist. 2); and for underground receptacles for agricultural produce (Vitruv. vi. 8, Varro, *De re rust.* i. 57). Tunnels, or galleries excavated in the living rock, were also called *cryptae*. Thus the tunnel to the north of Naples, through which the road passes to Puteoli, familiar to tourists as the "Grotto of Posilipo," was originally designated *crypta Neapolitana* (Seneca, Epist. 57). In early Christian times *crypta* was appropriately employed for the galleries of a catacomb, or for the catacomb itself. Jerome calls them by this name when describing his visits to them as a schoolboy, and the term is used by Prudentius (see CATACOMBS).

A crypt, as a portion of a church, had its origin in the subterranean chapels known as "confessiones," erected around the tomb of a martyr, or the place of his martyrdom. This is the origin of the spacious crypts, some of which may be called subterranean churches, of the Roman churches of S. Prisca, S. Prassede, S. Martino ai Monti, S. Lorenzo fuori le Mura, and above all of St Peter's—the crypt being thus the germ of the church or basilica subsequently erected above the hallowed spot. When the martyr's tomb was sunk in the surface of the ground, and not placed in a catacomb chapel, the original memorial-shrine would be only partially below the surface, and consequently the part of the church erected over it, which was always that containing the altar, would be elevated some height above the ground, and be approached by flights of steps. This fashion of raising the chancel or altar end of a church on a crypt was widely imitated long after the reason for adopting it ceased, and even where it never existed. The crypt under the altar at the basilica of St Maria Maggiore in Rome is merely imitative, and the same may be said of many of the crypts of the early churches in England. The original Saxon cathedral of Canterbury had a crypt beneath the eastern apse, containing the so-called body of St Dunstan, and other relics, "fabricated," according to Eadmer, "in the likeness of the confessionary of St Peter at Rome" (see BASILICA). St Wilfrid constructed crypts still existing beneath the churches erected by him in the latter part of the 7th century at Hexham and Ripon. These are peculiarly interesting from their similarity in form and arrangement to the catacomb chapels with which Wilfrid must have become familiar during his residence in Rome. The cathedral, begun by Æthelwold and finished by Alphege at Winchester, at the end of the 10th century, had spacious crypts "supporting the holy altar and the venerable relics of the saints" (Wulstan, *Life of St Æthelwold*), and they appear to have been common in the earlier churches in England. The arrangement was adopted by the Norman builders of the 11th and 12th centuries, and though far from universal is found in many of the cathedrals of that date. The object of the construction of these crypts was twofold,—to give the altar sufficient elevation to enable those below to witness the sacred mysteries, and to provide a place of burial for those holy men whose relics were the church's most precious possession. But the crypt was "a foreign fashion," derived, as has been said, from Rome, "which failed to take root in England, and indeed elsewhere barely outlasted the Romanesque period" (*Essays on Cathedrals*, ed. Howson, p. 331).

Of the crypts beneath English Norman cathedrals, that under the choir of Canterbury (*q.v.*) is by far the largest and most elaborate in its arrangements. It is, in fact, a subterranean church of vast size and considerable altitude. The whole crypt was dedicated to the Virgin Mary, and contained two chapels especially dedicated to her,—the central one beneath the high altar, enclosed with rich Gothic screen-work, and one under the south transept. This latter chapel was appropriated by Queen Elizabeth to the use of the French Huguenot refugees who had settled at Canterbury in the time of Edward VI. There were also in this crypt a large number of altars and chapels of other saints, some of whose hallowed bodies were buried here. At the extreme east end, beneath the Trinity chapel, the body of St Thomas (Becket) was buried the day after his martyrdom, and lay there till his translation, July 7, 1220.

The cathedrals of Winchester, Worcester and Gloucester have crypts of slightly earlier date (they may all be placed between 1080 and 1100), but of similar character, though less elaborate. They all contain piscinas and other evidences of the existence of altars in considerable numbers. They are all apsidal. The most picturesque is that of Worcester, the work of Bishop Wulfstan (1084), which is remarkable for the multiplicity of small pillars supporting its radiating vaults. Instead of having the air of a sepulchral vault like those of Winchester and Gloucester, this crypt is, in Professor Willis's words, "a complex and beautiful temple." Archbishop Roger's crypt at York, belonging to the next century (1154-1181), was filled up with earth when the present choir was built at the end of the 14th century, and its existence forgotten till its disinterment after the fire of 1829. The choir and presbytery at Rochester are supported by an extensive crypt, of which the western portion is Gundulf's work (1076-1107), but the eastern part, which displays slender cylindrical and octagonal shafts, with light vaulting springing from them, is of the same period as the superstructure, the first years of the 13th century. This crypt, and that beneath the Early English Lady chapel at Hereford, are the latest English existing cathedral crypts. That at Hereford was rendered necessary by the fall of the ground, and is an exceptional case. Later than any of these crypts was that of St Paul's, London. This was a really large and magnificent church of Decorated date, with a vaulted roof of rich and intricate character resting on a forest of clustered columns. Part of it served as the parish church of St Faith. A still more exquisite work of the Decorated period is the crypt of St Stephen's chapel at Westminster, than which it is difficult to conceive anything more perfect in design or more elaborate in ornamentation. Having happily escaped the conflagration of the Houses of Parliament in 1834—before which it was degraded to the purpose of the speaker's state dining-room—it has been restored to its former sumptuousness of decoration, and is now one of the most beautiful architectural gems in England.

Of Scottish cathedrals the only one that possesses a crypt is the cathedral of Glasgow, rendered celebrated by Sir Walter Scott in his novel of *Rob Roy* (ch. xx.). At the supposed date of the tale, and indeed till a comparatively recent period, this crypt was used as a place of worship by one of the three congregations among which the cathedral was partitioned, and was known as "the Laigh or Barony Kirk." It extends beneath the choir transepts and chapter-house; in consequence of the steep declivity on which the cathedral stands it is of unusual height and lightsomeness. It belongs to the 13th century, its style corresponding to Early English, and is simply constructional, the building being adapted to the locality. In architectural beauty it is quite unequalled by any crypt in the United Kingdom, and can hardly anywhere be surpassed. It is an unusually rich example of the style, the clustered piers and groining being exquisite in design and admirable in execution. The bosses of the roof and capitals of the piers are very elaborate, and the doors are much enriched with foliage. "There is a solidity in its architecture, a richness in its vaulting, and a variety of perspective in the spacing of its pillars, which make it one of the most perfect pieces of architecture in these kingdoms" (Fergusson).

In the centre of the main alley stands the mutilated effigy of St Mungo, the patron saint of Glasgow, and at the south-east corner is a well called after the same saint.

Crypts under parish churches are not very uncommon in England, but they are usually small and not characterized by any architectural beauty. A few of the earlier crypts, however, deserve notice. One of the earliest and most remarkable is that of the church of Lastingham near Pickering in Yorkshire, on the site of the monastery founded in 648 by Cedd, bishop of the East Saxons. The existing crypt, though exceedingly rude in structure, is of considerably later date than Bishop Cedd, forming part of the church erected by Abbot Stephen of Whitby in 1080, when he had been driven inland by the incursions of the northern pirates. This crypt is remarkable from its extending under the nave as well as the chancel of the upper church, the plan of which it accurately reproduces, with the exception of the westernmost bay. It forms a nave with side aisles of three bays, and an apsidal chancel, lighted by narrow deeply splayed slits. The roof of quadripartite vaulting is supported by four very dwarf thick cylindrical columns, the capitals of which and of the responds are clumsy imitations of classical work with rude volutes. Still more curious is the crypt beneath the chancel of the church of Repton in Derbyshire. This also consists of a centre and side aisles, divided by three arches on either side. The architectural character, however, is very different from that at Lastingham, and is in some respects almost unique, the piers being slender, and some of them of a singular spiral form, with a bead running in the sunken part of the spiral. Another very extensive and curious Norman crypt is that beneath the chancel of St Peter's-in-the-East at Oxford. This is five bays in length, the quadripartite vaulting being supported by eight low, somewhat slender, cylindrical columns with capitals bearing grotesque animal and human subjects. Its dimensions are 36 by 20 ft. and 10 ft. in height. This crypt has been commonly attributed to Grymboldt in the 9th century; but it is really not very early Norman. Under the church of St Mary-le-Bow in London there is an interesting Norman crypt not very dissimilar in character to that last described. Of a later date is the remarkably fine Early English crypt groined in stone, beneath the chancel of Hythe in Kent, containing a remarkable collection of skulls and bones, the history of which is quite uncertain. There is also a Decorated crypt beneath the chancel at Wimborne minster, and one of the same date beneath the southern chancel aisle at Grantham.

Among the more remarkable French crypts may be mentioned those of the cathedrals of Auxerre, said to date from the original foundation in 1085; of Bayeux, attributed to Odo, bishop of that see, uterine brother of William the Conqueror, where twelve columns with rude capitals support a vaulted roof; of Chartres, running under the choir and its aisles, frequently assigned to Bishop Fulbert in 1029, but more probably coeval with the superstructure; and of Bourges, where the crypt is in the Pointed style, extending beneath the choir. The church of the Holy Trinity attached to Queen Matilda's foundation—the "Abbaye aux Dames" at Caen—has a Norman crypt where the thirty-four pillars are as closely set as those at Worcester. The church of St Eutropius at Saintes has also a crypt of the 11th century, of very large dimensions, which deserves special notice; the capitals of the columns exhibit very curious carvings. Earlier than any already mentioned is that of St Gervase of Rouen, considered by E. A. Freeman "the oldest ecclesiastical work to be seen north of the Alps." It is apsidal, and in its walls are layers of Roman brick. It is said to contain the remains of two of the earliest apostles of Gaul—St Mello and St Avitian. There are numerous crypts in Germany. One at Göttingen may be mentioned, where cylindrical shafts with capitals of singular design support "vaulting of great elegance and lightness" (Fergusson), the curves being those of a horseshoe arch. The crypts of the cathedrals or churches at Halberstadt, Hildesheim and Naumburg also deserve to be noticed; that of Lübeck may be rather called a lower choir. It is 20 ft. high and vaulted.

The Italian crypts, when found, as a rule reproduce the "confessio" of the primitive churches. That beneath the chancel of S. Michele at Pavia is an excellent typical example, probably dating from the 10th century. It is apsidal and vaulted, and is seven bays in length. That at S. Zeno at Verona (*c.* 1138) is still more remarkable; its vaulted roof is upborne by forty columns, with curiously carved capitals. It is approached from the west by a double flight of steps and contains many ancient monuments. S. Miniato at Florence, begun in 1013, has a very spacious crypt at the east end, forming virtually a second choir. It is seven bays in length and vaulted. The most remarkable crypt in Italy, however, is perhaps that of St Mark's, Venice. The plan of this is almost a Greek cross. Four rows of nine columns each run from end to end, and two rows of three each occupy the arms of the cross, supporting low stunted arches on which rests the pavement of the church above. This also constitutes a lower church, containing a *chorus cantorum* formed by a low stone screen, not unlike that of S. Clemente at Rome (see BASILICA), enclosing a massive stone altar with four low columns. This crypt is reasonably supposed to belong to the church founded by the doge P. Orseolo in 977. There are also crypts deserving notice at the cathedrals of Brescia, Fiesole and Modena, and the churches of S. Ambrogio and S. Eustorgio at Milan. The former was unfortunately modernized by St Charles Borromeo. The crypt at Assisi is really a second church at a lower level, and being built on the steep side of a hill is well lighted. The whole fabric is a beautiful specimen of Italian Gothic, and both the lower and upper churches are covered with rich frescoes.

Domestic crypts are of frequent occurrence. Medieval houses had as a rule their chief rooms raised above the level of the ground upon vaulted substructures, which were used as cellars and storerooms. These were sometimes partially underground, sometimes entirely above it. The underground vaults often remain when all the superstructure has been swept away, and from their Gothic character are frequently mistaken for ecclesiastical buildings. The older English towns are full of crypts of this character, now used as cellars. They occur in Oxford and Rochester, are very abundant in the older parts of Bristol, and, according to J. H. Parker, "nearly the whole city of Chester is built upon a series of them with the Rows or passages made on the top of the vaults" (*Domestic Architecture*, iii. 91). The crypt of Gerard's Hall in London, destroyed in the construction of New Cannon Street, figured by Parker (*Dom. Arch.* ii. 185), was a beautiful example of the lower storey of the residence of a wealthy merchant of the time of Edward I. It was divided down the middle by a row of four slender cylindrical columns supporting a very graceful vault. The finest example of a secular crypt now remaining in England is that beneath the Guildhall of London. The date of this is early in the 15th century—1411. It is a large and lofty apartment, divided into four alleys by two rows of clustered shafts supporting a rich lierne vault with ribs of considerable intricacy. There is a fine vaulted crypt of the same date and of similar character beneath St Mary's Hall, the Guildhall of the city of Coventry.

(E. V.)

CRYPTEIA

(Gr. κρύπτειν, to hide), a kind of secret police in ancient Sparta, founded, according to Aristotle, by Lycurgus; there is, however, no real evidence as to the date of its origin. The institution was under the supervision of the ephors, who, on entering office, annually proclaimed war against the helots (serf-class) and thus absolved from the guilt of murder any Spartan who should slay a helot. It was instituted primarily as a precaution against the ever-present danger of a helot revolt, and secondarily perhaps as a training for young Spartans, who were sent out by the ephors to keep watch on the helots and assassinate any who might appear dangerous. Plato (*Laws*, i. p. 633) emphasizes the former aspect, but there can be little doubt that, at all events after the revolt of 464 (see Cimon), its more sinister purpose was predominant, as we may gather from the secret massacre of 2000 helots who, on the invitation of the ephors, claimed to have rendered distinguished service (Thuc. iv. 80).

See HELOTS; EPHOR; also A. H. J. Greenidge, *Handbook of Gk. Const. Hist.* (London, 1896); G. Gilbert, *Gk. Const. Antiq.* (Eng. trans., London, 1895).

CRYPTOBRANCHUS

A genus of thoroughly aquatic, but lung-breathing tailed Batrachia, of the family *Amphiumidae*, characterized by a heavy, flattened build, a very porous tubercular skin, with a frilled fold along each side, short stout limbs with four very short fingers and five very short toes, and minute eyes without lids. The vertebrae are biconcave, and although the gills are lost in the adult, ossified gill-arches, two to four in number, persist. A strong series of vomerine teeth extends across the palate. Three species of this genus are known. One is the well-known fossil of Oeningen first described as *Homo diluvii testis* and shown by Cuvier to be nearly related to the gigantic salamander of Japan, *Cryptobranchus maximus*, which has since been found to inhabit China also; the third is the hellbender, mud-puppy or water-dog of North America, *C. alleghaniensis*, also known under the name of *Menopoma*. Both the fossil *C. scheuchzeri* and *C. maximus* grow to a length of over 5 ft. and are by far the largest Urodeles known, whilst *C. alleghaniensis* reaches the respectable length of 18 in.

The eggs are laid in rosary-like strings. They have been found, in Japan, deposited in deep holes in the water, where they form large clumps (70 to 80 eggs) round which the female coils herself. The gigantic salamander has also bred in the Amsterdam zoological gardens, the eggs numbering upwards of 500; the male, it is stated, took charge of the eggs, and for the ten weeks which elapsed before the release of the last larva, he kept close to them, at times crawling among the coiled mass of egg-strings or lifting them up, evidently for the purpose of aeration. The larva on leaving the egg is about an inch long, provided with three branched external gills on each side, and showing mere rudiments of the four limbs.

CRYPTOGRAPHY

(from Gr. κρύπτος, hidden, and γράφειν, to write), or writing in cipher, called also steganography (from Gr. στεγάνη, a covering), the art of writing in such a way as to be incomprehensible except to those who possess the key to the system employed. The unravelling of the writing is called deciphering. Cryptography having become a distinct art, Bacon (Lord Verulam) classed it (under the name *ciphers*) as a part of grammar. Secret modes of communication have been in use from the earliest times. The Lacedemonians had a method called the *scytale*, from the staff (σκυτάλη) employed in constructing and deciphering the message. When the Spartan ephors wished to forward their orders to their commanders abroad, they wound slantwise a narrow strip of parchment upon the σκυτάλη so that the edges met close together, and the message was then added in such a way that the centre of the line of writing was on the edges of the parchment. When unwound the scroll consisted of broken letters; and in that condition it was despatched to its destination, the general to whose hands it came deciphering it by means of a σκυτάλη exactly corresponding to that used by the ephors. Polybius has enumerated other methods of cryptography.

The art was in use also amongst the Romans. Upon the revival of letters methods of secret correspondence were introduced into private business, diplomacy, plots, &c.; and as the study of this art has always presented attractions to the ingenious, a curious body of literature has been the result.

John Trithemius (d. 1516), the abbot of Spanheim, was the first important writer on cryptography. His *Polygraphia*, published in 1518, has passed through many editions, and has supplied the basis upon which subsequent writers have worked. It was begun at the desire of the duke of Bavaria; but Trithemius did not at first intend to publish it, on the ground that it would be injurious to public interests. A *Steganographia* published at Lyons (? 1551) and later at Frankfort (1606), is also attributed to him. The next treatises of importance were those of Giovanni Battista della Porta, the Neapolitan mathematician, who wrote *De furtivis litterarum notis*, 1563; and of Blaise de Vigenere, whose *Traité des chiffres* appeared in Paris, 1587. Bacon proposed an ingenious system of cryptography on the plan of what is called the double cipher; but while thus lending to the art the influence of his great name, he gave an intimation as to the general opinion formed of it and as to the classes of men who used it. For when prosecuting the earl of Somerset in the matter of the poisoning of Overbury, he urged it as an aggravation of the crime that the earl and Overbury "had cyphers and jargons for the king and queen and all the great men,—things seldom used but either by princes and their ambassadors and ministers, or by such as work or practise against or, at least, upon princes."

Other eminent Englishmen were afterwards connected with the art. John Wilkins, subsequently bishop of Chester, published in 1641 an anonymous treatise entitled *Mercury, or The Secret and Swift Messenger*,—a small but comprehensive work on the subject, and a timely gift to the diplomatists and leaders of the Civil War. The deciphering of many of the royalist papers of that period, such as the letters that fell into the hands of the parliament at the battle of Naseby, has by Henry Stubbe been charged on the celebrated mathematician Dr John Wallis (*Athen. Oxon.* iii. 1072), whose connexion with the subject of cipher-writing is referred to by himself in the Oxford edition of his mathematical works, 1689, p. 659; as also by John Davys. Dr Wallis elsewhere states that this art, formerly scarcely known to any but the secretaries of princes, &c., had grown very common and familiar during the civil commotions, "so that now there is scarce a person of quality but is more or less acquainted with it, and doth, as there is occasion, make use of it." Subsequent writers on the subject are John Falconer (*Cryptomenysis patefacta*), 1685; John Davys (*An Essay on the Art of Decyphering: in which is inserted a*

Discourse of Dr Wallis), 1737; Philip Thicknesse (*A Treatise on the Art of Decyphering and of Writing in Cypher*), 1772; William Blair (the writer of the comprehensive article "Cipher" in Rees's *Cyclopaedia*), 1819; and G. von Marten (Cours *diplomatique*), 1801 (a fourth edition of which appeared in 1851). Perhaps the best modern work on this subject is the *Kryptographik* of J. L. Klüber (Tübingen, 1809), who was drawn into the investigation by inclination and official circumstances. In this work the different methods of cryptography are classified. Amongst others of lesser merit who have treated of this art may be named Gustavus Selenus (*i.e.* Augustus, duke of Brunswick), 1624; Cospi, translated by Niceron in 1641; the marquis of Worchester, 1659; Kircher, 1663; Schott, 1665; Ludwig Heinrich Hiller, 1682; Comiers; 1690; Baring, 1737; Conrad, 1739, &c. See also a paper on *Elizabethan Cipher-books* by A. J. Butler in the Bibliographical Society's *Transactions*, London, 1901.

Schemes of cryptography are endless in their variety. Bacon lays down the following as the "virtues" to be looked for in them:—"that they be not laborious to write and read; that they be impossible to decipher; and, in some cases, that they be without suspicion." These principles are more or less disregarded by all the modes that have been advanced, including that of Bacon himself, which has been unduly extolled by his admirers as "one of the most ingenious methods of writing in cypher, and the most difficult to be decyphered, of any yet contrived" (Thicknesse, p. 13).

The simplest and commonest of all the ciphers is that in which the writer selects in place of the proper letters certain other letters in regular advance. This method of transposition was used by Julius Caesar. He, "per quartam elementorum literam," wrote *d* for *a*, *e* for *b*, and so on. There are instances of this arrangement in the Jewish rabbis, and even in the sacred writers. An illustration of it occurs in Jeremiah (xxv. 26), where the prophet, to conceal the meaning of his prediction from all but the initiated, writes *Sheshak* instead of Babel (Babylon), the place meant; *i.e.* in place of using the second and twelfth letters of the Hebrew alphabet (*b*, *b*, *l*) from the beginning, he wrote the second and twelfth (*sh*, *sh*, *k*) from the end. To this kind of cipher-writing Buxtorf gives the name Athbash (from *a* the first letter of the Hebrew alphabet, and *th* the last; *b* the second from the beginning, and *h* the second from the end). Another Jewish cabalism of like nature was called Albam; of which an example is in Isaiah vii. 6, where Tabeal is written for Remaliah. In its adaptation to English this method of transposition, of which there are many modifications, is comparatively easy to decipher. A rough key may be derived from an examination of the respective quantities of letters in a type-founder's bill, or a printer's "case." The decipherer's first business is to classify the letters of the secret message in the order of their frequency. The letter that occurs oftenest is *e*; and the next in order of frequency is *t*. The following groups come after these, separated from each other by degrees of decreasing recurrence:—*a*, *o*, *n*, *i*; *r*, *s*, *h*; *d*, *l*; *c*, *w*, *u*, *m*; *f*, *y*, *g*, *p*, *b*; *v*, *k*; *x*, *q*, *j*, *z*. All the single letters must be *a*, *I* or *O*. Letters occurring together are *ee*, *oo*, *ff*, *ll*, *ss*, &c. The commonest words of two letters are (roughly arranged in the order of their frequency) *of*, *to*, *in*, *it*, *is*, *be*, *he*, *by*, *or*, *as*, *at*, *an*, *so*, &c. The commonest words of three letters are *the* and *and* (in great excess), *for*, *are*, *but*, *all*, *not*, &c.; and of four letters—*that*, *with*, *from*, *have*, *this*, *they*, &c. Familiarity with the composition of the language will suggest numerous other points that are of value to the decipherer. He may obtain other hints from Poe's tale called *The Gold Bug*. As to messages in the continental languages constructed upon this system of transposition, rules for deciphering may be derived from Breithaupt's *Ars decifratoria* (1737), and other treatises.

Bacon remarks that though ciphers were commonly in letters and alphabets yet they might be in words. Upon this basis codes have been constructed, classified words taken from dictionaries being made to represent complete ideas. In recent years such codes have been adapted by merchants and others to communications by telegraph, and have served the purpose not only of keeping business affairs private, but also of reducing the excessive cost of telegraphic messages to distant markets.

Obviously this class of ciphers presents greater difficulties to the skill of the decipherer.

Figures and other characters have been also used as letters; and with them ranges of numerals have been combined as the representatives of syllables, parts of words, words themselves, and complete phrases. Under this head must be placed the despatches of Giovanni Michael, the Venetian ambassador to England in the reign of Queen Mary, documents which have only of late years been deciphered. Many of the private letters and papers from the pen of Charles I. and his queen, who were adepts in the use of ciphers, are of the same description. One of that monarch's letters, a document of considerable interest, consisting entirely of numerals purposely complicated, was in 1858 deciphered by Professor Wheatstone, the inventor of the ingenious crypto-machine, and printed by the Philobiblon Society. Other letters of the like character have been published in the *First Report of the Royal Commission on Historical Manuscripts* (1870). In the second and subsequent reports of the same commission several keys to ciphers have been catalogued, which seem to refer themselves to the methods of cryptography under notice. In this connexion also should be mentioned the "characters," which the diarist Pepys drew up when clerk to Sir George Downing and secretary to the earl of Sandwich and to the admiralty, and which are frequently mentioned in his journal. Pepys describes one of them as "a great large character," over which he spent much time, but which was at length finished, 25th April 1660; "it being," says he, "very handsomely done and a very good one in itself, but that not truly alphabetical."

Shorthand marks and other arbitrary characters have also been largely imported into cryptographic systems to represent both letters and words, but more commonly the latter. This plan is said to have been first put into use by the old Roman poet Ennius. It formed the basis of the method of Cicero's freedman, Tiro, who seems to have systematized the labours of his predecessors. A large quantity of these characters have been engraved in Gruter's *Inscriptiones*. The correspondence of Charlemagne was in part made up of marks of this nature. In Rees's *Cyclopaedia* specimens were engraved of the cipher used by Cardinal Wolsey at the court of Vienna in 1524, of that used by Sir Thomas Smith at Paris in 1563, and of that of Sir Edward Stafford in 1586; in all of which arbitrary marks are introduced. The first English system of shorthand—Bright's *Characterie*, 1588—almost belongs to the same category of ciphers. A favourite system of Charles I., used by him during the year 1646, was one made up of an alphabet of twenty-four letters, which were represented by four simple strokes varied in length, slope and position. This alphabet is engraved in Clive's *Linear System of Shorthand* (1830), having been found amongst the royal manuscripts in the British Museum. An interest attaches to this cipher from the fact that it was employed in the well-known letter addressed by the king to the earl of Glamorgan, in which the former made concessions to the Roman Catholics of Ireland.

Complications have been introduced into ciphers by the employment of "dummy" letters,—"nulls and insignificants," as Bacon terms them. Other devices have been introduced to perplex the decipherer, such as spelling words backwards, making false divisions between words, &c. The greatest security against the decipherer has been found in the use of elaborate tables of letters, arranged in the form of the multiplication table, the message being constructed by the aid of preconcerted key-words. Details of the working of these ciphers may be found in the treatises named in this article. The deciphering of them is one of the most difficult of tasks. A method of this kind is explained in the Latin and English lives of Dr John Barwick, whose correspondence with Hyde, afterwards earl of Clarendon, was carried on in cryptography. In a letter dated 20th February 1659/60, Hyde, alluding to the skill of his political opponents in deciphering, says that "nobody needs to fear them, if they write carefully in good cyphers." In his next he allays his correspondent's apprehensiveness as to the deciphering of their letters.

"I confess to you, as I am sure no copy could be gotten of any of my cyphers from hence, so I did not think it probable that they could be got on your side the water. But I was as confident, till you tell me you believe it, that the devil himself cannot decypher a letter that is well written, or find that 100 stands for Sir H. Vane. I have heard of many of the pretenders to that skill, and have spoken with some of them, but have found them all to be mountebanks; nor did I ever hear that more of the King's letters that were found at Naseby, than those which they found decyphered, or found the cyphers in which they were writ, were decyphered. And I very well remember that in the volume they published there was much left in cypher which could not be understood, and which I believe they would have explained if it had been in their power."

An excellent modification of the key-word principle was constructed by Admiral Sir Francis Beaufort.

Ciphers have been constructed on the principle of altering the places of the letters without changing their powers. The message is first written Chinese-wise, upward and downward, and the letters are then combined in given rows from left to right. In the celebrated cipher used by the earl of Argyll when plotting against James II., he altered the positions of the words. Sentences of an indifferent nature were constructed, but the real meaning of the message was to be gathered from words, placed at certain intervals. This method, which is connected with the name of Cardan, is sometimes called the trellis or cardboard cipher.

The wheel-cipher, which is an Italian invention, the string-cipher, the circle-cipher and many others are fully explained, with the necessary diagrams, in the authorities named above—more particularly by Klüber in his *Kryptographik*.

(J. E. B.)

CRYPTOMERIA

Or Japanese Cedar, a genus of conifers, containing a single species, *C. japonica*, native of China and Japan, which was introduced into Great Britain by the Royal Horticultural Society in 1844. It is described as one of the finest trees in Japan, reaching a height of 100 or more feet, usually divested of branches along the lower part of the trunk and crowned with a conical head. The narrow, pointed leaves are spirally arranged and persist for four or five years; the cones are small, globose and borne at the ends of the branchlets, the scales are thickened at the extremity and divided into sharply pointed lobes, three to five seeds are borne on each scale. *Cryptomeria* is extensively used in Japan for reafforesting denuded lands, as it is a valuable timber tree; it is also planted to form avenues along the public roads. In Veitch's *Manual of Coniferae* (ed. 2, 1900, p. 265) reference is made to "an avenue of Cryptomerias 7 m. in extent near Lake Hakone" in which "the trees are more than 100 ft. high, with perfectly straight trunks crowned with conical heads of foliage." Professor C. S. Sargent, in his *Forest Flora of Japan*, says, "Japan owes much of the beauty of its groves and gardens to the *Cryptomeria*. Nowhere is there a more solemn and impressive group of trees than that which surrounds the temples and tombs at Nikko where they rise to a height of 100 to 125 ft.; it is a stately tree with no rival except in the sequoias of California." Many curious varieties have been obtained by Japanese horticulturists, including some dwarf shrubby forms not exceeding a few feet in height. When grown in Great Britain *Cryptomeria* requires a deep, well-drained soil with plenty of moisture, and protection from cold winds.

CRYPTO-PORTICUS

(Gr. κρυπτός, concealed, and Lat. *porticus*), an architectural term for a concealed or covered passage, generally underground, though lighted and ventilated from the open air. One of the best-known examples is the crypto-porticus under the palaces of the Caesars in Rome. In Hadrian's villa in Rome they formed the principal private intercommunication between the several buildings.

CRYSTAL-GAZING

Or Scrying, the term commonly applied to the induction of visual hallucinations by concentrating the gaze on any clear deep, such as a crystal or a ball of polished rock crystal. Some persons do not even find a clear deep necessary, and are content to gaze at the palm of the hand, for example, when hallucinatory pictures, as they declare, emerge. Among objects used are a pool of ink in the hand (Egypt), the liver of an animal (tribes of the North-West Indian frontier), a hole filled with water (Polynesia), quartz crystals (the Apaches and the Euahlayi tribe of New South Wales), a smooth slab of polished black stone (the Huille-che of South America), water in a vessel (Zulus and Siberians), a crystal (the Incas), a mirror (classical Greece and the middle ages), the finger-nail, a sword-blade, a ring-stone, a glass of sherry, in fact almost anything. Much depends on what the "seer" is accustomed to use, and some persons who can "scry" in a glass ball or a glass water-bottle cannot "scry" in ink.

The practice of inducing pictorial hallucinations by such methods as these has been traced among the natives of North and South America, Asia, Australia, Africa, among the Maoris, who sometimes use a drop of blood, and in Polynesia, and is thus practically of world-wide diffusion. This fact was not observed (that is, the collections of examples were not made) till recently, when experiments in private non-spiritualist circles drew attention to crystal-gazing, a practice always popular among peasants, and known historically to have survived through classical and medieval times, and, as in the famous case of Dr Dee, after the Reformation.

The early church condemned *specularii* (mirror-gazers), and Aubrey and the *Memoirs* of Saint-Simon contain "scrying" anecdotes of the 17th and 18th centuries, while Sir Walter Scott's story, *My Aunt Margaret's Mirror*, is based on a tradition of about 1750 in a noble Scottish family. The practice, in all times and countries, was used for purposes of divination. The gazer detected unknown criminals, or described remote events, or even professed to foretell things future. Sometimes the supposed magician or medicine man himself did the scrying; occasionally he enabled his client to see for himself; often a child was selected as the scryer. The process was usually explained as the result of the action of a spirit, angel or devil, and many unessential formulae, invocations, "calls," written charms with cabbalistic signs, and fumigations, were employed. These things may have had some effect by way of suggestion; the scryer may have been brought by them into an appropriate frame of mind; but, as a whole, they are tedious and superfluous.

A person can either induce the pictorial hallucinations (he may discover his capacity by accident, like George Sand, as she tells in her *Memoirs*—and other cases are known), or he cannot induce them, though he stare till his eyes water. It is almost universally found, in cases of successful experiment, that the glass ball, for example, takes a milky or misty aspect, that it then grows black, reflections disappearing, and that then the pictures emerge. Some people arrive at seeing the glass ball milky or misty, and can go no further. Others see pictures of persons or landscapes, only in black and white, and motionless. Others see in the glass coloured figures of men, women and animals in motion; while in rarer cases the ball disappears from view, and the scryer finds himself apparently looking at an actual scene. In a few attested cases two persons have shared the same vision. In experiments with magnifying glasses, and through spars, the ordinary effects of magnifying and of alteration of view are sometimes produced; sometimes they are not. The evidence, of course, is necessarily only that of the scryers themselves, but repeated experiments by persons of probity, and unfamiliar with the topic, combined with the world-wide existence of the practice, prove that hallucinatory pictures are really induced.

It has not been found possible to determine, before experiment, whether any given man or woman will prove capable of the hallucinatory experiences. Many subjects with strong powers of "visualization," or seeing things "in the mind's eye," cannot scry; others are successful in various degrees. We might expect persons who have experienced spontaneous visual hallucinations, of the kind vulgarly styled "ghosts" or "wraiths," to succeed in inducing pictures in a glass ball. As a matter of fact such persons sometimes can and sometimes cannot see pictures in the way of crystal-gazing; while many who can see in the crystal have had no spontaneous hallucinations. It is useless to make experiments with hysterical and visionary people, "whose word no man relies on"; they may have the hallucinatory experiences, but they would say that they had in any case.

The nearest analogy to crystal visions, as described, is the common experience of "hypnagogic illusions" (cf. Alfred Maury. *Les Rêves et le sommeil*). With closed eyes, between sleeping and waking, many people see faces, landscapes and other things flash upon their view, pictures often brilliant, but of very brief duration and rapid mutation. Sometimes the subject opens his eyes to get rid of an unpleasant vision of this kind. People who cannot scry may have these hypnagogic illusions, and, so far, may partly understand the experience of the scryer who is wide awake. But the visions of the scryer often endure for a considerable time. He or she may put the glass down and converse, and may find the picture still there when the ball is taken up again. New figures may join the figure first seen, as when one enters a room. In these respects, and in the awakeness of the scryer, crystal pictures differ from hypnagogic illusions. In other ways the experiences coincide, the pictures are either fanciful, like illustrations of some unread history or romance, or are revivals of remembered places and faces.

Occasionally, in hypnagogic illusions, the observer can see the picture develop rapidly out of a blot of light or colour, beheld by the closed eyes. One or two scryers think that they, too, can trace the picture as it develops on the suggestion of some passage of light, colour or shadow in the glass or crystal. But, as a rule, the scryer cannot detect any process of development from such *points de mire*; though this may be the actual process.

On the whole there seems little doubt that successful crystal-gazing is the exertion of a not uncommon though far from universal faculty, like those of "chromatic audition"—the vivid association of certain sounds with certain colours—and the mental seeing of figures arranged in coloured diagrams (Galton, *Inquiry into Human Faculty*, pp. 114-154). The experience of hypnagogic illusions also seems far more rare than ordinary dreaming in sleep. Unfortunately, while these phenomena have been carefully studied by officially scientific characters, in England orthodox *savants* have disdained to observe crystal-gazing, while in France psychologists have too commonly experimented with subjects professionally hysterical and quite untrustworthy. Our remarks are therefore based mainly on considerable personal study of "scrying" among normal British subjects of both sexes, to whom the topic was previously unknown.

The superstitious associations of crystal-gazing, as of hypnotism, appear to bar the way to official scientific investigation, and the fluctuating proficiency of the seers, who cannot command success, or determine the causes and conditions of success and failure, tends in the same direction. The existence, too, of paid professionals who lead astray silly women, encourages the natural scientific contempt for the study of the faculty.

The seeing of the pictures, as far as we have spoken of it, appears to be a thing unusual, but in no way abnormal, any more than dreams or hypnagogic illusions are abnormal. Crystal pictures, however, are commonly dismissed as mere results of "imagination," a theory which, of course, is of no real assistance to psychology. Persons of recognized "imaginativeness," such as novelists and artists, do not seem more or less capable of the hallucinatory experiences than their sober neighbours; while persons not otherwise recognizably "imaginative" (we could quote a singularly accurate historian) are capable of the experiences. It is unfortunate, as it awakens prejudice, but in the present writer's opinion it is true, that crystal-gazing sometimes is rewarded with results which may be styled "supra-normal." In addition to the presentation of revived memories, and of "objectivation of ideas or images consciously or unconsciously in the mind of the percipient," there occur "visions, possibly telepathic or clairvoyant, implying acquirement of knowledge by supra-normal means."[1]

A number of examples occurring during experiments made by the present writer and by his acquaintances in 1897 were carefully recorded and attested by the signatures of all concerned The cases, or rather a selection of the cases, are printed in A. Lang's book, *The Making of Religion* (2nd ed., London, 1902, pp. 87-104). Others are chronicled in A. Lang's Introduction to Mr N. W. Thomas's work, *Crystal Gazing* (1905). The experiments took this form: any person might ask the scryer (a lady who had never previously heard of crystal-gazing) "to see what he was thinking of." The scryer, who was a stranger in a place which she had not visited before, gave, in a long series of cases, a description of the person or place on which the inquirer's thoughts were fixed. The descriptions, though three or four entire failures occurred, were of remarkable accuracy as a rule, and contained facts and incidents unknown to the inquirers, but confirmed as accurate. In fact, some Oriental scenes and descriptions of incidents were corroborated by a letter from India which arrived just after the experiment; and the same thing happened when the events described were occurring in places less remote. On one occasion a curious set of incidents were described, which happened to be vividly present to the mind of a sceptical stranger who chanced to be in the room during the experiment; events unknown to the inquirer in this instance. As an example of the minuteness of description, an inquirer, thinking of a brother in India, an officer in the army, whose hair had suffered in an encounter with a tiger, had described to her an officer in undress uniform, with bald scars through the hair on his temples, such as he really bore. The number and proportion of successes was too high to admit of explanation by chance coincidence, but success was not invariable. On one occasion the scryer could see nothing, "the crystal preserved its natural diaphaneity," as Dr Dee says; and there were failures with two or three inquirers. On the other hand no record was kept in several cases of success.

Whoever can believe that the successes were numerous and that descriptions were given correctly—not only of facts present to the minds of inquirers, and of other persons present who were not consciously taking a share in the experiments, but also of facts necessarily unknown to all concerned—must of course be most impressed by the latter kind of success. If the process commonly styled "telepathy" exists (see TELEPATHY), that may account for the scryer's power of seeing facts which are in the mind of the inquirer. But when the scryers see details of various sorts, which are unknown to the inquirer, but are verified on inquiry, then telepathy perhaps fails to provide an explanation. We seem to be confronted with actual clairvoyance (*q.v.*), or *vue à distance*. It would be vain to form hypotheses as to the conditions or faculties which make *vue à distance* possible. This way lie metaphysics, with Hegel's theory of the Sensitive Soul, or Myers' theory of the Subliminal Self. "The intuitive soul," says Hegel, "oversteps the conditions of time and space; it beholds things remote, things long past, and things to come."[2]

What we need, if any progress is to be made in knowledge of the subject, is not a metaphysical hypothesis, but a large, carefully tested, and well-recorded collection of examples, made by *savants* of recognized standing. At present we are where we were in electrical science, when Newton produced curious sparks while rubbing glass with paper. By way of facts, we have only a large body of unattested anecdotes of supra-normal successes in crystal-gazing, in many lands and ages; and the scanty records of modern amateur investigators, like the present writer. Even from these, if the honesty of all concerned be granted (and even clever dishonesty could not have produced many of the results), it would appear that we are investigating a strange and important human faculty. The writer is acquainted with no experiments in which it was attempted to discern the future (except in trivial cases as to events on the turf, when chance coincidence might explain the successes), and only with two or three cases in which there was an attempt to help historical science and discern the past by aid of psychical methods. The results were interesting and difficult to explain, but the experiments were few. Ordinary scryers of fancy pictures are common enough, but scryers capable of apparently supra-normal successes are apparently rare. Perhaps something depends on the inquirer as well as the scryer.

The method of scrying, as generally practised, is simple. It is usual to place a glass ball on a dark ground, to sit with the back to the light, to focus the gaze on the ball (disregarding reflections, if these cannot be excluded), and to await results. Perhaps from five to ten minutes is a long enough time for the experiment. The scryer may let his consciousness play freely, but should not be disturbed by lookers-on. As a rule, if a person has the faculty he "sees" at the first attempt; if he fails in the first three or four efforts he need not persevere. Solitude is advisable at first, but few people can find time amounting to ten minutes for solitary studies of this sort, so busy and so gregarious is mankind. The writer has no experience of trance, sleep or auto-hypnotization produced in such experiments; scryers have always seemed to retain their full normal consciousness. As regards scepticism concerning the faculty we may quote what Mr Galton says about the faculty of visualization: "Scientific men as a class have feeble power of visual reproduction.... They had a mental deficiency of which they were unconscious, and, naturally enough, supposed that those who affirmed *they* were possessed of it were romancing."

Authorities.—A useful essay is that of "Miss X" (Miss Goodrich Freer) in the *Proceedings of the Society for Psychical Research*, v. The history of crystal-gazing is here traced, and many examples of the author's own experiments are recorded. A. Lang's *The Making of Religion*, ch. v., contains anthropological examples and a series of experiments. In N. W. Thomas's *Crystal Gazing* the history and anthropology of the subject are investigated, with modern instances. For Egypt, see Lane's *Modern Egyptians*, and the *Journal* of Sir Walter Scott, xi. 419-421, with *Quarterly Review*, No. 117, pp. 196-208. These Egyptian experiments of 1830 were vitiated by their method, the scryer being asked to see and describe a given person, named. He ought not, of course, to be told more than that he is to descry the inquirer's thoughts, and there ought never to be physical contact, as in holding hands, between the inquirer and the scryer during the experiment. There is a chapter on crystal-gazing in *Les Névroses et les idées fixes* of Dr Janet (1898). His statements are sometimes demonstrably inaccurate (see *Making of Religion*, Appendix C). A curious passage on the subject, by Ibn Khaldun, an Arabian medieval *savant*, is quoted by Mr Thomas from the printed Extracts of MSS. in the Bibliothèque Nationale. There is also a chapter on crystal-gazing in Myers' *Human Personality*.

(A. L.)

---

1 *Proceedings of the Society for Psychical Research*, v. 486.

2 "Philosophie der Geistes," Hegel's *Werke*, vii. 179, 406, 408 (Berlin, 1845). Cf. Wallace's translation (Oxford, 1894).

CRYSTALLITE

In media which, on account of their viscosity, offer considerable resistance to those molecular movements which are necessary for the building and growth of crystals, rudimentary or imperfect forms of crystallization very frequently occur. Such media are the volcanic rocks when they are rapidly cooled, producing various kinds of pitchstone, obsidian, &c. When examined under the microscope these rocks consist largely of a perfectly amorphous or glassy base, through which are scattered great numbers of very minute crystals (microliths), and other bodies, termed crystallites, which seem to be stages in the formation of crystals. Crystallites may also be produced by allowing a solution of sulphur in carbon disulphide mixed with Canada balsam to evaporate slowly, and their development may be watched on a microscopic slide. Small globules appear (globulites), spherical and non-crystalline (so far as can be ascertained). They may coalesce or may arrange themselves into rows like strings of beads—margarites—(Gr. μαργαρίτης, a pearl) or into groups with a somewhat radiate arrangement—globospherites. Occasionally they take elongated shapes—longulites and baculites (Lat. *baculus*, a staff). The largest may become crystalline, changing suddenly into polyhedral bodies with evident double refraction and the optical properties belonging to crystals. Others become long and thread-like—trichites (Gr. θρίξ, τριχός, hair)—and these are often curved, and a group of them may be implanted on the surface of a small crystal. All these forms are found in vitreous igneous rocks. H. P. J. Vogelsang, who was the first to direct much attention to them, believes that the globulites are preliminary stages in the formation of crystals.

Microliths, as distinguished from crystallites, have crystalline properties, and evidently belong to definite minerals or salts. When sufficiently large they are often recognizable, but usually they are so small, so opaque, or so densely crowded together that this is impossible. In igneous rocks they are usually felspar, augite, enstatite, and iron oxides, and are found in abundance only where there is much uncrystallized glassy base; in contact-altered sediments, slags, &c., microlithic forms of garnet, spinel, sillimanite, cordierite, various lime silicates, and many other substances have been observed. Their form varies greatly, *e.g.* thin fibres (sillimanite, augite), short prisms or rods (felspar, enstatite, cordierite), or equidimensional grains (augite, spinel, magnetite). Occasionally they are perfectly shaped though minute crystals; more frequently they appear rounded (magnetite, &c.), or have brush-like terminations (augite, felspar, &c.). The larger microliths may contain enclosures of glass, and it is very common to find that the prisms have hollow, funnel-shaped ends, which are filled with vitreous material. These microliths, under the influence of crystalline forces, may rank themselves side by side to make up skeleton crystals and networks, or feathery and arborescent forms, which obey more or less closely the laws of crystallization of the substance to which they belong. They bear a very close resemblance to the arborescent frost flowers seen on window panes in winter, and to the stellate snow crystals. In magnetite the growths follow three axes at right angles to one another; in augite this is nearly, though not exactly, the case; in hornblende an angle of 57° may frequently be observed, corresponding to the prism angle of the fully-developed crystal. The interstices of the network may be partly filled up by a later growth. In other cases the crystalline arrangement of the microliths is less perfect, and branching, arborescent or feathery groupings are produced (*e.g.* felspar, augite, hornblende). Spherulites may be regarded as radiate aggregates of such microliths (mostly felspar mixed with quartz or tridymite). If larger porphyritic crystals occur in the rock, the microliths of the vitreous base frequently grow outwards from their faces; in some cases a definite parallelism exists between the two, but more frequently the early crystal has served merely as a centre, or nucleus, from which the microliths and spherulites have spread in all directions.

(J. S. F.)

CRYSTALLIZATION

The art of obtaining a substance in the form of crystals; it is an important process in chemistry since it permits the purification of a substance, or the separation of the constituents of a mixture. Generally a substance is more soluble in a solvent at a high temperature than at a low, and consequently, if a boiling concentrated solution be allowed to cool, the substance will separate in virtue of the diminished solubility, and the slower the cooling the larger and more perfect will be the crystals formed. If, as sometimes appears, such a solution refuses to crystallize, the expedient of inoculating the solution with a minute crystal of the same substance, or with a similar substance, may be adopted; shaking the solution, or the addition of a drop of another solvent, may also occasion the desired result. "Fractional crystallization" consists in repeatedly crystallizing a salt so as to separate the substances of different solubilities. Examples are especially presented in the study of the rare-earths. Other conditions under which crystals are formed are given in the article CRYSTALLOGRAPHY.

CRYSTALLOGRAPHY

(from the Gr. κρύσταλλος, ice, and γράφειν, to write), the science of the forms, properties and structure of crystals. Homogeneous solid matter, the physical and chemical properties of which are the same about every point, may be either amorphous or crystalline. In amorphous matter all the properties are the same in every direction in the mass; but in crystalline matter certain of the physical properties vary with the direction. The essential properties of crystalline matter are of two kinds, viz. the general properties, such as density, specific heat, melting-point and chemical composition, which do not vary with the direction; and the directional properties, such as cohesion and elasticity, various optical, thermal and electrical properties, as well as external form. By reason of the homogeneity of crystalline matter the directional properties are the same in all parallel directions in the mass, and there may be a certain symmetrical repetition of the directions along which the properties are the same.

When the crystallization of matter takes place under conditions free from outside influences the peculiarities of internal structure are expressed in the external form of the mass, and there results a solid body bounded by plane surfaces intersecting in straight edges, the directions of which bear an intimate relation to the internal structure. Such a polyhedron (πολύς, many, ἕδρα, base or face) is known as a crystal. An example of this is sugar-candy, of which a single isolated crystal may have grown freely in a solution of sugar. Matter presenting well-defined and regular crystal forms, either as a single crystal or as a group of individual crystals, is said to be crystallized. If, on the other hand, crystallization has taken place about several centres in a confined space, the development of plane surfaces may be prevented, and a crystalline aggregate of differently orientated crystal-individuals results. Examples of this are afforded by loaf sugar and statuary marble.

After a brief historical sketch, the more salient principles of the subject will be discussed under the following sections:—

Most chemical elements and compounds are capable of assuming the crystalline condition. Crystallization may take place when solid matter separates from solution (*e.g.* sugar, salt, alum), from a fused mass (*e.g.* sulphur, bismuth, felspar), or from a vapour (*e.g.* iodine, camphor, haematite; in the last case by the interaction of ferric chloride and steam). Crystalline growth may also take place in solid amorphous matter, for example, in the devitrification of glass, and the slow change in metals when subjected to alternating stresses. Beautiful crystals of many substances may be obtained in the laboratory by one or other of these methods, but the most perfectly developed and largest crystals are those of mineral substances found in nature, where crystallization has continued during long periods of time. For this reason the physical science of crystallography has developed side by side with that of mineralogy. Really, however, there is just the same connexion between crystallography and chemistry as between crystallography and mineralogy, but only in recent years has the importance of determining the crystallographic properties of artificially prepared compounds been recognized.

History.—The word "crystal" is from the Gr. κρύσταλλος, meaning clear ice (Lat. *crystallum*), a name which was also applied to the clear transparent quartz ("rock-crystal") from the Alps, under the belief that it had been formed from water by intense cold. It was not until about the 17th century that the word was extended to other bodies, either those found in nature or obtained by the evaporation of a saline solution, which resembled rock-crystal in being bounded by plane surfaces, and often also in their clearness and transparency.

The first important step in the study of crystals was made by Nicolaus Steno, the famous Danish physician, afterwards bishop of Titiopolis, who in his treatise *De solido intra solidum naturaliter contento* (Florence, 1669; English translation, 1671) gave the results of his observations on crystals of quartz. He found that although the faces of different crystals vary considerably in shape and relative size, yet the angles between similar pairs of faces are always the same. He further pointed out that the crystals must have grown in a liquid by the addition of layers of material upon the faces of a nucleus, this nucleus having the form of a regular six-sided prism terminated at each end by a six-sided pyramid. The thickness of the layers, though the same over each face, was not necessarily the same on different faces, but depended on the position of the faces with respect to the surrounding liquid; hence the faces of the crystal, though variable in shape and size, remained parallel to those of the nucleus, and the angles between them constant. Robert Hooke in his *Micrographia* (London, 1665) had previously noticed the regularity of the minute quartz crystals found lining the cavities of flints, and had suggested that they were built up of spheroids. About the same time the double refraction and perfect rhomboidal cleavage of crystals of calcite or Iceland-spar were studied by Erasmus Bartholinus (*Experimenta crystalli Islandici disdiaclastici*, Copenhagen, 1669) and Christiaan Huygens (*Traité de la lumière*, Leiden, 1690); the latter supposed, as did Hooke, that the crystals were built up of spheroids. In 1695 Anton van Leeuwenhoek observed under the microscope that different forms of crystals grow from the solutions of different salts. Andreas Libavius had indeed much earlier, in 1597, pointed out that the salts present in mineral waters could be ascertained by an examination of the shapes of the crystals left on evaporation of the water; and Domenico Guglielmini (*Riflessioni filosofiche dedotte dalle figure de' sali*, Padova, 1706) asserted that the crystals of each salt had a shape of their own with the plane angles of the faces always the same.

The earliest treatise on crystallography is the *Prodromus Crystallographiae* of M. A. Cappeller, published at Lucerne in 1723. Crystals were mentioned in works on mineralogy and chemistry; for instance, C. Linnaeus in his *Systema Naturae* (1735) described some forty common forms of crystals amongst minerals. It was not, however, until the end of the 18th century that any real advances were made, and the French crystallographers Romé de l'Isle and the abbé Haüy are rightly considered as the founders of the science. J. B. L. de Romé de l'Isle (*Essai de cristallographie*, Paris, 1772; *Cristallographie, ou description des formes propres à tous les corps du règne minéral*, Paris, 1783) made the important discovery that the various shapes of crystals of the same natural or artificial substance are all intimately related to each other; and further, by measuring the angles between the faces of crystals with the goniometer (*q.v.*), he established the fundamental principle that these angles are always the same for the same kind of substance and are characteristic of it. Replacing by single planes or groups of planes all the similar edges or solid angles of a figure called the "primitive form" he derived other related forms. Six kinds of primitive forms were distinguished, namely, the cube, the regular octahedron, the regular tetrahedron, a rhombohedron, an octahedron with a rhombic base, and a double six-sided pyramid. Only in the last three can there be any variation in the angles: for example, the primitive octahedron of alum, nitre and sugar were determined by Romé de l'Isle to have angles of 110°, 120° and 100° respectively. René Just Haüy in his *Essai d'une théorie sur la structure des crystaux* (Paris, 1784; see also his Treatises on Mineralogy and Crystallography, 1801, 1822) supported and extended these views, but took for his primitive forms the figures obtained by splitting crystals in their directions of easy fracture of "cleavage," which are aways the same in the same kind of substance. Thus he found that all crystals of calcite, whatever their external form (see, for example, figs. 1-6 in the article Calcite), could be reduced by cleavage to a rhombohedron with interfacial angles of 75°. Further, by stacking together a number of small rhombohedra of uniform size he was able, as had been previously done by J. G. Gahn in 1773, to reconstruct the various forms of calcite crystals. Fig. 1 shows a scalenohedron (σκαληνός, uneven) built up in this manner of rhombohedra; and fig. 2 a regular octahedron built up of cubic elements, such as are given by the cleavage of galena and rock-salt.

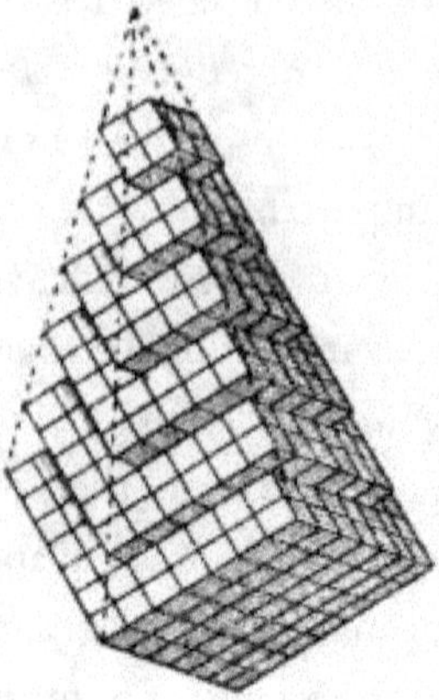

Fig. 1.—Scalenohedron built up of Rhombohedra.

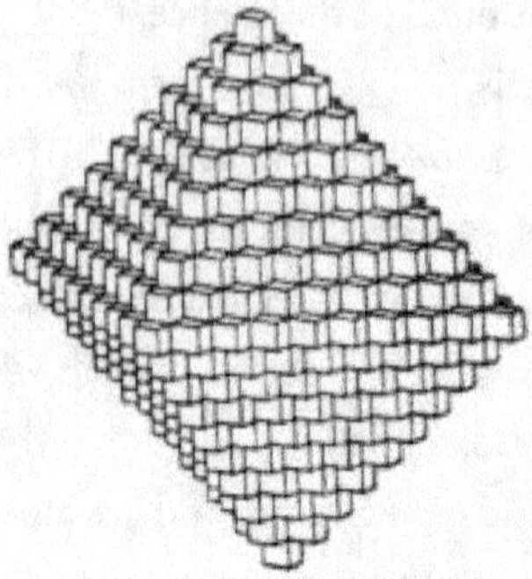

Fig. 2.—Octahedron built up of Cubes.

The external surfaces of such a structure, with their step-like arrangement, correspond to the plane faces of the crystal, and the bricks may be considered so small as not to be separately visible. By making the steps one, two or three bricks in width and one, two or three bricks in height the various secondary faces on the crystal are related to the primitive form or "cleavage nucleus" by a law of whole numbers, and the angles between them can be arrived at by mathematical calculation. By measuring with the goniometer the inclinations of the secondary faces to those of the primitive form Haüy found that the secondary forms are always related to the primitive form on crystals of numerous substances in the manner indicated, and that the width and the height of a step are always in a simple ratio, rarely exceeding that of 1 : 6. This laid the foundation of the important "law of rational indices" of the faces of crystals.

The German crystallographer C. S. Weiss (*De indagando formarum crystallinarum charactere geometrico principali dissertatio*, Leipzig, 1809; *Übersichtliche Darstellung der verschiedenen natürlichen Abtheilungen der Krystallisations-Systeme*, Denkschrift der Berliner Akad. der Wissensch., 1814-1815) attacked the problem of crystalline form from a purely geometrical point of view, without reference to primitive forms or any theory of structure. The faces of crystals were considered by their intercepts on co-ordinate axes, which were drawn joining the opposite corners of certain forms; and in this way the various primitive forms of Haüy were grouped into four classes, corresponding to the four systems described below under the names cubic, tetragonal, hexagonal and orthorhombic. The same result was arrived at independently by F. Mohs, who further, in 1822, asserted the existence of two additional systems with oblique axes. These two systems (the monoclinic and anorthic) were, however, considered by Weiss to be only hemihedral or tetartohedral modifications of the orthorhombic system, and they were not definitely established until 1835, when the optical characters of the crystals were found to be distinct. A system of notation to express the relation of each face of a crystal to the co-ordinate axes of reference was devised by Weiss, and other notations were proposed by F. Mohs, A. Lévy (1825), C. F. Naumann (1826), and W. H. Miller (*Treatise on Crystallography*, Cambridge, 1839). For simplicity and utility in calculation the Millerian notation, which was first suggested by W. Whewell in 1825, surpasses all others and is now generally adopted, though those of Lévy and Naumann are still in use.

Although the peculiar optical properties of Iceland-spar had been much studied ever since 1669, it was not until much later that any connexion was traced between the optical characters of crystals and their external form. In 1818 Sir David Brewster found that crystals could be divided optically into three classes, viz. isotropic, uniaxial and biaxial, and that these classes corresponded with Weiss's four systems (crystals belonging to the cubic system being isotropic, those of the tetragonal and hexagonal being uniaxial, and the orthorhombic being biaxial). Optically biaxial crystals were afterwards shown by J. F. W. Herschel and F. E. Neumann in 1822 and 1835 to be of three kinds, corresponding with the orthorhombic, monoclinic and anorthic systems. It was, however, noticed by Brewster himself that there are many apparent exceptions, and the "optical anomalies" of crystals have been the subject of much study. The intimate relations existing between various other physical properties of crystals and their external form have subsequently been gradually traced.

The symmetry of crystals, though recognized by Romé de l'Isle and Haüy, in that they replaced all similar edges and corners of their primitive forms by similar secondary planes, was not made use of in defining the six systems of crystallization, which depended solely on the lengths and inclinations of the axes of reference. It was, however, necessary to recognize that in each system there are certain forms which are only partially symmetrical, and these were described as hemihedral and tetartohedral forms (*i.e.* ἡμι-, half-faced, and τέταρτος, quarter-faced forms).

As a consequence of Haüy's law of rational intercepts, or, as it is more often called, the law of rational indices, it was proved by J. F. C. Hessel in 1830 that thirty-two types of symmetry are possible in crystals. Hessel's work remained overlooked for sixty years, but the same important result was independently arrived at by the same method by A. Gadolin in 1867. At the present day, crystals are considered as belonging to one or other of thirty-two classes, corresponding with these thirty-two types of symmetry, and are grouped in six systems. More recently, theories of crystal structure have attracted attention, and have been studied as purely geometrical problems of the homogeneous partitioning of space.

The historical development of the subject is treated more fully in the article CRYSTALLOGRAPHY in the 9th edition of this work. Reference may also be made to C. M. Marx, *Geschichte der Crystallkunde* (Karlsruhe and Baden, 1825); W. Whewell, *History of the Inductive Sciences*, vol. iii. (3rd ed., London, 1857); F. von Kobell, *Geschichte der Mineralogie von 1650-1860* (München, 1864); L. Fletcher, *An Introduction to the Study of Minerals* (British Museum Guide-Book); L. Fletcher, *Recent Progress in Mineralogy and Crystallography* [1832-1894] (Brit. Assoc. Rep., 1894).

I. CRYSTALLINE FORM

The fundamental laws governing the form of crystals are:—

1. Law of the Constancy of Angle.

2. Law of Symmetry.

3. Law of Rational Intercepts or Indices.

According to the first law, the angles between corresponding faces of all crystals of the same chemical substance are always the same and are characteristic of the substance.

(*a*) *Symmetry of Crystals.*

Crystals may, or may not, be symmetrical with respect to a point, a line or axis, and a plane; these "elements of symmetry" are spoken of as a centre of symmetry, an axis of symmetry, and a plane of symmetry respectively.

Centre of Symmetry.—Crystals which are centro-symmetrical have their faces arranged in parallel pairs; and the two parallel faces, situated on opposite sides of the centre (*O* in fig. 3) are alike in surface characters, such as lustre, striations, and figures of corrosion. An octahedron (fig. 3) is bounded by four pairs of parallel faces. Crystals belonging to many of the hemihedral and tetartohedral classes of the six systems of crystallization are devoid of a centre of symmetry.

Axes of Symmetry.—Consider the vertical axis joining the opposite corners a_3 and $\bar{a}_3$ of an octahedron (fig. 3) and passing through its centre *O*: by rotating the crystal about this axis through a right angle (90°) it reaches a position such that the orientation of its faces is the same as before the rotation; the face $\bar{a}_1\bar{a}_2\bar{a}_3$, for example, coming into the position of $a_1\bar{a}_2a_3$.

During a complete rotation of 360° (= 90° × 4), the crystal occupies four such interchangeable positions. Such an axis of symmetry is known as a tetrad axis of symmetry. Other tetrad axes of the octahedron are $a_2\bar{a}_2$ and a_1a_1.

An axis of symmetry of another kind is that which passing through the centre *O* is normal to a face of the octahedron. By rotating the crystal about such an axis *Op* (fig. 3) through an angle of 120° those faces which are not perpendicular to the axis occupy interchangeable positions; for example, the face $a_1a_3a_2$ comes into the position of $\bar{a}_2a_1\bar{a}_3$, and $\bar{a}_2a_1\bar{a}_3$ to $a_3\bar{a}_2\bar{a}_1$. During a complete rotation of 360° (= 120° × 3) the crystal occupies similar positions three times. This is a triad axis of symmetry; and there being four pairs of parallel faces on an octahedron, there are four triad axes (only one of which is drawn in the figure).

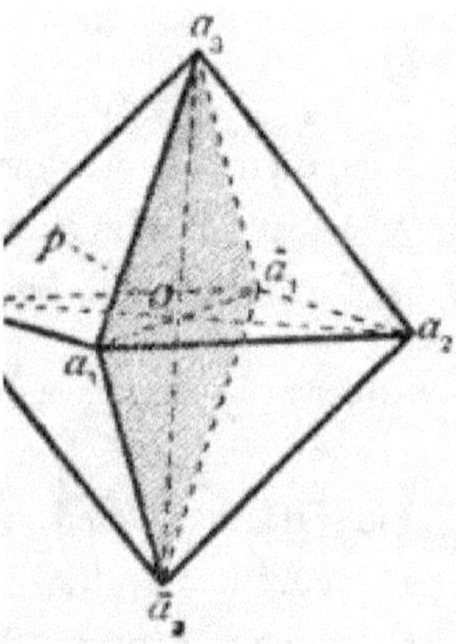

FIG. 3.

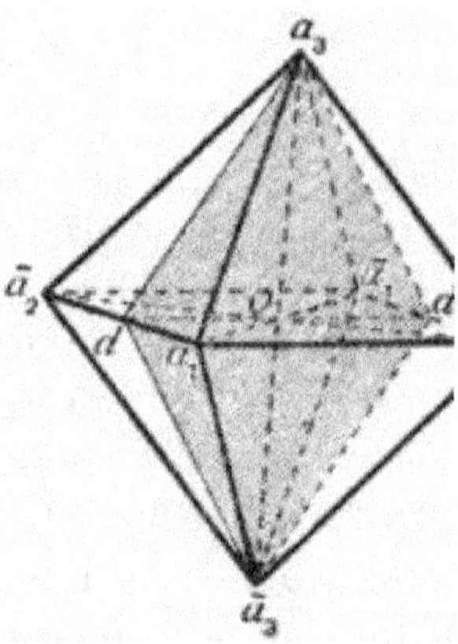

FIG. 4.

Axes and Planes of Symmetry of an Octahedron.

An axis passing through the centre *O* and the middle points d of two opposite edges of the octahedron (fig. 4), *i.e.* parallel to the edges of the octahedron, is a dyad axis of symmetry. About this axis there may be rotation of 180°, and only twice in a complete revolution of 360° (= 180° × 2) is the crystal brought into interchangeable positions. There being six pairs of parallel edges on an octahedron, there are consequently six dyad axes of symmetry.

A regular octahedron thus possesses thirteen axes of symmetry (of three kinds), and there are the same number in the cube. Fig. 5 shows the three tetrad (or tetragonal) axes (*aa*), four triad (or trigonal) axes (*pp*), and six dyad (diad or diagonal) axes (*dd*).

Although not represented in the cubic system, there is still another kind of axis of symmetry possible in crystals. This is the hexad axis or hexagonal axis, for which the angle of rotation is 60°, or one-sixth of 360°. There can be only one hexad axis of symmetry in any crystal (see figs. 77-80).

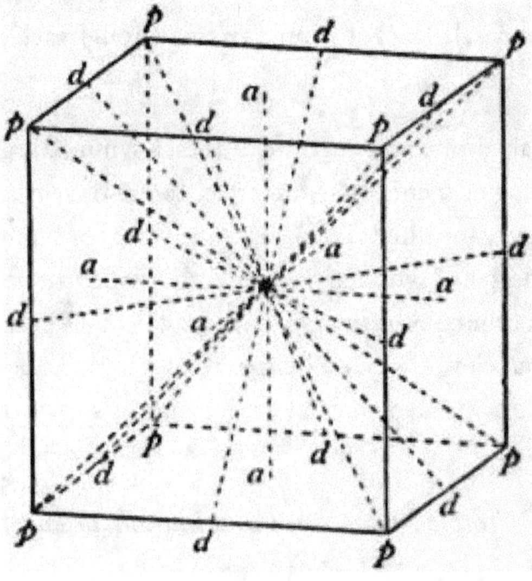

FIG. 5.—Axes of Symmetry of a Cube.

Planes of Symmetry.—A regular octahedron can be divided into two equal and similar halves by a plane passing through the corners $a_1a_3\bar{a}_1\bar{a}_3$ and the centre *O* (fig. 3). One-half is the mirror reflection of the other in this plane, which is called a plane of symmetry. Corresponding planes on either side of a plane of symmetry are inclined to it at equal angles. The octahedron can also be divided by similar planes of symmetry passing through the corners $a_1a_2\bar{a}_1\bar{a}_2$ and $a_2a_3\bar{a}_2\bar{a}_3$. These three similar planes of symmetry are called the cubic planes of symmetry, since they are parallel to the faces of the cube (compare figs. 6-8, showing combinations of the octahedron and the cube).

A regular octahedron can also be divided symmetrically into two equal and similar portions by a plane passing through the corners a_3 and $\bar{a}_3$, the middle points *d* of the edges $a_1\bar{a}_2$ and $\bar{a}_1a_2$, and the centre *O* (fig. 4). This is called a dodecahedral plane of symmetry, being parallel to the face of the rhombic dodecahedron which truncates the edge a_1a_2 (compare fig. 14, showing a combination of the octahedron and rhombic dodecahedron). Another similar plane of symmetry is that passing through the corners $a_3\bar{a}_3$ and the middle points of the edges a_1a_2 and $\bar{a}_1\bar{a}_2$, and altogether there are six dodecahedral planes of symmetry, two through each of the corners a_1, a_2, a_3 of the octahedron.

A regular octahedron and a cube are thus each symmetrical with respect to the following elements of symmetry: a centre of symmetry, thirteen axes of symmetry (of three kinds), and nine planes of symmetry (of two kinds). This degree of symmetry, which is the type corresponding to one of the classes of the cubic system, is the highest possible in crystals. As will be pointed out below, it is possible, however, for both the octahedron and the cube to be associated with fewer elements of symmetry than those just enumerated.

(b) Simple Forms and Combinations of Forms.

A single face $a_1a_2a_3$ (figs. 3 and 4) may be repeated by certain of the elements of symmetry to give the whole eight faces of the octahedron. Thus, by rotation about the vertical tetrad axis $a_3\bar{a}_3$ the four upper faces are obtained; and by rotation of these about one or other of the horizontal tetrad axes the eight faces are derived. Or again, the same repetition of the faces may be arrived at by reflection across the three cubic planes of symmetry. (By reflection across the six dodecahedral planes of symmetry a tetrahedron only would result, but if this is associated with a centre of symmetry we obtain the octahedron.) Such a set of similar faces, obtained by symmetrical repetition, constitutes a "simple form." An octahedron thus consists of eight similar faces, and a cube is bounded by six faces all of which have the same surface characters, and parallel to each of which all the properties of the crystal are identical.

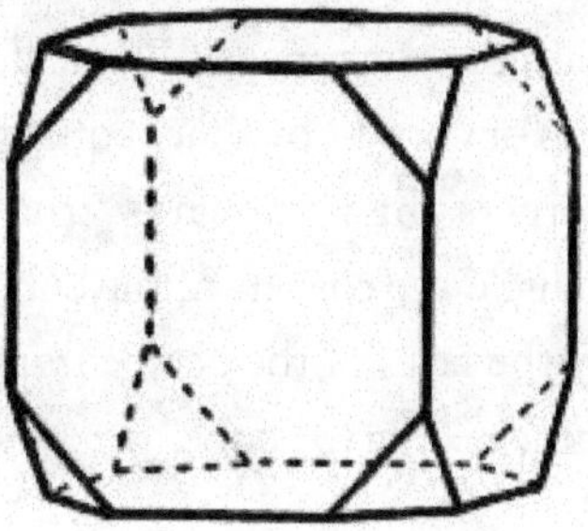

Fig. 6.—Cube in combination with Octahedron.

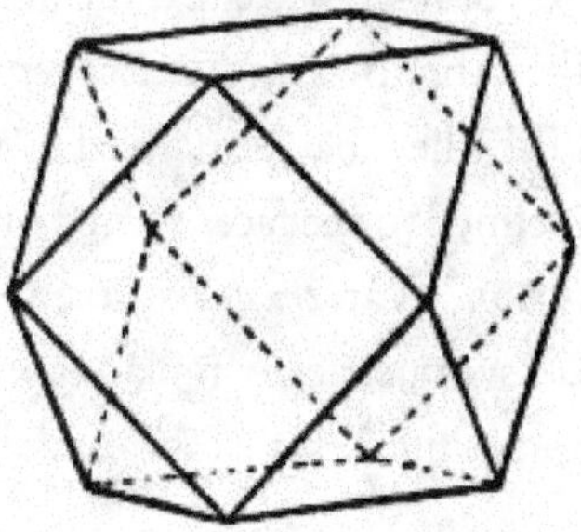

Fig. 7.—Cubo-octahedron.

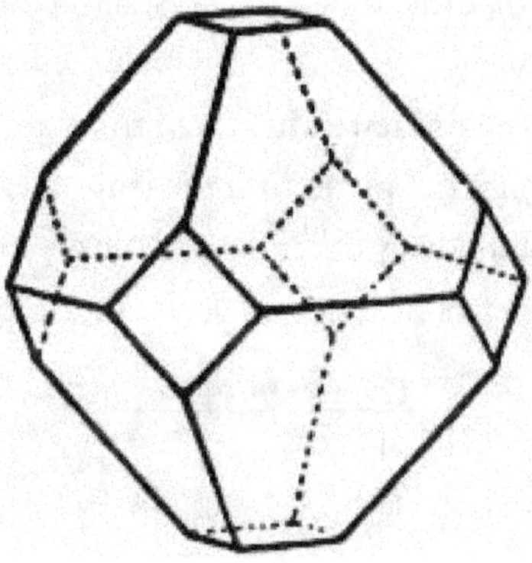

Fig. 8.—Octahedron in combination with Cube.

Examples of simple forms amongst crystallized substances are octahedra of alum and spinel and cubes of salt and fluorspar. More usually, however, two or more forms are present on a crystal, and we then have a combination of forms, or simply a "combination." Figs. 6, 7 and 8 represent combinations of the octahedron and the cube; in the first the faces of the cube predominate, and in the third those of the octahedron; fig. 7

with the two forms equally developed is called a cubo-octahedron. Each of these combined forms has all the elements of symmetry proper to the simple forms.

The simple forms, though referable to the same type of symmetry and axes of reference, are quite independent, and cannot be derived one from the other by symmetrical repetition, but, after the manner of Romé de l'Isle, they may be derived by replacing edges or corners by a face equally inclined to the faces forming the edges or corners; this is known as "truncation" (Lat. *truncare*, to cut off). Thus in fig. 6 the corners of the cube are symmetrically replaced or truncated by the faces of the octahedron, and in fig. 8 those of the octahedron are truncated by the cube.

(c) Law of Rational Intercepts.

For axes of reference, *OX*, *OY*, *OZ* (fig. 9), take any three edges formed by the intersection of three faces of a crystal. These axes are called the crystallographic axes, and the planes in which they lie the axial planes. A fourth face on the crystal intersecting these three axes in the points *A*, *B*, *C* is taken as the parametral plane, and the lengths *OA* : *OB* : *OC* are the parameters of the crystal. Any other face on the crystal may be referred to these axes and parameters by the ratio of the intercepts

$$\frac{.OC:OB:OA}{lkh}$$

Thus for a face parallel to the plane A Be the intercepts are in the ratio *OA* : *OB* : *Oe*, or

$$\frac{OA}{1}:\frac{OB}{1}:\frac{OC}{2}$$

and for a plane *fgC* they are *Of* : *Og* : *OC* or

$$\frac{OA}{2}:\frac{OB}{3}:\frac{OC}{1}.$$

Now the important relation existing between the faces of a crystal is that the denominators *h*, *k* and *l* are always rational whole numbers, rarely exceeding 6, and usually 0, 1, 2 or 3. Written in the form (*hkl*), *h* referring to the axis *OX*, *k* to *OY*, and *l* to *OZ*, they are spoken of as the indices (Millerian indices) of the face. Thus of a face parallel to the plane *ABC* the indices are (111), of *ABe* they are (112), and of *fgC* (231). The indices are thus inversely proportional to the intercepts, and the law of rational intercepts is often spoken of as the "law of rational indices."

The angular position of a face is thus completely fixed by its indices; and knowing the angles between the axial planes and the parametral plane all the angles of a crystal can be calculated when the indices of the faces are known.

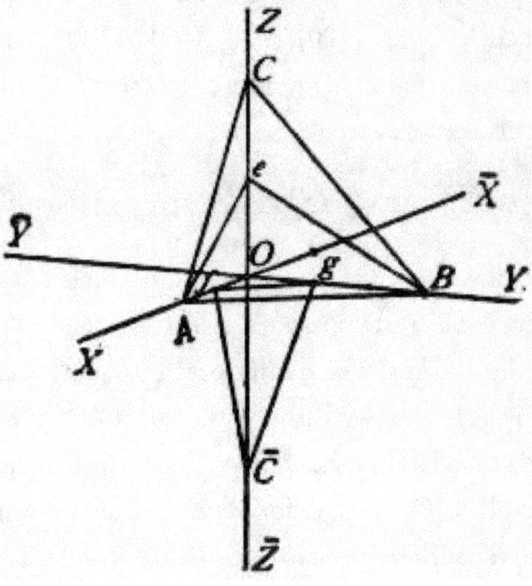

FIG. 9.—Crystallographic axes of reference.

Although any set of edges formed by the intersection of three planes may be chosen for the crystallographic axes, it is in practice usual to select certain edges related to the symmetry of the crystal, and usually coincident with axes of symmetry; for then the indices will be simpler and all faces of the same simple form will have a similar set of indices. The angles between the axes and the ratio of the lengths of the parameters *OA* : *OB* : *OC* (usually given as *a* : *b* : *c*) are spoken of as the "elements" of a crystal, and are constant for and characteristic of all crystals of the same substance.

The six systems of crystal forms, to be enumerated below, are defined by the relative inclinations of the crystallographic axes and the lengths of the parameters. In the cubic system, for example, the three crystallographic axes are taken parallel to the three tetrad axes of symmetry, *i.e.* parallel to the edges of the cube (fig. 5) or joining the opposite corners of the octahedron (fig. 3), and they are therefore all at right angles; the parametral plane (111) is a face of the octahedron, and the parameters are all of equal length. The indices of the eight faces of the octahedron will then be (111), (111), (111), (111), (111), (111), (111), (111). The symbol {111} indicates all the faces belonging to this simple form. The indices of the six faces of the cube are (100), (010), (001), (100), (010), (001); here each face is parallel to two axes, *i.e.* intercepts them at infinity, so that the corresponding indices are zero.

(d) Zones.

An important consequence of the law of rational intercepts is the arrangement of the faces of a crystal in zones. All faces, whether they belong to one or more simple forms, which intersect in parallel edges are said to lie in the same zone. A line drawn through the centre *O* of the crystal parallel to these edges is called a zone-axis, and a plane perpendicular to this axis is called a zone-plane. On a cube, for example, there are three zones each containing four faces, the zone-axes being coincident with the three tetrad axes of symmetry. In the crystal of zircon (fig. 88) the eight prism-faces *a, m*, &c. constitute a zone, denoted by [*a, m, a'*, &c.], with the vertical tetrad axis of symmetry as zone-axis. Again the faces [*a, x, p, e', p', x''', a''*] lie in another zone, as may be seen by the parallel edges of intersection of the faces in figs. 87 and 88; three other similar zones may be traced on the same crystal.

The direction of the line of intersection (*i.e.* zone-axis) of any two planes (*hkl*) and (*h1k1l1*) is given by the zone-indices [**uvw**], where $\mathbf{u} = kl_1 - lk_1$, $\mathbf{v} = lh_1 - hl_1$, and $\mathbf{w} = hk_1 - kh_1$, these being obtained from the face-indices by cross multiplication as follows:—

$$
\begin{array}{cccccc}
h & k & l & h & k & l \\
 & \times & \times & \times & & \\
h_1 & k_1 & l_1 & h_1 & k_1 & l_1.
\end{array}
$$

Any other face (*h2k2l2*) lying in this zone must satisfy the equation

$$h_2\mathbf{u} + k_2\mathbf{v} + l_2\mathbf{w} = 0.$$

This important relation connecting the indices of a face lying in a zone with the zone-indices is known as Weiss's zone-law, having been first enunciated by C. S. Weiss. It may be pointed out that the indices of a face may be arrived at by adding together the indices of faces on either side of it and in the same zone; thus, (311) in fig. 12 lies at the intersections of the three zones [210, 101], [201, 110] and [211, 100], and is obtained by adding together each set of indices.

(e) Projection and Drawing of Crystals.

The shapes and relative sizes of the faces of a crystal being as a rule accidental, depending only on the distance of the faces from the centre of the crystal and not on their angular relations, it is often more convenient to consider only the directions of the normals to the faces. For this purpose projections are drawn, with the aid of which the zonal relations of a crystal are more readily studied and calculations are simplified.

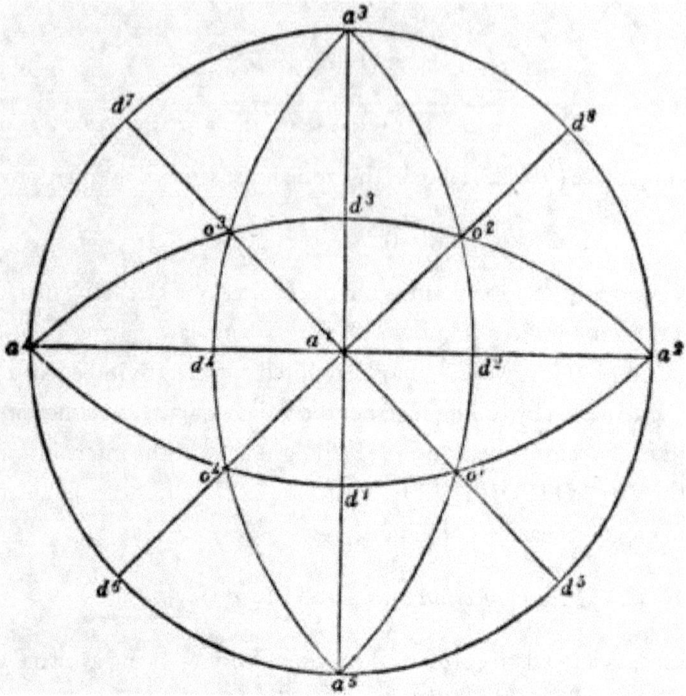

FIG. 10.—Stereographic Projection of a Cubic Crystal.

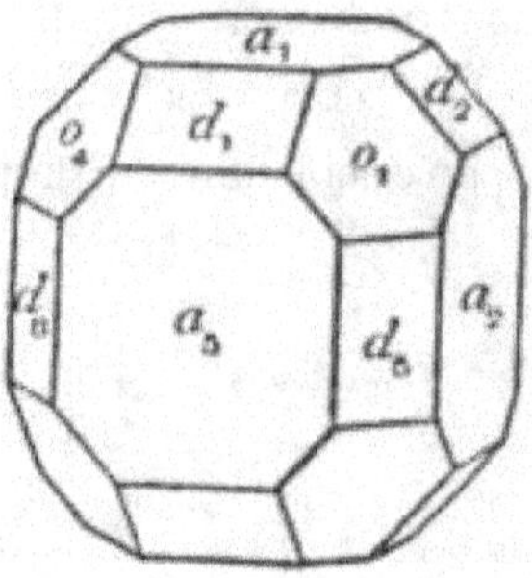

FIG. 11.—Clinographic Drawing of a Cubic Crystal.

The kind of projection most extensively used is the "stereographic projection." The crystal is considered to be placed inside a sphere from the centre of which normals are drawn to all the faces of the crystal. The points at which these normals intersect the surface of the sphere are called the poles of the faces, and by these poles the positions of the faces are fixed. The poles of all faces in the same zone on the crystal will lie on a great circle of the sphere, which are therefore called zone-circles. The calculation of the angles between the normals of faces and between zone-circles is then performed by the ordinary methods of spherical trigonometry. The stereographic projection, however, represents the poles and zone-circles on a plane surface and not on a spherical surface. This is achieved by drawing lines joining all the poles of the faces with the north or south pole of the sphere and finding their points of intersection with the plane of the equatorial great circle, or primitive circle, of the sphere, the projection being represented on this plane. In fig. 10 is shown the stereographic projection, or stereogram, of a cubic crystal; a_1, a_2, &c. are the poles of the faces of the cube. o_1, o_2, &c. those of the octahedron, and d_1, d_2, &c. those of the rhombic dodecahedron. The straight lines and circular arcs are the projections on the equatorial plane of the great circles in which the nine planes of symmetry intersect the sphere. A drawing of a crystal showing a combination of the cube, octahedron and rhombic dodecahedron is shown in fig. 11, in which the faces are lettered the same as the corresponding poles in the projection. From the zone-circles in the projection and the parallel edges in the drawing the zonal relations of the faces are readily seen: thus [$a_1o_1d_5$], [$a_1d_1a_5$], [$a_5o_1d_2$], &c. are zones. A stereographic projection of a rhombohedral crystal is given in fig. 72.

Another kind of projection in common use is the "gnomonic projection" (fig. 12). Here the plane of projection is tangent to the sphere, and normals to all the faces are drawn from the centre of the sphere to intersect the plane of projection. In this case all zones are represented by straight lines. Fig. 12 is the gnomonic projection of a cubic crystal, the plane of projection being tangent to the sphere at the pole of an octahedral face (111), which is therefore in the centre of the projection. The indices of the several poles are given in the figure.

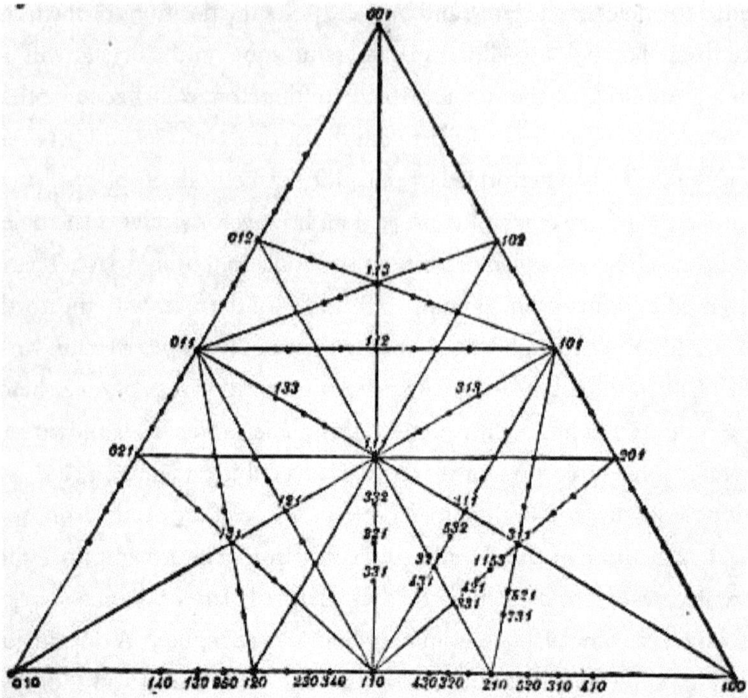

FIG. 12.—Gnomonic Projection of a Cubic Crystal.

In drawing crystals the simple plans and elevations of descriptive geometry (*e.g.* the plans in the lower part of figs. 87 and 88) have sometimes the advantage of showing the symmetry of a crystal, but they give no idea of solidity. For instance, a cube would be represented merely by a square, and an octahedron by a square with lines joining the opposite corners. True perspective drawings are never used in the representation of crystals, since for showing the zonal relations it is important to preserve the parallelism of the edges. If, however, the eye, or point of vision, is regarded as being at an infinite distance from the object all the rays will be parallel, and edges which are parallel on the crystal will be represented by parallel lines in the drawing. The plane of the drawing, in which the parallel rays joining the corners of the crystals and the eye intersect, may be either perpendicular or oblique to the rays; in the former case we have an "orthographic" (ὀρθός, straight; γράφειν, to draw) drawing, and in the latter a "clinographic" (κλίνειν, to incline) drawing. Clinographic drawings are most frequently used for representing crystals. In representing, for example, a cubic crystal (fig. 11) a cube face a_5 is first placed parallel to the plane on which the crystal is to be projected and with one set of edges vertical; the crystal is then turned through a small angle about a vertical axis until a second cube face a_2 comes into view, and the eye is then raised so that a third cube face a_1 may be seen.

(f) Crystal Systems and Classes.

According to the mutual inclinations of the crystallographic axes of reference and the lengths intercepted on them by the parametral plane, all crystals fall into one or other of six groups or systems, in each of which there are several classes depending on the degree of symmetry. In the brief description which follows of these six systems and thirty-two classes of crystals we shall proceed from those in which the symmetry is most complex to those in which it is simplest.

1. CUBIC SYSTEM

(Isometric; Regular; Octahedral; Tesseral).

In this system the three crystallographic axes of reference are all at right angles to each other and are equal in length. They are parallel to the edges of the cube, and in the different classes coincide either with tetrad or dyad axes of symmetry. Five classes are included in this system, in all of which there are, besides other elements of symmetry, four triad axes.

In crystals of this system the angle between any two faces *P* and *Q* with the indices (*hkl*) and (*pqr*) is given by the equation

$$\mathrm{COS\ PQ} = \frac{hp + kq + lr}{\sqrt{(h^2 + k^2 + l^2)(p^2 + q^2 + r^2)}}.$$

The angles between faces with the same indices are thus the same in all substances which crystallize in the cubic system: in other systems the angles vary with the substance and are characteristic of it.

Holosymmetric Class

(Holohedral (ὅλος, whole); Hexakis-octahedral).

Crystals of this class possess the full number of elements of symmetry already mentioned above for the octahedron and the cube, viz. three cubic planes of symmetry, six dodecahedral planes, three tetrad axes of symmetry, four triad axes, six dyad axes, and a centre of symmetry.

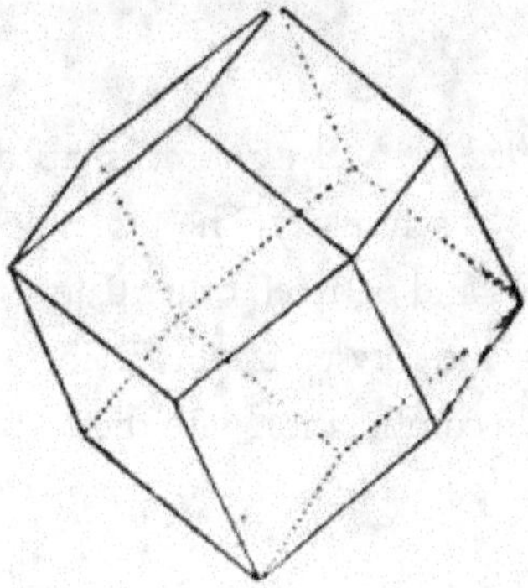

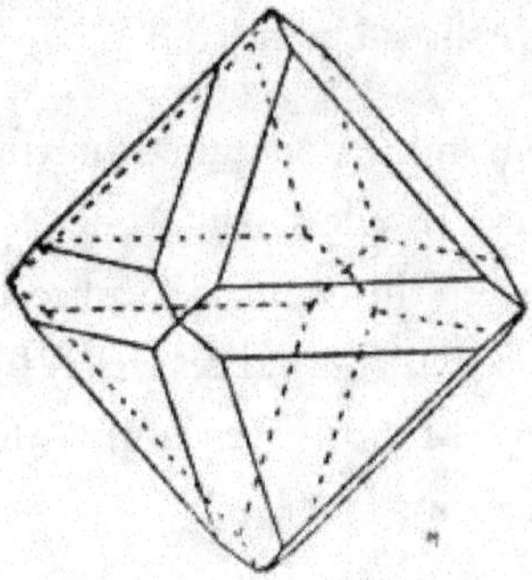

FIG. 13.—Rhombic Dodecahedron.

FIG. 14.—Combination of Rhombic Dodecahedron and Octahedron.

There are seven kinds of simple forms, viz.:—

Cube (fig. 5). This is bounded by six square faces parallel to the cubic planes of symmetry; it is known also as the hexahedron. The angles between the faces are 90°, and the indices of the form are {100}. Salt, fluorspar and galena crystallize in simple cubes.

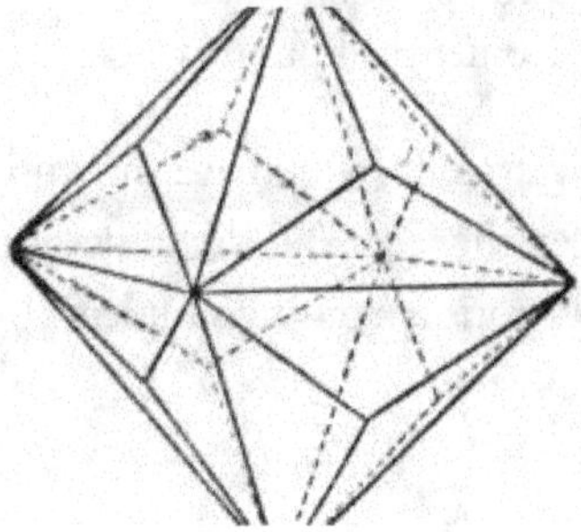

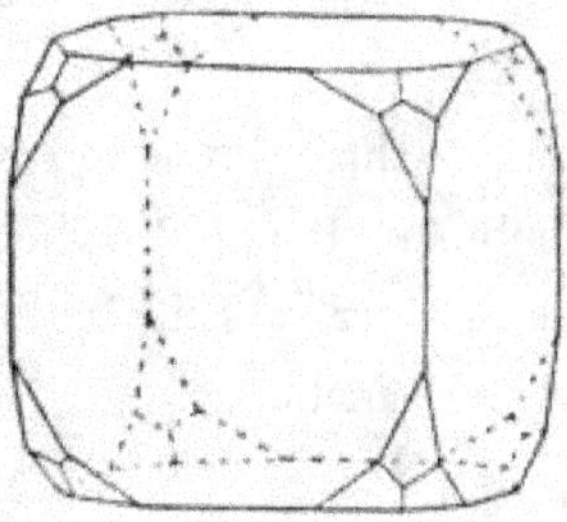

FIG. 15.—Triakis-octahedron.

FIG. 16.—Combination of Triakis-octahedron and Cube.

Octahedron (fig. 3). Bounded by eight equilateral triangular faces perpendicular to the triad axes of symmetry. The angles between the faces are 70° 32′ and 109° 28′, and the indices are {111}. Spinel, magnetite and gold crystallize in simple octahedra. Combinations of the cube and octahedron are shown in figs. 6-8.

Rhombic dodecahedron (fig. 13). Bounded by twelve rhomb-shaped faces parallel to the six dodecahedral planes of symmetry. The angles between the normals to adjacent faces are 60°, and between other pairs of faces 90°; the indices are {110}. Garnet frequently crystallizes in this form. Fig. 14 shows the rhombic dodecahedron in combination with the octahedron.

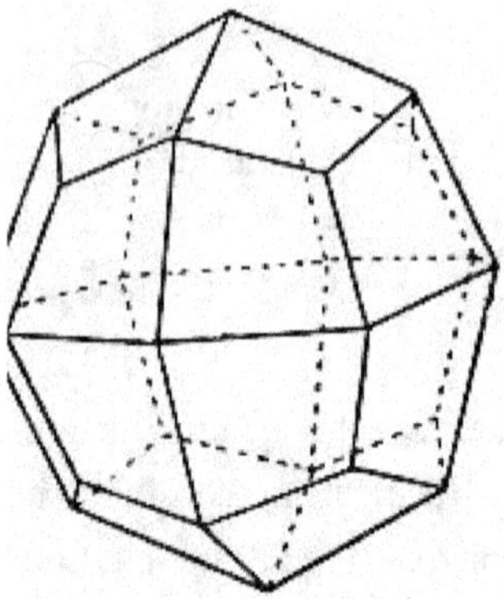

FIG. 17.—Icositetrahedron.

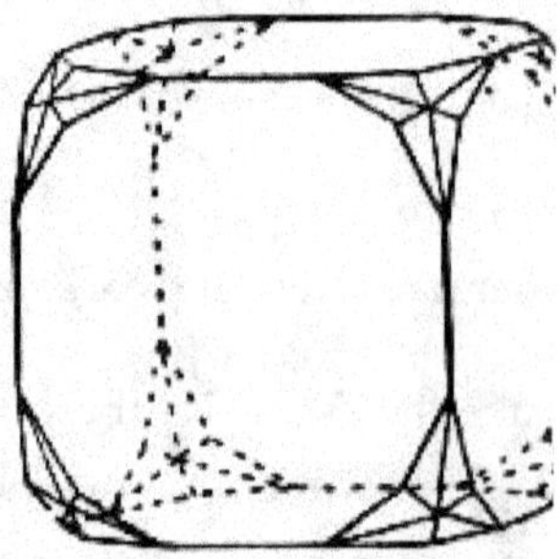

FIG. 18.—Combination of Icositetrahedron and Cube.

In these three simple forms of the cubic system (which are shown in combination in fig. 11) the angles between the faces and the indices are fixed and are the same in all crystals; in the four remaining simple forms they are variable.

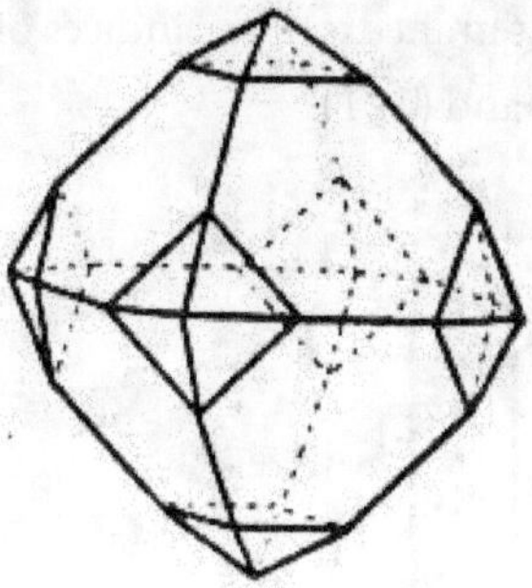

FIG. 19.—Combination of Icositetrahedron and Octahedron.

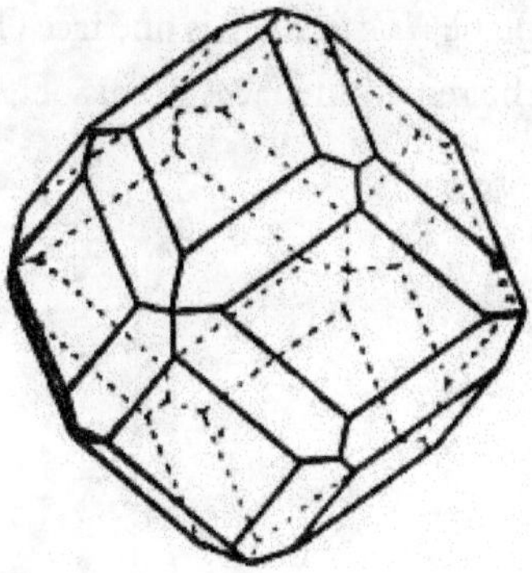

FIG. 20.—Combination of Icositetrahedron {211} and Rhombic Dodecahedron.

Triakis-octahedron (three-faced octahedron) (fig. 15). This solid is bounded by twenty-four isosceles triangles, and may be considered as an octahedron with a low triangular pyramid on each of its faces. As the inclinations of the faces may vary there is a series of these forms with the indices {221}, {331}, {332}, &c. or in general {*hhk*}.

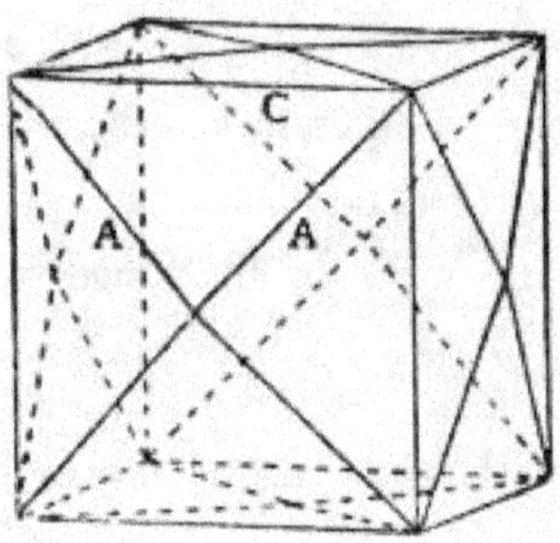

FIG. 21.—Tetrakis-hexahedron.

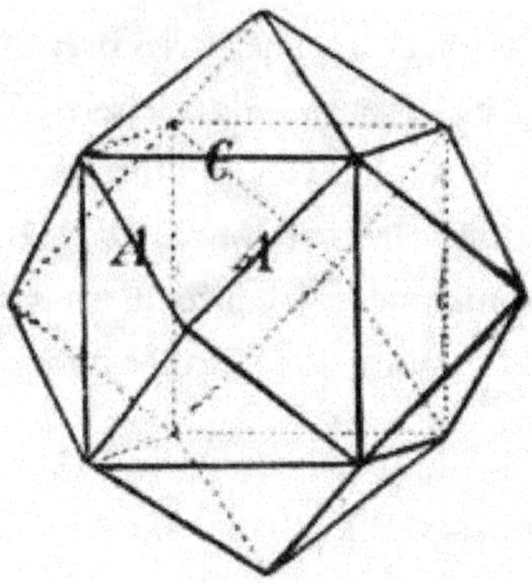

FIG. 22.—Tetrakis-hexahedron.

Icositetrahedron (fig. 17). Bounded by twenty-four trapezoidal faces, and hence sometimes called a "trapezohedron." The indices are {211}, {311}, {322}, &c., or in general {*hkk*}. Analcite, leucite and garnet often crystallize in the simple form {211}. Combinations are shown in figs. 18-20. The plane *ABe* in fig. 9 is one face (112) of an icositetrahedron; the indices of the remaining faces in this octant being (211) and (121).

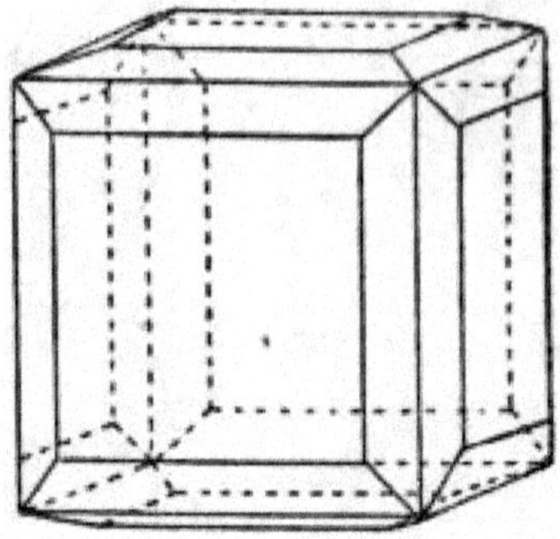

FIG. 23.—Combination of Tetrakis-hexahedron and Cube.

Tetrakis-hexahedron (four-faced cube) (figs. 21 and 22). Like the triakis-octahedron this solid is also bounded by twenty-four isosceles triangles, but here grouped in fours over the cubic faces. The two figures show how, with different inclinations of the faces, the form may vary, approximating in fig. 21 to the cube and in fig. 22 to the rhombic dodecahedron. The angles over the edges lettered *A* are different from the angles over the edges lettered *C*. Each face is parallel to one of the crystallographic axes and intercepts the two others in different lengths; the indices are therefore {210}, {310}, {320}, &c., in general {*hko*}. Fluorspar sometimes crystallizes in the simple form {310}; more usually, however, in combination with the cube (fig. 23).

Hexakis-octahedron (fig. 24). Here each face of the octahedron is replaced by six scalene triangles, so that altogether there are forty-eight faces. This is the greatest number of faces possible for any simple form in crystals. The faces are all oblique to the planes and axes of symmetry, and they intercept the three crystallographic axes in different lengths, hence the indices are all unequal, being in general {*hkl*}, or in particular cases {321}, {421}, {432}, &c. Such a form is known as the "general form" of the class. The interfacial angles over the three edges of each triangle are all different.

These forms usually exist only in combination with other cubic forms (for example, fig. 25), but {421} has been observed as a simple form on fluorspar.

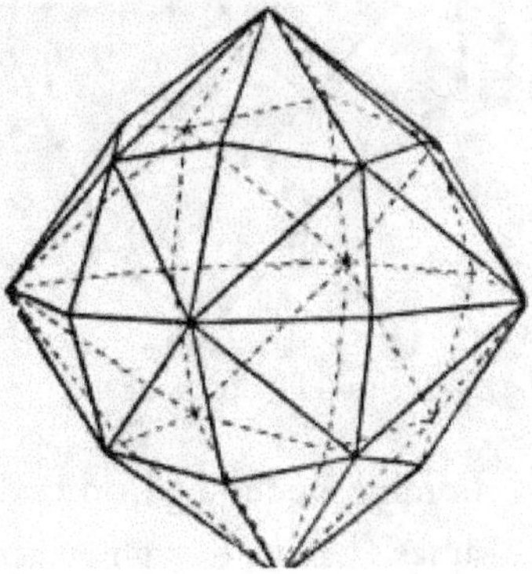

FIG. 24.—Hexakis-octahedron.

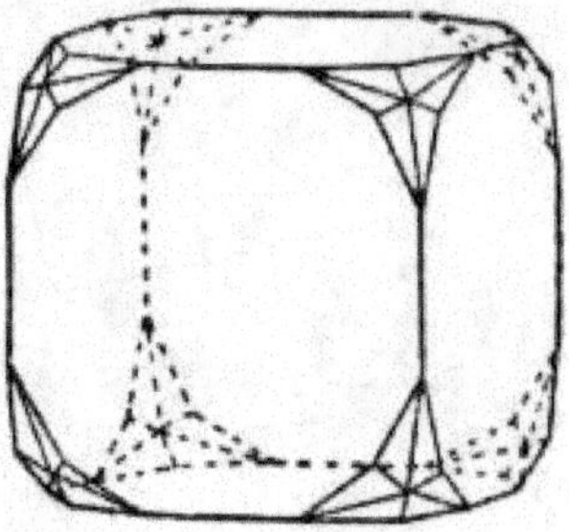

FIG. 25.—Combination of Hexakis-octahedron and Cube.

Several examples of substances which crystallize in this class have been mentioned above under the different forms; many others might be cited—for instance, the metals iron, copper, silver, gold, platinum, lead, mercury, and the non-metallic elements silicon and phosphorus.

Tetrahedral Class

(Tetrahedral-hemihedral; Hexakis-tetrahedral).

In this class there is no centre of symmetry nor cubic planes of symmetry; the three tetrad axes become dyad axes of symmetry, and the four triad axes are polar, *i.e.* they are associated with different faces at their two ends. The other elements of symmetry (six dodecahedral planes and six dyad axes) are the same as in the last class.

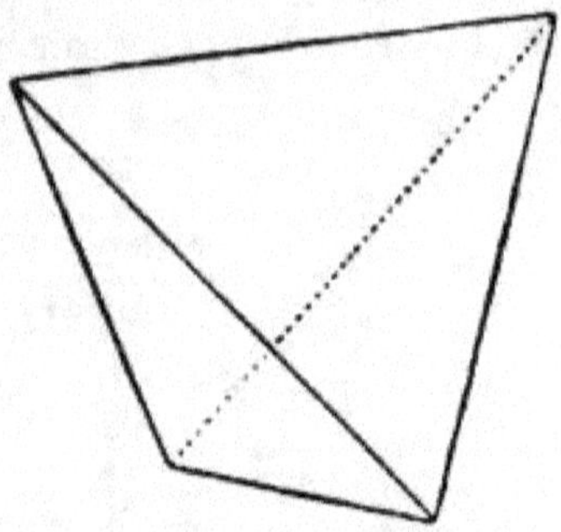

FIG. 26.—Tetrahedron.

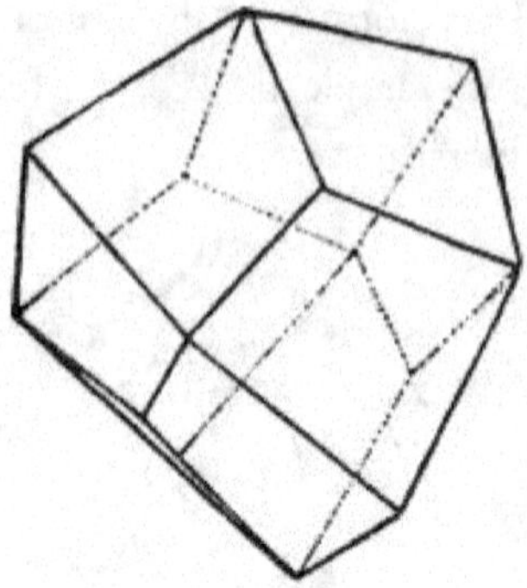

FIG. 27.—Deltoid Dodecahedron.

Of the seven simple forms, the cube, rhombic dodecahedron and tetrakis-hexahedron are geometrically the same as before, though on actual crystals the faces will have different surface characters. For instance, the cube faces will be striated parallel to only one of the diagonals (fig. 90), and etched figures on this face will be symmetrical with respect to two lines, instead of four as in the last class. The remaining simple forms have, however, only half the number of faces as the corresponding form in the last class, and are spoken of as "hemihedral with inclined faces."

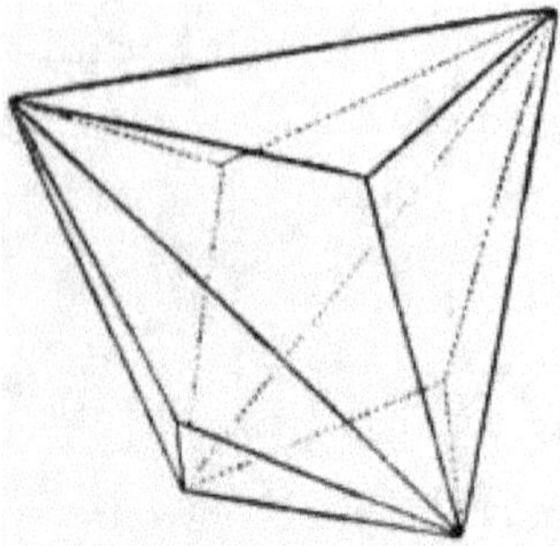

FIG. 28.—Triakis-tetrahedron.

FIG. 29.—Hexakis-tetrahedron.

Tetrahedron (fig. 26). This is bounded by four equilateral triangles and is identical with the regular tetrahedron of geometry. The angles between the normals to the faces are 109° 28′. It may be derived from the octahedron by suppressing the alternate faces.

Deltoid[1] dodecahedron (fig. 27). This is the hemihedral form of the triakis-octahedron; it has the indices {*hhk*} and is bounded by twelve trapezoidal faces.

Triakis-tetrahedron (fig. 28). The hemihedral form {*hkk*} of the icositetrahedron; it is bounded by twelve isosceles triangles arranged in threes over the tetrahedron faces.

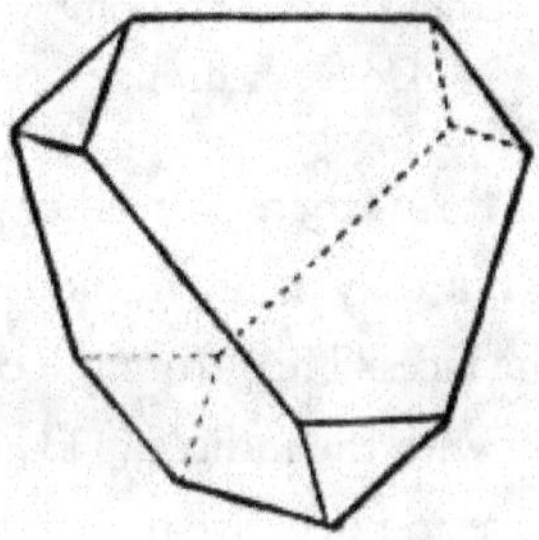

FIG. 30.—Combination of two Tetrahedra.

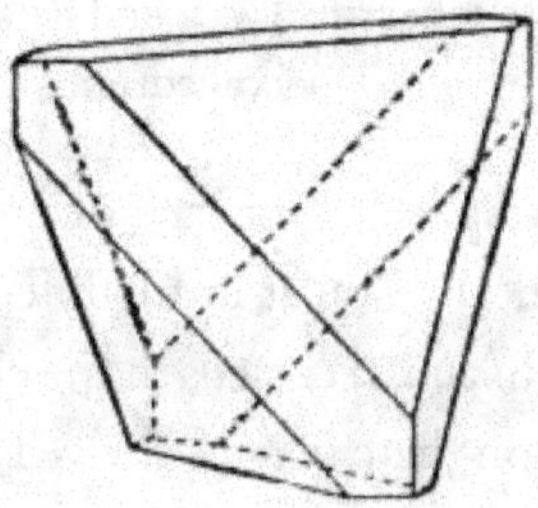

FIG. 31.—Combination of Tetrahedron and Cube.

Hexakis-tetrahedron (fig. 29). The hemihedral form {*hkl*} of the hexakis-octahedron; it is bounded by twenty-four scalene triangles and is the general form of the class.

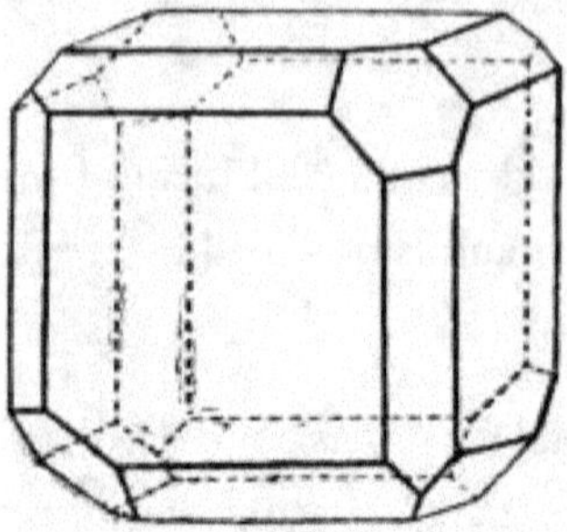

FIG. 32.—Combination of Tetrahedron, Cube and Rhombic Dodecahedron.

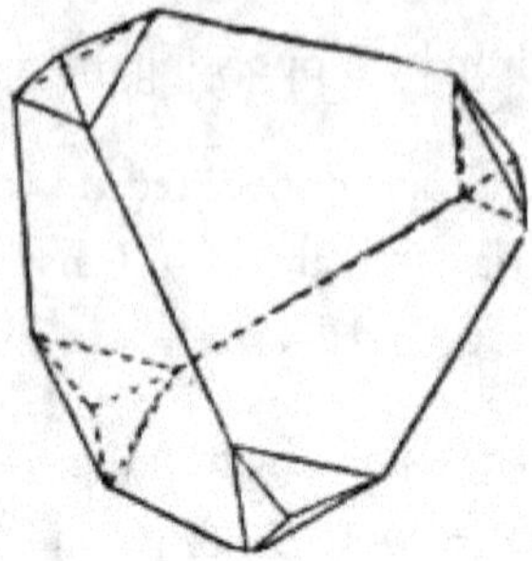

FIG. 33.—Combination of Tetrahedron and Rhombic Dodecahedron.

Corresponding to each of these hemihedral forms there is another geometrically similar form, differing, however, not only in orientation, but also in actual crystals in the characters of the faces. Thus from the octahedron there may be derived two tetrahedra with the indices {111} and {111}, which may be distinguished as positive and negative respectively. Fig. 30 shows a combination of these two tetrahedra, and represents a crystal of blende, in which the four larger faces are dull and striated, whilst the four smaller are bright and smooth. Figs. 31-33 illustrate other tetrahedral combinations.

Tetrahedrite, blende, diamond, boracite and pharmacosiderite are substances which crystallize in this class.

Pyritohedral[2] Class

(Parallel-faced hemihedral; Dyakis-dodecahedral).

Crystals of this class possess three cubic planes of symmetry but no dodecahedral planes. There are only three dyad axes of symmetry, which coincide with the crystallographic axes; in addition there are three triad axes and a centre of symmetry.

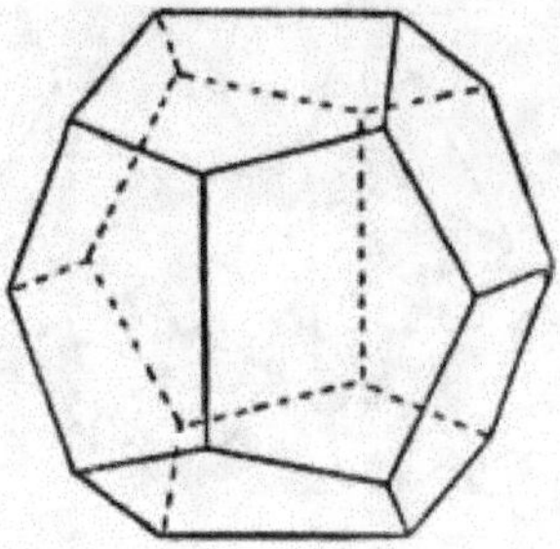

FIG. 34. Pentagonal Dodecahedron.

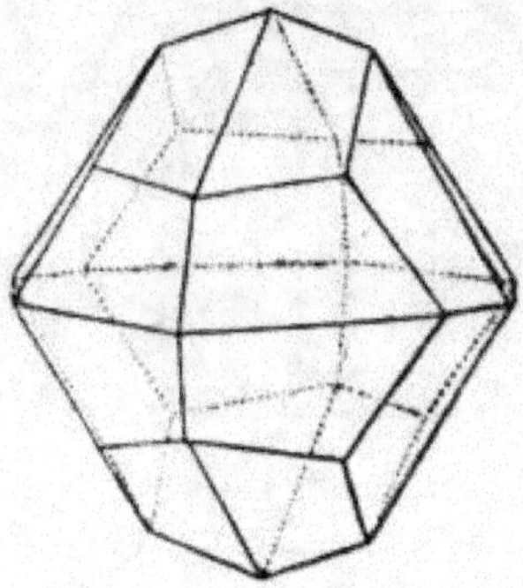

FIG. 35. Dyakis-dodecahedron.

Here the cube, octahedron, rhombic dodecahedron, triakis-octahedron and icositetrahedron are geometrically the same as in the first class. The characters of the faces will, however, be different; thus the cube faces will be striated parallel to one edge only (fig. 89), and triangular markings on the octahedron faces will be placed obliquely to the edges. The remaining simple forms are "hemihedral with parallel faces," and from the corresponding holohedral forms two hemihedral forms, a positive and a negative, may be derived.

Pentagonal dodecahedron (fig. 34). This is bounded by twelve pentagonal faces, but these are not regular pentagons, and the angles over the three sets of different edges are different. The regular dodecahedron of geometry, contained by twelve regular pentagons, is not a possible form in crystals. The indices are {*hko*}: as a simple form {210} is of very common occurrence in pyrites.

Dyakis-dodecahedron (fig. 35). This is the hemihedral form of the hexakis-octahedron and has the indices {*hkl*}; it is bounded by twenty-four faces. As a simple form {321} is met with in pyrites.

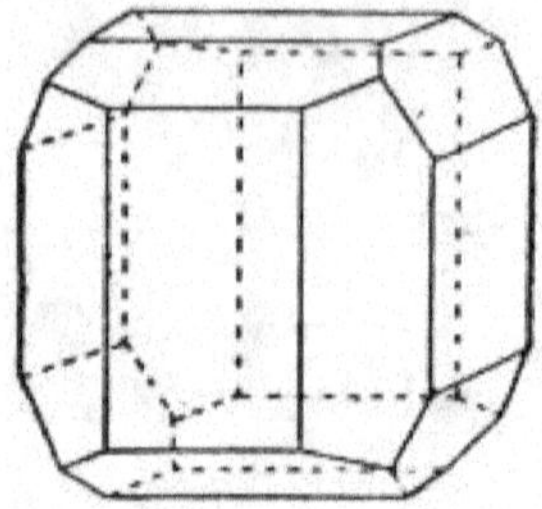

FIG. 36.—Combination of Pentagonal Dodecahedron and Cube.

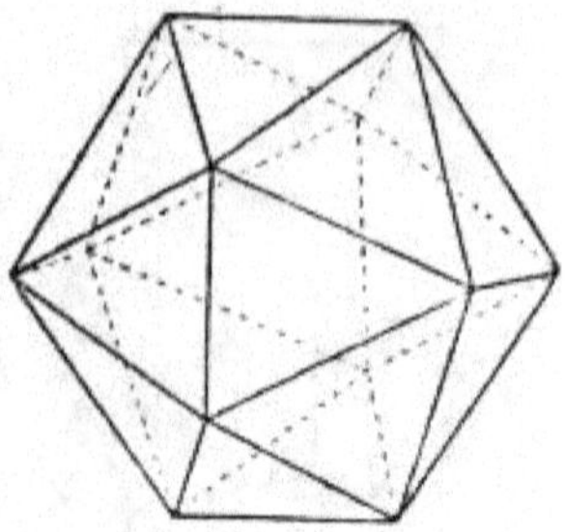

FIG. 37.—Combination of Pentagonal Dodecahedron and Octahedron.

Combinations (figs. 36-39) of these forms with the cube and the octahedron are common in pyrites. Fig. 37 resembles in general appearance the regular icosahedron of geometry, but only eight of the faces are equilateral triangles. Cobaltite, smaltite and other sulphides and sulpharsenides of the pyrites group of minerals crystallize in these forms. The alums also belong to this class; from an aqueous solution they crystallize as simple octahedra, sometimes with subordinate faces of the cube and rhombic dodecahedron, but from an acid solution as octahedra combined with the pentagonal dodecahedron {210}.

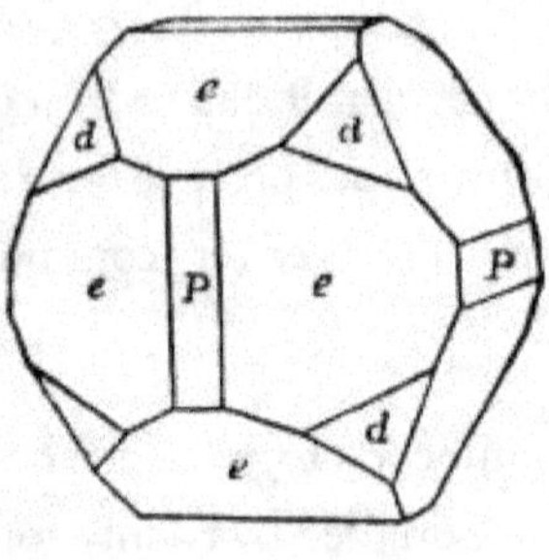

FIG. 38.—Combination of Pentagonal Dodecahedron, Cube and Octahedron.

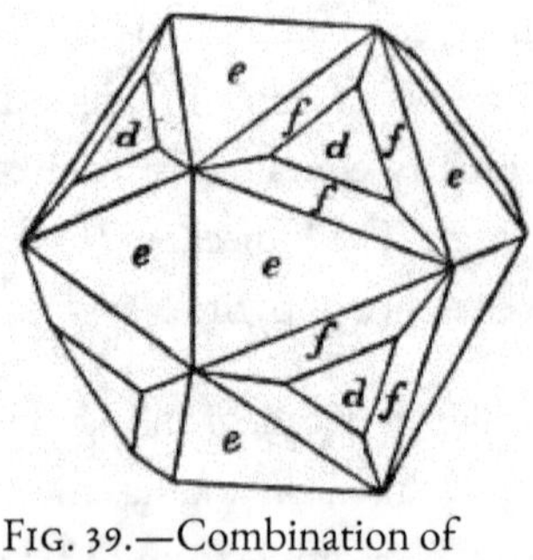

FIG. 39.—Combination of Pentagonal Dodecahedron *e* {210}, Dyakis-dodecahedron *f* {321}, and Octahedron *d* {111}.

Plagihedral[3] Class

(Plagihedral-hemihedral; Pentagonal icositetrahedral; Gyroidal[4]).

In this class there are the full number of axes of symmetry (three tetrad, four triad and six dyad), but no planes of symmetry and no centre of symmetry.

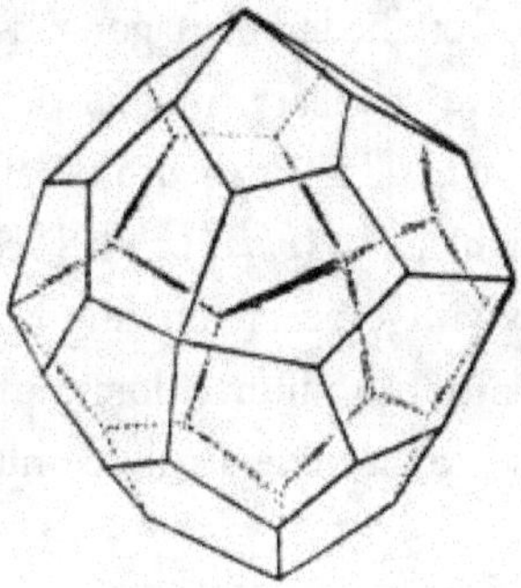

FIG. 40.—Pentagonal Icositetrahedron.

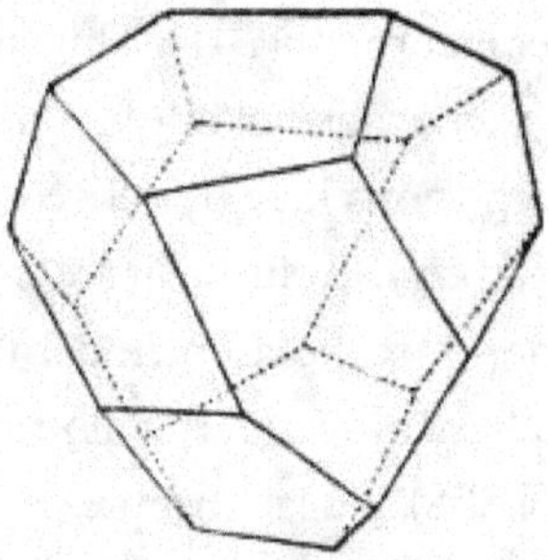

FIG. 41.—Tetrahedral Pentagonal Dodecahedron.

Pentagonal icositetrahedron (fig. 40). This is the only simple form in this class which differs geometrically from those of the holosymmetric class. By suppressing either one or other set of alternate faces of the hexakis-octahedron two pentagonal icositetrahedra {*hkl*} and {*khl*} are derived. These are each bounded by twenty-four irregular pentagons, and although similar to each other they are respectively right- and left-handed, one being the mirror image of the other; such similar but nonsuperposable forms are said to be enantiomorphous (ἐναντίος, opposite, and μορφή, form), and crystals showing such forms sometimes rotate the plane of polarization of plane-polarized light. Faces of a pentagonal icositetrahedron with high indices have been very rarely observed on crystals of cuprite, potassium chloride and ammonium chloride, but none of these are circular polarizing.

Tetartohedral Class

(Tetrahedral pentagonal dodecahedral).

Here, in addition to four polar triad axes, the only other elements of symmetry are three dyad axes, which coincide with the crystallographic axes. Six of the simple forms, the cube, tetrahedron, rhombic dodecahedron, deltoid dodecahedron, triakis-tetrahedron and pentagonal dodecahedron, are geometrically the same in this class as in either the tetrahedral or pyritohedral classes. The general form is the Tetrahedral pentagonal dodecahedron (fig. 41). This is bounded by twelve irregular pentagons, and is a tetartohedral or quarter-faced form of the hexakis-octahedron. Four such forms may be derived, the indices of which are $\{hkl\}$, $\{khl\}$, $\{hkl\}$ and $\{khl\}$; the first pair are enantiomorphous with respect to one another, and so are the last pair. Barium nitrate, lead nitrate, sodium chlorate and sodium bromate crystallize in this class, as also do the minerals ullmannite (NiSbS) and langbeinite ($K_2Mg_2(SO_4)_3$).

2. TETRAGONAL SYSTEM

(Pyramidal; Quadratic; Dimetric).

In this system the three crystallographic axes are all at right angles, but while two are equal in length and interchangeable the third is of a different length. The unequal axis is spoken of as the principal axis or morphological axis of the crystal, and it is always placed in a vertical position; in five of the seven classes of this system it coincides with the single tetrad axis of symmetry.

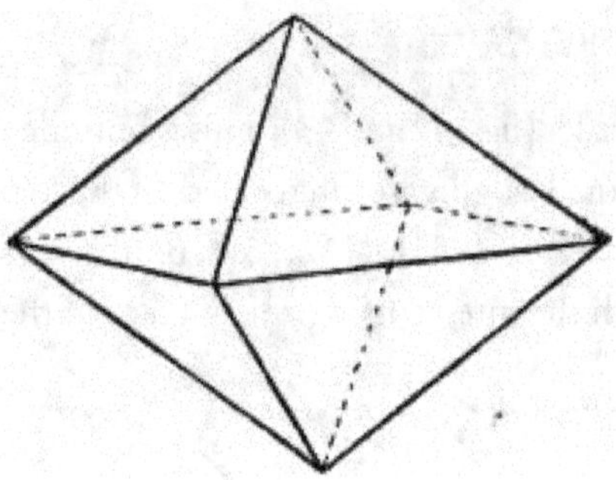

FIG. 42.

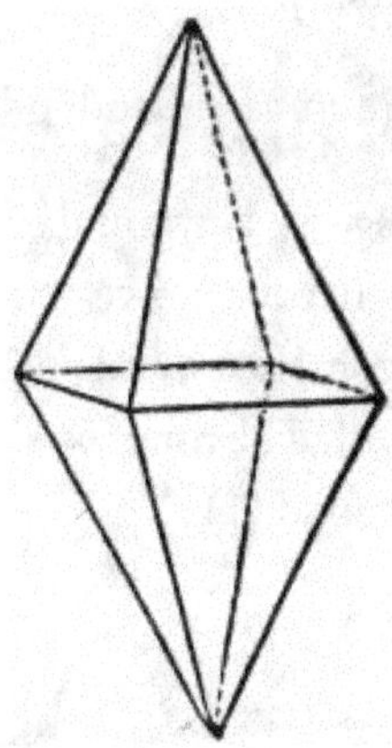

FIG. 43.

Tetragonal Bipyramids.

The parameters are $a : a : c$, where a refers to the two equal horizontal axes, and c to the vertical axis; c may be either shorter (as in fig. 42) or longer (fig. 43) than a. The ratio $a : c$ is spoken of as the axial ratio of a crystal, and it is dependent on the angles between the faces. In all crystals of the same substance this ratio is constant, and is characteristic of the substance; for other substances crystallizing in the tetragonal system it will be different.

For example, in cassiterite it is given as $a : c = 1 : 0.67232$ or simply as $c = 0.67232$, a being unity; and in anatase as $c = 1.7771$.

Holosymmetric Class

(Holohedral; Ditetragonal bipyramidal).

Crystals of this class are symmetrical with respect to five planes, which are of three kinds; one is perpendicular to the principal axis, and the other four intersect in it; of the latter, two are perpendicular to the equal crystallographic axes, while the two others bisect the angles between them. There are five axes of symmetry, one tetrad and two pairs of dyad, each perpendicular to a plane of symmetry. Finally, there is a centre of symmetry.

There are seven kinds of simple forms, viz.:—

Tetragonal bipyramid of the first order (figs. 42 and 43). This is bounded by eight equal isosceles triangles. Equal lengths are intercepted on the two horizontal axes, and the indices are {111}, {221}, {112}, &c., or in general {*hhl*}. The parametral plane with the intercepts $a : a : c$ is a face of the bipyramid {111}.

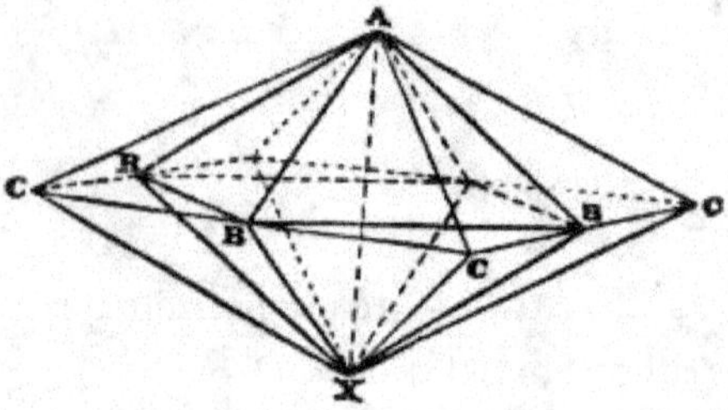

FIG. 44.

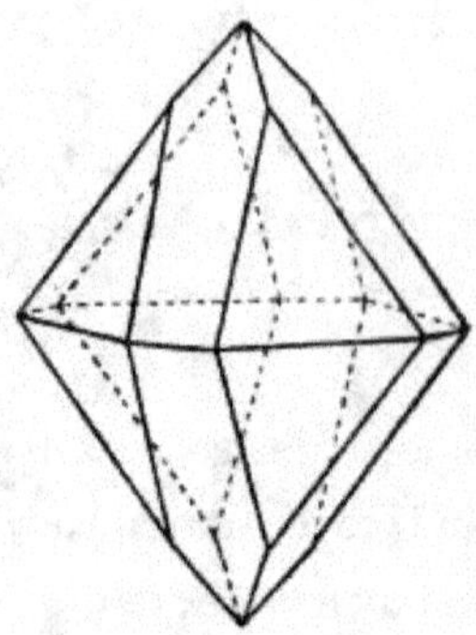

FIG. 45.

Tetragonal Bipyramids of the first and second orders.

Tetragonal bipyramid of the second order. This is also bounded by eight equal isosceles triangles, but differs from the last form in its position, four of the faces being parallel to each of the horizontal axes; the indices are therefore {101}, {201}, {102}, &c., or {*hol*}.

Fig. 44 shows the relation between the tetragonal bipyramids of the first and second orders when the indices are {111} and {101} respectively: *ABB* is the face (111), and *ACC* is (101). A combination of these two forms is shown in fig. 45.

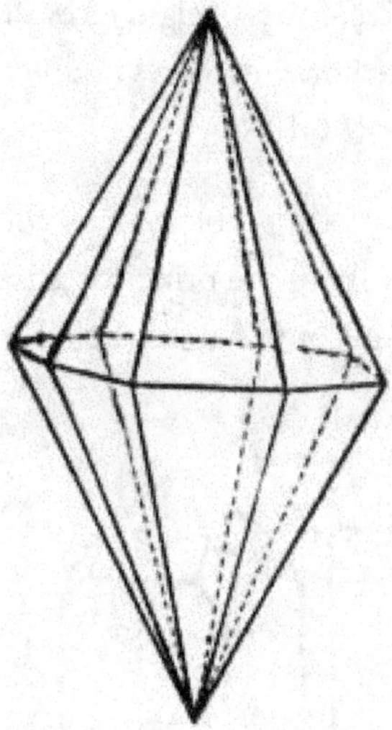

FIG. 46.—Ditetragonal Bipyramid.

Ditetragonal bipyramid (fig. 46). This is the general form; it is bounded by sixteen scalene triangles, and all the indices are unequal, being {321}, &c., or {*hkl*}.

Tetragonal prism of the first order. The four faces intersect the horizontal axes in equal lengths and are parallel to the principal axis; the indices are therefore {110}. This form does not enclose space, and is therefore called an "open form" to distinguish it from a "closed form" like the tetragonal bipyramids and all the forms of the cubic system. An open form can exist only in combination with other forms; thus fig. 47 is a combination of the tetragonal prism {110} with the basal pinacoid {001}. If the faces (110) and (001) are of equal size such a figure will be geometrically a cube, since all the angles are right angles; the variety of apophyllite known as tesselite crystallizes in this form.

Tetragonal prism of the second order. This has the same number of faces as the last prism, but differs in position; each face being parallel to the vertical axis and one of the horizontal axes; the indices are {100}.

Ditetragonal prism. This consists of eight faces all parallel to the principal axis and intercepting the horizontal axes in different lengths; the indices are {210}, {320}, &c., or {*hko*}.

Basal pinacoid (from πίναξ, a tablet). This consists of a single pair of parallel faces perpendicular to the principal axis. It is therefore an open form and can exist only in combination (fig. 47).

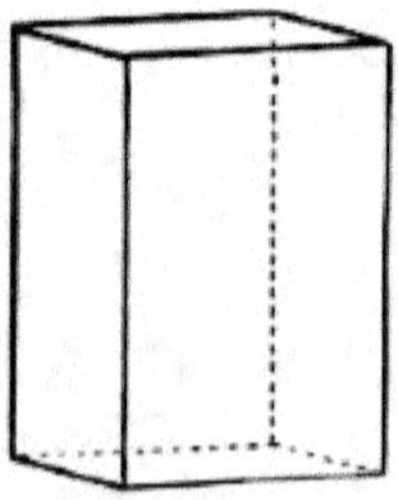

FIG. 47.
Combination of
Tetragonal Prism
and Basal Pinacoid.

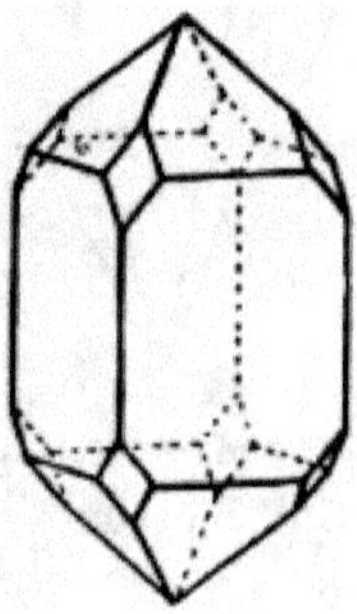

FIG. 48.

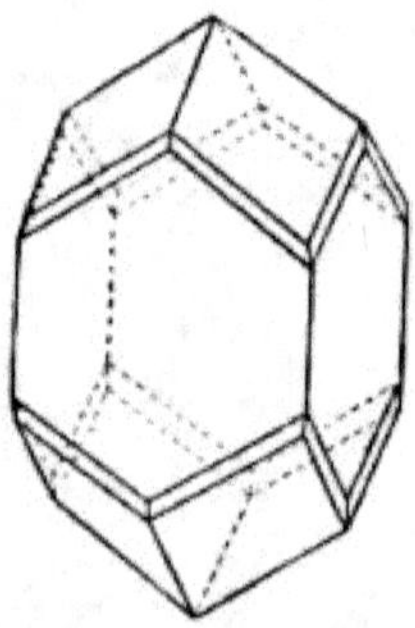

FIG. 49.

Combinations of Tetragonal Prisms and Pyramids.

Combinations of holohedral tetragonal forms are shown in figs. 47-49; fig. 48 is a combination of a bipyramid of the first order with one of the second order and the prism of the first order; fig. 49 a combination of a bipyramid of the first order with a ditetragonal bipyramid and the prism of the second order. Compare also figs. 87 and 88.

Examples of substances which crystallize in this class are cassiterite, rutile, anatase, zircon, thorite, vesuvianite, apophyllite, phosgenite, also boron, tin, mercuric iodide.

Scalenohedral Class

(Bisphenoidal-hemihedral).

Here there are only three dyad axes and two planes of symmetry, the former coinciding with the crystallographic axes and the latter bisecting the angles between the horizontal pair. The dyad axis of symmetry, which in this class coincides with the principal axis of the crystal, has certain of the characters of a tetrad axis, and is sometimes called a tetrad axis of "alternating symmetry"; a face on the upper half of the crystal if rotated through 90° about this axis and reflected across the equatorial plane falls into the position of a face on the lower half of the crystal. This kind of symmetry, with simultaneous rotation about an axis and reflection across a plane, is also called "composite symmetry."

In this class all except two of the simple forms are geometrically the same as in the holosymmetric class.

Bisphenoid (σφήν, a wedge) (fig. 50). This is a double wedge-shaped solid bounded by four equal isosceles triangles; it has the indices {111}, {211}, {112}, &c., or in general {*hhl*}. By suppressing either one or other set of alternate faces of the tetragonal bipyramid of the first order (fig. 42) two bisphenoids are derived, in the same way that two tetrahedra are derived from the regular octahedron.

Tetragonal scalenohedron or ditetragonal bisphenoid (fig. 51). This is bounded by eight scalene triangles and has the indices {*hkl*}. It may be considered as the hemihedral form of the ditetragonal bipyramid.

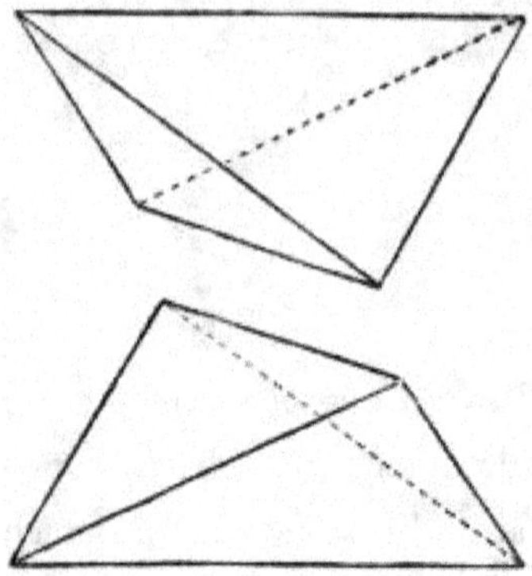

FIG. 50.—Tetragonal Bisphenoids.

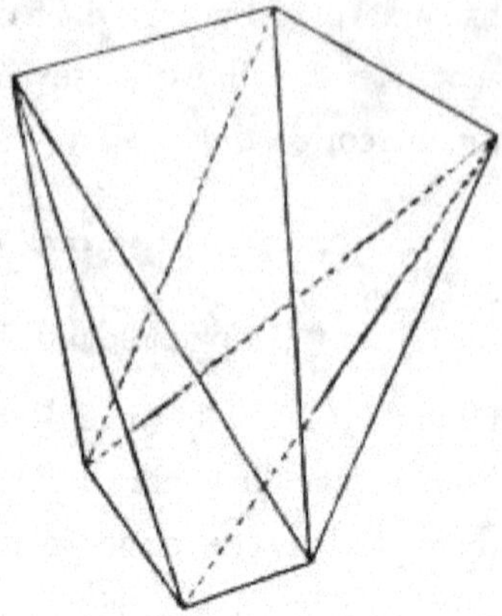

FIG. 51.—Tetragonal Scalenohedron.

The crystal of chalcopyrite ($CuFeS_2$) represented in fig. 52 is a combination of two bisphenoids (*P* and *P'*), two bipyramids of the second order (*b* and *c*), and the basal pinacoid (*a*). Stannite (Cu_2FeSnS_4), acid potassium phosphate (H_2KPO_4), mercuric cyanide, and urea ($CO(NH_2)_2$) also crystallize in this class.

Bipyramidal Class

(Parallel-faced hemihedral).

The elements of symmetry are a tetrad axis with a plane perpendicular to it, and a centre of symmetry. The simple forms are the same here as in the holosymmetric class, except the prism {*hko*}, which has only four faces, and the bipyramid {*hkl*}, which has eight faces and is distinguished as a "tetragonal pyramid of the third order."

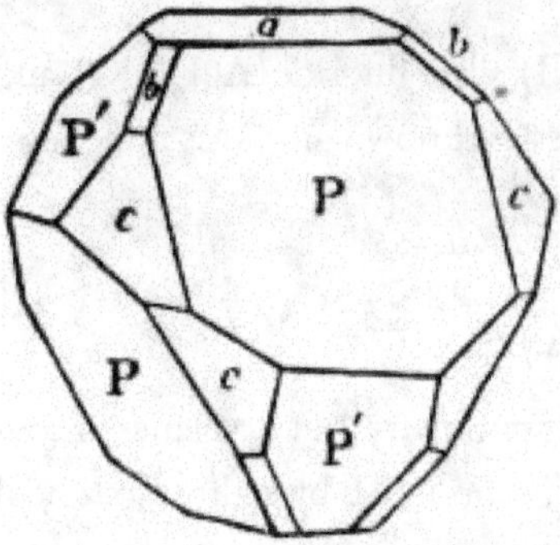

FIG. 52.—Crystal of Chalcopyrite.

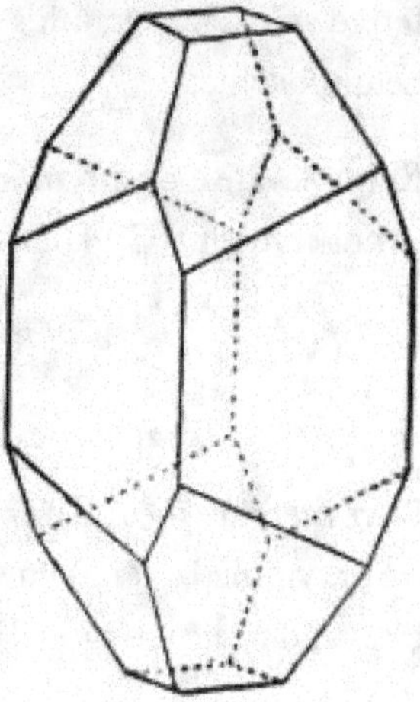

FIG. 53.—Crystal of Fergusonite.

Fig. 53 shows a combination of a tetragonal prism of the first order with a tetragonal bipyramid of the third order and the basal pinacoid, and represents a crystal of fergusonite. Scheelite (*q.v.*), scapolite (*q.v.*), and erythrite ($C_4H_{10}O_4$) also crystallize in this class.

Pyramidal Class

(Hemimorphic-tetartohedral).

Here the only element of symmetry is the tetrad axis. The pyramids of the first {*hhl*}, second {*hol*} and third {*hkl*} orders have each only four faces at one or other end of the crystal, and are hemimorphic. All the simple forms are thus open forms.

Examples are wulfenite ($PbMoO_4$) and barium antimonyl dextro-tartrate ($(Ba(SbO)_2(C_4H_4O_6)\cdot H_2O)$).

Ditetragonal Pyramidal Class

(Hemimorphic-hemihedral).

Here there are two pairs of vertical planes of symmetry intersecting in the tetrad axis. The pyramids {*hhl*} and {*hol*} and the bipyramid {*hkl*} are all hemimorphic.

Examples are iodosuccimide ($C_4H_4O_2NI$), silver fluoride ($AgF{\cdot}H_2O$), and penta-erythrite ($C_5H_{12}O_4$). No examples are known amongst minerals.

Trapezohedral Class

(Trapezohedral-hemihedral).

Here there are the full number of axes of symmetry, but no planes or centre of symmetry. The general form {*hkl*} is bounded by eight trapezoidal faces and is the tetragonal trapezohedron.

Examples are nickel sulphate ($NiSO_4{\cdot}6H_2O$), guanidine carbonate ($((CH_5N_3)_2H_2CO_3)$), strychnine sulphate ($((C_{21}H_{22}N_2O_2)_2{\cdot}H_2SO_4{\cdot}6H_2O)$).

Bisphenoidal Class

(Bisphenoidal-tetartohedral).

Here there is only a single dyad axis of symmetry, which coincides with the principal axis. All the forms, except the prisms and basal pinacoid, are sphenoids. Crystals possessing this type of symmetry have not yet been observed.

3. ORTHORHOMBIC SYSTEM

(Rhombic; Prismatic; Trimetric).

In this system the three crystallographic axes are all at right angles, but they are of different lengths and not interchangeable. The parameters, or axial ratios, are $a : b : c$, these referring to the axes OX, OY and OZ respectively. The choice of a vertical axis, $OZ = c$, is arbitrary, and it is customary to place the longer of the two horizontal axes from left to right ($OY = b$) and take it as unity: this is called the "macro-axis" or "macro-diagonal" (from μακρός, long), whilst the shorter horizontal axis ($OX = a$) is called the "brachy-axis" or "brachy-diagonal" (from βραχύς, short). The axial ratios are constant for crystals of any one substance and are characteristic of it; for example, in barytes ($BaSO_4$), $a : b : c = 0.8152 : 1 : 1.3136$; in anglesite ($PbSO_4$), $a : b : c = 0.7852: 1 : 1.2894$; in cerussite ($PbCO_3$), $a : b : c = 0.6100 : 1 : 0.7230$.

There are three symmetry-classes in this system:—

Holohedral Class

(Holohedral; Bipyramidal).

Here there are three dissimilar dyad axes of symmetry, each coinciding with a crystallographic axis; perpendicular to them are three dissimilar planes of symmetry; there is also a centre of symmetry. There are seven kinds of simple forms:—

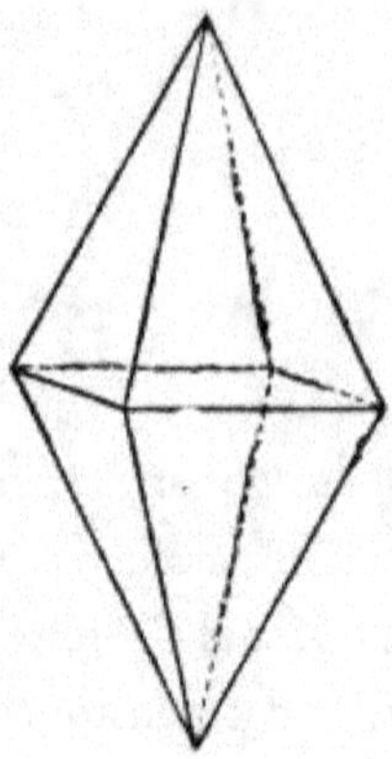

FIG. 54.

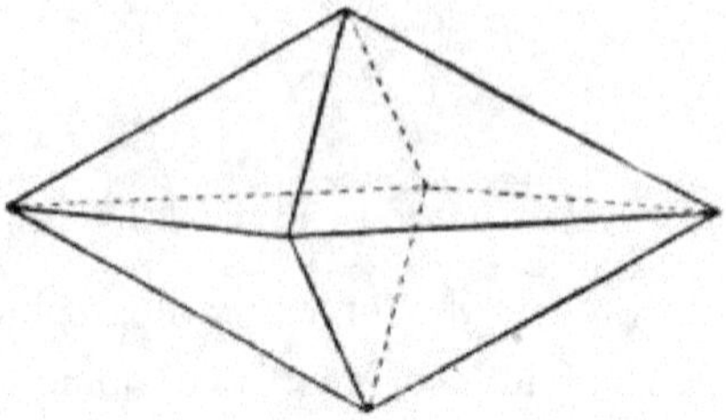

FIG. 55.

Orthorhombic Bipyramids.

Bipyramid (figs. 54 and 55). This is the general form and is bounded by eight scalene triangles; the indices are {111}, {211}, {221}, {112}, {321}, {123}, &c., or in general {*hkl*}. The crystallographic axes join opposite corners of these pyramids and in the fundamental bipyramid {111} the parametral plane has the intercepts $a : b : c$. This is the only closed form in this class; the others are open forms and can exist only in combination. Sulphur often crystallizes in simple bipyramids.

Prism. This consists of four faces parallel to the vertical axis and intercepting the horizontal axes in the lengths a and b or in any multiples of these; the indices are therefore {110}, {210}, {120} or {*hko*}.

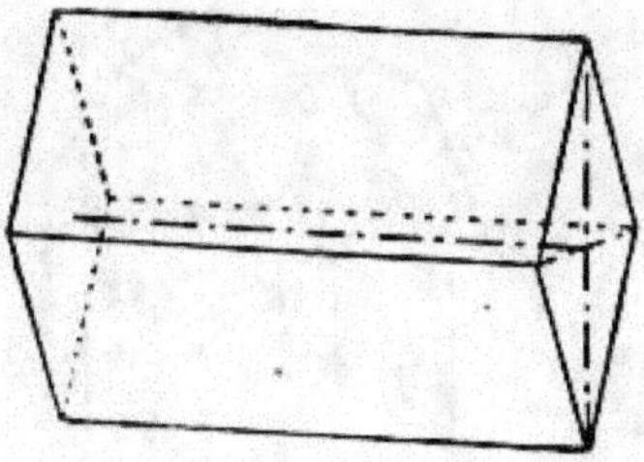

FIG. 56.—Macro-prism and Brachy-pinacoid.

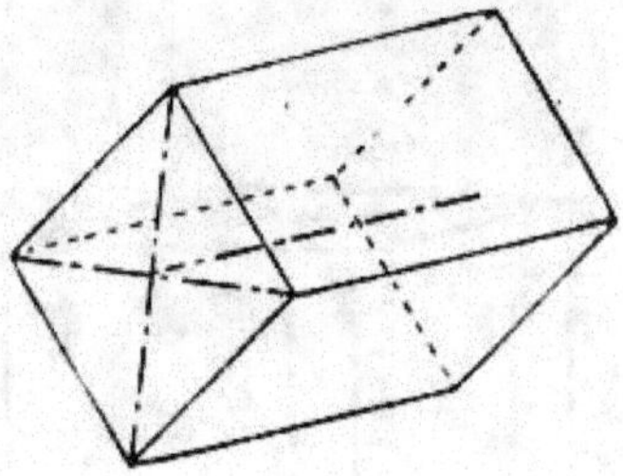

FIG. 57.—Brachy-prism and Macro-pinacoid.

Macro-prism. This consists of four faces parallel to the macro-axis, and has the indices {101}, {201} ... or {*hol*}.

Brachy-prism. This consists of four faces parallel to the brachy-axis, and has the indices {011}, {021} ... {*okl*}. The macro- and brachy-prisms are often called "domes."

Basal pinacoid, consisting of a pair of parallel faces perpendicular to the vertical axis; the indices are {001}. The macro-pinacoid {100} and the brachy-pinacoid {010} each consist of a pair of parallel faces respectively parallel to the macro- and the brachy-axis.

Figs. 56-58 show combinations of these six open forms, and fig. 59 a combination of the macro-pinacoid (*a*), brachy-pinacoid (*b*), a prism (*m*), a macro-prism (*d*), a brachy-prism (*k*), and a bipyramid (*u*).

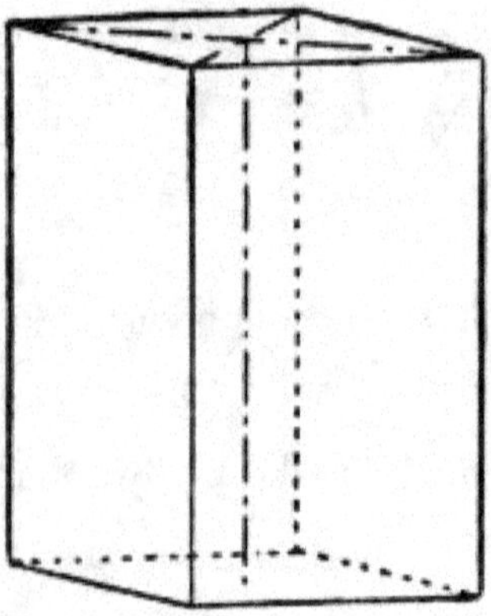

FIG. 58.—Prism and Basal Pinacoid.

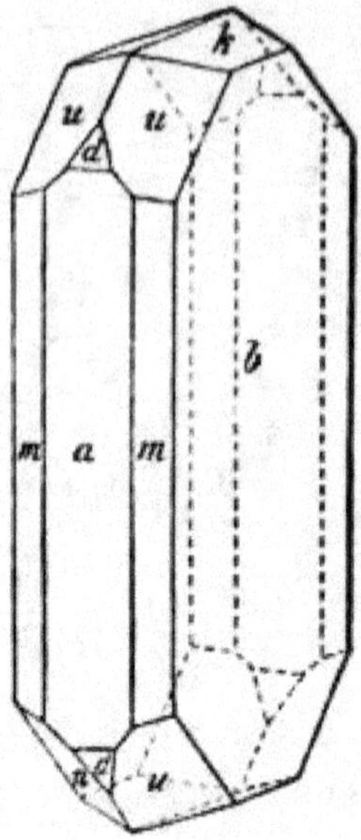

FIG. 59.—Crystal of Hypersthene.

Holohedral Orthorhombic Combinations.

Examples of substances crystallizing in this class are extremely numerous; amongst minerals are sulphur, stibnite, cerussite, chrysoberyl, topaz, olivine, nitre, barytes, columbite and many others; and amongst artificial products iodine, potassium permanganate, potassium sulphate, benzene, barium formate, &c.

Pyramidal Class

(Hemimorphic).

Here there is only one dyad axis in which two planes of symmetry intersect. The crystals are usually so placed that the dyad axis coincides with the vertical crystallographic axis, and the planes of symmetry are also vertical.

The pyramid $\{hkl\}$ has only four faces at one end or other of the crystal. The macro-prism and the brachy-prism of the last class are here represented by the macro-dome and brachy-dome respectively, so called because of the resemblance of the pair of equally sloped faces to the roof of a house. The form {001} is a single plane at the top of the crystal, and is called a "pedion"; the parallel pedion {001}, if present at the lower end of the crystal, constitutes a different form. The prisms $\{hko\}$ and the macro- and brachy-pinacoids are geometrically the same in this class as in the last. Crystals of this class are therefore differently developed at the two ends and are said to be "hemimorphic."

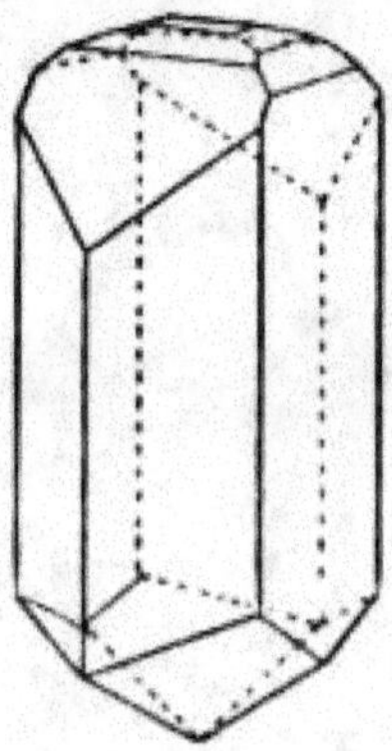

FIG. 60.—Crystal of Hemimorphite.

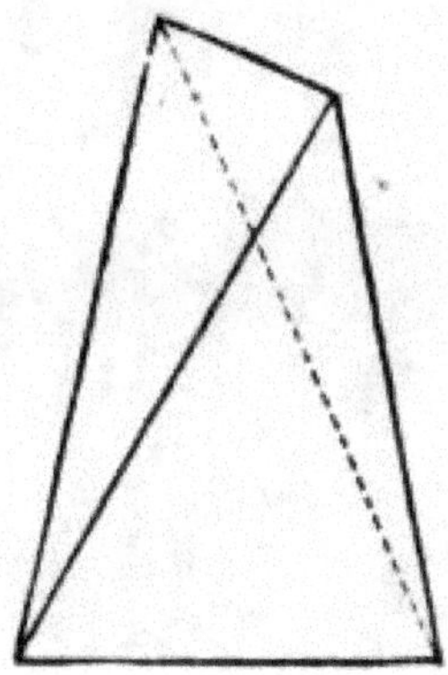

FIG. 61.—Orthorhombic Bisphenoid.

Fig. 60 shows a crystal of the mineral hemimorphite ($H_2Zn_2SiO_5$) which is a combination of the brachy-pinacoid {010} and a prism, with the pedion (001), two brachy-domes and two macro-domes at the upper end, and a pyramid at the lower end. Examples of other substances belonging to this class are struvite ($NH_4MgPO_4{\cdot}6H_2O$), bertrandite ($H_2Be_4Si_2O_9$), resorcin, and picric acid.

Bisphenoidal Class

(Hemihedral).

Here there are three dyad axes, but no planes of symmetry and no centre of symmetry. The general form {*hkl*} is a bisphenoid (fig. 61) bounded by four scalene triangles. The other simple forms are geometrically the same as in the holosymmetric class.

Examples: epsomite (Epsom salts, $MgSO_4 \cdot 7H_2O$), goslarite ($ZnSO_4 \cdot 7H_2O$), silver nitrate, sodium potassium dextro-tartrate (seignette salt, $NaKC_4H_4O_6 \cdot 4H_2O$), potassium antimonyl dextro-tartrate (tartar-emetic, $K(SbO)C_4H_4O_6$), and asparagine ($C_4H_8N_2O_8 \cdot H_2O$).

4. MONOCLINIC5 SYSTEM

(Oblique; Monosymmetric).

In this system two of the angles between the crystallographic axes are right angles, but the third angle is oblique, and the axes are of unequal lengths. The axis which is perpendicular to the other two is taken as $OY = b$ (fig. 62) and is called the ortho-axis or ortho-diagonal. The choice of the other two axes is arbitrary; the vertical axis ($OZ = c$) is usually taken parallel to the edges of a prominently developed prismatic zone, and the clino-axis or clino-diagonal ($OX = a$) parallel to the zone-axis of some other prominent zone on the crystal. The acute angle between the axes OX and OZ is usually denoted as β, and it is necessary to know its magnitude, in addition to the axial ratios $a : b : c$, before the crystal is completely determined. As in other systems, except the cubic, these elements, $a : b : c$ and β, are characteristic of the substance. Thus for gypsum $a : b : c = 0.6899 : 1 : 0.4124$; $\beta = 80° 42'$; for orthoclase $a : b : c = 0.6585 : 1 : 0.5554$; $\beta = 63° 57'$; and for cane-sugar $a : b : c = 1.2595 : 1 : 0.8782$; $\beta = 76° 30'$.

Holosymmetric Class

(Holohedral; Prismatic).

Here there is a single plane of symmetry perpendicular to which is a dyad axis; there is also a centre of symmetry. The dyad axis coincides with the ortho-axis OY, and the vertical axis OZ and the clino-axis OX lie in the plane of symmetry.

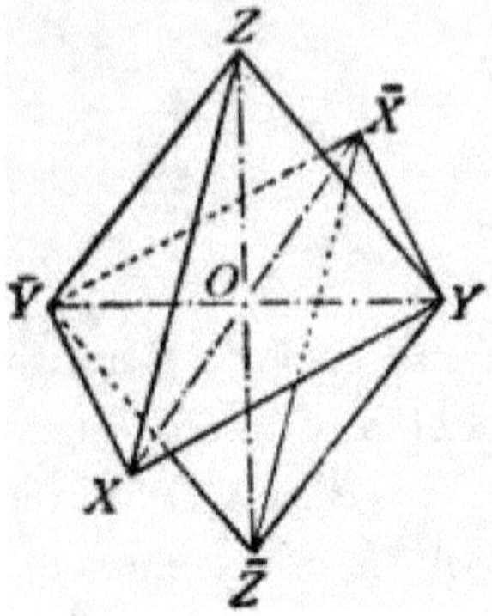

FIG. 62.—Monoclinic Axes and Hemi-pyramid.

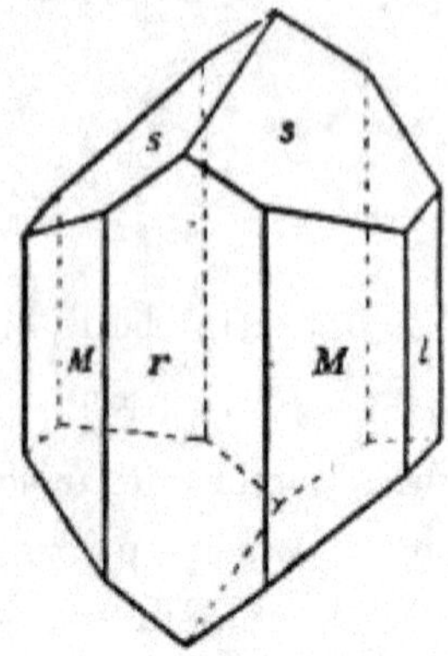

FIG. 63.—Crystal of Augite.

All the forms are open, being either pinacoids or prisms; the former consisting of a pair of parallel faces, and the latter of four faces intersecting in parallel edges and with a rhombic cross-section. The pair of faces parallel to the plane of symmetry is distinguished as the "clino-pinacoid" and has the indices {010}. The other pinacoids are all perpendicular to the plane of symmetry (and parallel to the ortho-axis); the one parallel to the vertical axis is called the "ortho-pinacoid" {100}, whilst that parallel to the clino-axis is the "basal pinacoid" {001}; pinacoids not parallel to the arbitrarily chosen clino- and vertical axes may have the indices {101}, {201}, {102} ... {*hol*} or {101}, {201}, {102} ... {*hol*}, according to whether they lie in the obtuse or the acute axial angle. Of the prisms, those with edges (zone-axis) parallel to the clino-axis, and having indices {011}, {021}, {012} ... {*okl*}, are called "clino-prisms"; those with edges parallel to the vertical axis, and with the indices {110}, {210}, {120} ... {*hko*}, are called simply "prisms." Prisms with edges parallel to neither of the axes *OX* and *OY* have the indices {111}, {221}, {211}, {321} ... {*hkl*} or {111} ... {*hkl*}, and are usually called "hemi-pyramids" (fig. 62); they are distinguished as negative or positive according to whether they lie in the obtuse or the acute axial angle β.

Fig. 63 represents a crystal of augite bounded by the clino-pinacoid (*l*), the ortho-pinacoid (*r*), a prism (*M*), and a hemi-pyramid (*s*).

The substances which crystallize in this class are extremely numerous: amongst minerals are gypsum, orthoclase, the amphiboles, pyroxenes and micas, epidote, monazite, realgar, borax, mirabilite ($Na_2SO_4 \cdot 10H_2O$), melanterite ($FeSO_4 \cdot 7H_2O$) and many others; amongst artificial products are monoclinic sulphur, barium chloride ($BaCl_2 \cdot 2H_2O$), potassium chlorate, potassium ferrocyanide ($K_4Fe(CN)_6 \cdot 3H_2O$), oxalic acid ($C_2O_4H_2 \cdot 2H_2O$), sodium acetate ($NaC_2H_3O_2 \cdot 3H_2O$) and naphthalene.

Hemimorphic Class

(Sphenoidal).

In this class the only element of symmetry is a single dyad axis, which is polar in character, being dissimilar at the two ends.

The form {010} perpendicular to the axis of symmetry consists of a single plane or pedion; the parallel face is dissimilar in character and belongs to the pedion {010}. The pinacoids {100}, {001}, {*hol*} and {*hol*} parallel to the axis of symmetry are geometrically the same in this class as in the holosymmetric class. The remaining forms consist each of only two planes on the same side of the axial plane *XOZ* and equally inclined to the dyad axis (*e.g.* in fig. 62 the two planes *XYZ* and *XYZ*); such a wedge-shaped form is sometimes called a sphenoid.

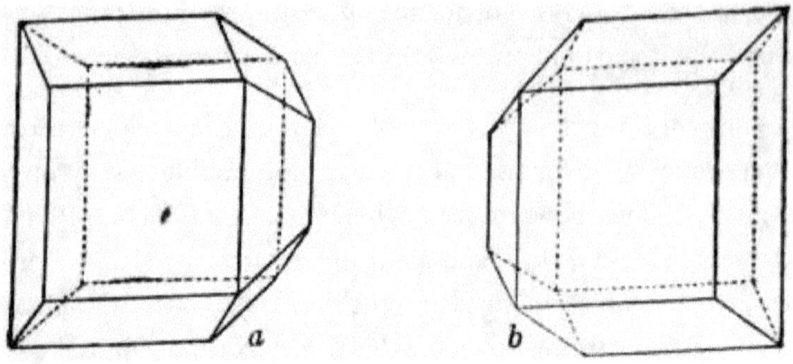

FIG. 64.—Enantiomorphous Crystals of Tartaric Acid.

Fig. 64 shows two crystals of tartaric acid, *a* a right-handed crystal of dextro-tartaric acid, and *b* a left-handed crystal of laevo-tartaric acid. The two crystals are enantiomorphous, *i.e.* although they have the same interfacial angles they are not superposable, one being the mirror image of the other. Other examples are potassium dextro-tartrate, cane-sugar, milk-sugar, quercite, lithium sulphate ($Li_2SO_4{\cdot}H_2O$); amongst minerals the only example is the hydrocarbon fichtelite (C_5H_8).

Clinohedral Class

(Hemihedral; Domatic).

Crystals of this class are symmetrical only with respect to a single plane. The only form which is here geometrically the same as in the holosymmetric class is the clino-pinacoid {010}. The forms perpendicular to the plane of symmetry are all pedions, consisting of single planes with the indices {100}, {100}, {001}, {001}, {*hol*}, &c. The remaining forms, {*hko*}, {*okl*} and {*hkl*}, are domes or "gonioids" (γωνία, an angle, and εἶδος, form), consisting of two planes equally inclined to the plane of symmetry.

Examples are potassium tetrathionate ($K_2S_4O_6$), hydrogen trisodium hypophosphate ($HNa_3P_2O_6{\cdot}9H_2O$); and amongst minerals, clinohedrite ($H_2ZnCaSiO_4$) and scolectite.

5. ANORTHIC SYSTEM

(Triclinic).

In the anorthic (from ἀν, privative, and ὀρθός, right) or triclinic system none of the three crystallographic axes are at right angles, and they are all of unequal lengths. In addition to the parameters a : b : c, it is necessary to know the angles, α, β, and γ, between the axes. In anorthite, for example, these elements are $a : b : c = 0.6347 : 1 : 0.5501$; $\alpha = 93° 13'$, $\beta = 115° 55'$, $\gamma = 91° 12'$.

Holosymmetric Class

(Holohedral; Pinacoidal).

Here there is only a centre of symmetry. All the forms are pinacoids, each consisting of only two parallel faces. The indices of the three pinacoids parallel to the axial planes are {100}, {010} and {001}; those of pinacoids parallel to only one axis are {*hko*}, {*hol*} and {*okl*}; and the general form is {*hkl*}.

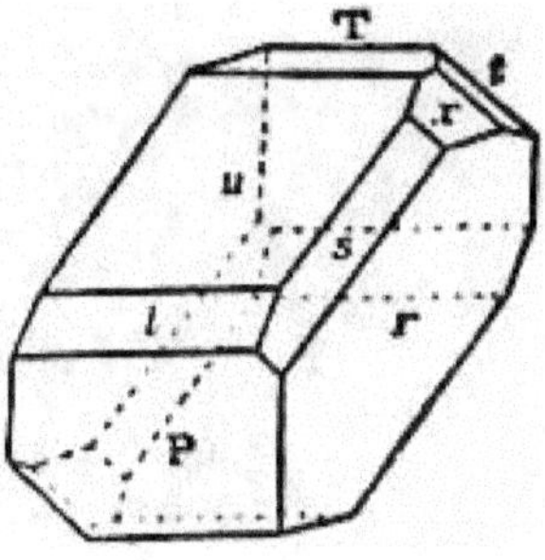

Fig. 65.—Crystal of Axinite.

Several minerals crystallize in this class; for example, the plagioclastic felspars, microcline, axinite (fig. 65), cyanite, amblygonite, chalcanthite ($CuSO_4 \cdot 5H_2O$), sassolite (H_3BO_3); among artificial substances are potassium bichromate, racemic acid ($C_4H_6O_6 \cdot 2H_2O$), dibrom-para-nitrophenol, &c.

Asymmetric Class

(Hemihedral, Pediad).

Crystals of this class are devoid of any elements of symmetry. All the forms are pedions, each consisting of a single plane; they are thus hemihedral with respect to crystals of the last class. Although there is a total absence of symmetry, yet the faces are arranged in zones on the crystals.

Examples are calcium thiosulphate ($CaS_2O_3 \cdot 6H_2O$) and hydrogen strontium dextro-tartrate ($(C_4H_4O_6H)_2Sr \cdot 5H_2O$); there is no example amongst minerals.

6. HEXAGONAL SYSTEM

Crystals of this system are characterized by the presence of a single axis of either triad or hexad symmetry, which is spoken of as the "principal" or "morphological" axis. Those with a triad axis are grouped together in the rhombohedral or trigonal division, and those with a hexad axis in the hexagonal division. By some authors these two divisions are treated as separate systems; or again the rhombohedral forms may be considered as hemihedral developments of the hexagonal. On the other hand, hexagonal forms may be considered as a combination of two rhombohedral forms.

Owing to the peculiarities of symmetry associated with a single triad or hexad axis, the crystallographic axes of reference are different in this system from those used in the five other systems of crystals. Two methods of axial representation are in common use; rhombohedral axes being usually used for crystals of the rhombohedral division, and hexagonal axes for those of the hexagonal division; though sometimes either one or the other set is employed in both divisions.

Rhomobohedral axes are taken parallel to the three sets of edges of a rhombohedron (fig. 66). They are inclined to one another at equal oblique angles, and they are all equally inclined to the principal axis; further, they are all of equal length and are interchangeable. With such a set of axes there can be no statement of an axial ratio, but the angle between the axes (or some other angle which may be calculated from this) may be given as a constant of the substance. Thus in calcite the rhombohedral angle (the angle between two faces of the fundamental rhombohedron) is 74° 55′, or the angle between the normal to a face of this rhombohedron and the principal axis is 44° 36½′.

Hexagonal axes are four in number, viz. a vertical axis coinciding with the principal axis of the crystal, and three horizontal axes inclined to one another at 60° in a plane perpendicular to the principal axis. The three horizontal axes, which are taken either parallel or perpendicular to the faces of a hexagonal prism (fig. 71) or the edge of a hexagonal bipyramid (fig. 70), are equal in length (*a*) but the vertical axis is of a different length (*c*). The indices of planes referred to such a set of axes are four in number; they are written as $\{hikl\}$, the first three ($h + i + k = 0$) referring to the horizontal axes and the last to the vertical axis. The ratio $a : c$ of the parameters, or the axial ratio, is characteristic of all the crystals of the same substance. Thus for beryl (including emerald) $a : c = 1 : 0.4989$ (often written $c = 0.4989$); for zinc $c = 1.3564$.

Rhombohedral Division.

In the rhomobohedral or trigonal division of the hexagonal system there are seven symmetry-classes, all of which possess a single triad axis of symmetry.

Holosymmetric Class

(Holohedral; Ditrigonal scalenohedral).

In this class, which presents the commonest type of symmetry of the hexagonal system, the triad axis is associated with three similar planes of symmetry inclined to one another at 60° and intersecting in the triad axis; there are also three similar dyad axes, each perpendicular to a plane of symmetry, and a centre of symmetry. The seven simple forms are:—

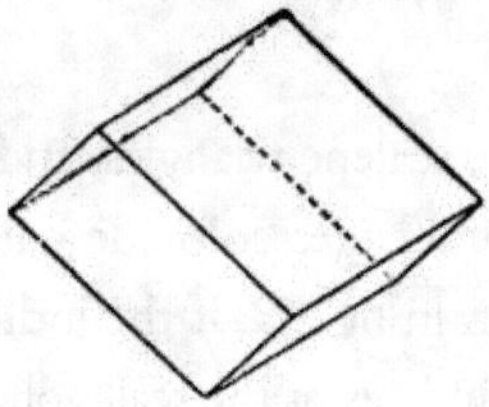

FIG. 66.

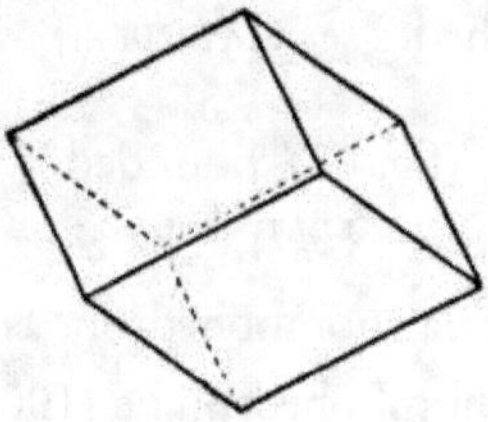

FIG. 67.

Direct and Inverse Rhombohedra.

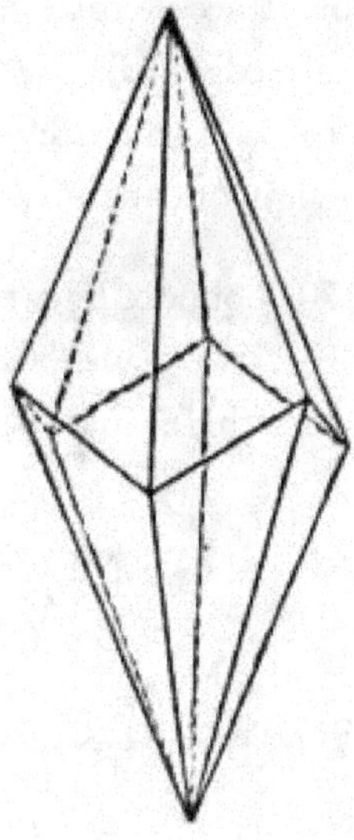

FIG. 68.—Scalenohedron.

Rhombohedron (figs. 66 and 67), consisting of six rhomb-shaped faces with the edges all of equal lengths: the faces are perpendicular to the planes of symmetry. There are two sets of rhombohedra, distinguished respectively as direct and inverse; those of one set (fig. 66) are brought into the orientation of the other set (fig. 67) by a rotation of 60° or 180° about the principal axis. For the fundamental rhombohedron, parallel to the edges of which are the crystallographic axes of reference, the indices are {100}. Other rhombohedra may have the indices {211}, {411}, {110},

{221}, {111}, &c., or in general {*hkk*}. (Compare fig. 72; for figures of other rhombohedra see CALCITE.)

Scalenohedron (fig. 68), bounded by twelve scalene triangles, and with the general indices {*hkl*}. The zig-zag lateral edges coincide with the similar edges of a rhombohedron, as shown in fig. 69; if the indices of the inscribed rhombohedron be {100}, the indices of the scalenohedron represented in the figure are {201}. The scalenohedron {201} is a characteristic form of calcite, which for this reason is sometimes called "dog-tooth-spar." The angles over the three edges of a face of a scalenohedron are all different; the angles over three alternate polar edges are more obtuse than over the other three polar edges. Like the two sets of rhombohedra, there are also direct and inverse scalenohedra, which may be similar in form and angles, but different in orientation and indices.

Hexagonal bipyramid (fig. 70), bounded by twelve isosceles triangles each of which are equally inclined to two planes of symmetry. The indices are {210}, {412}, &c., or in general (hkl), where $h - 2k + l = 0$.

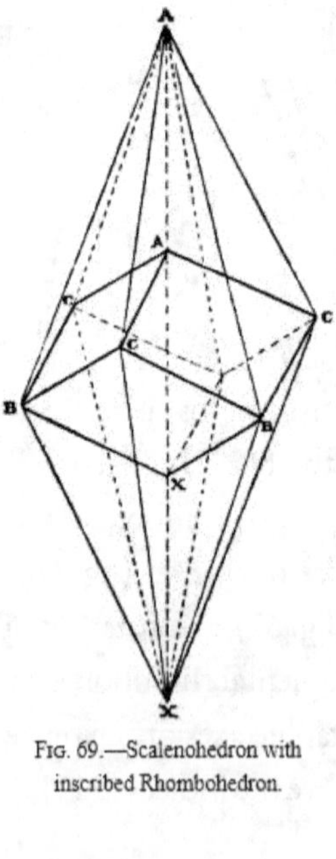

FIG. 69.—Scalenohedron with inscribed Rhombohedron.

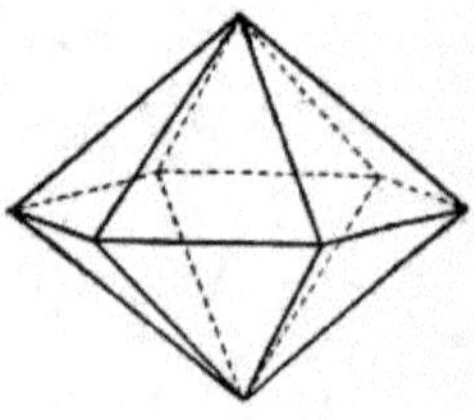

FIG. 70.—Hexagonal Bipyramid.

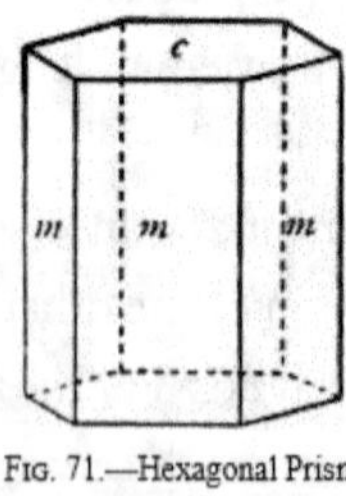

FIG. 71.—Hexagonal Prism and Basal Pinacoid.

Hexagonal prism of the first order (211), consisting of six faces parallel to the principal axis and perpendicular to the planes of symmetry; the angles between (the normals to) the faces are 60°.

Hexagonal prism of the second order (101), consisting of six faces parallel to the principal axis and parallel to the planes of symmetry. The faces of this prism are inclined to 30° to those of the last prism.

Dihexagonal prism, consisting of twelve faces parallel to the principal axis and inclined to the planes of symmetry. There are two sets of angles between the faces. The indices are {321}, {532} ... {*hkl*}, where $h + k + l = 0$.

Basal pinacoid {111}, consisting of a pair of parallel faces perpendicular to the principal axis.

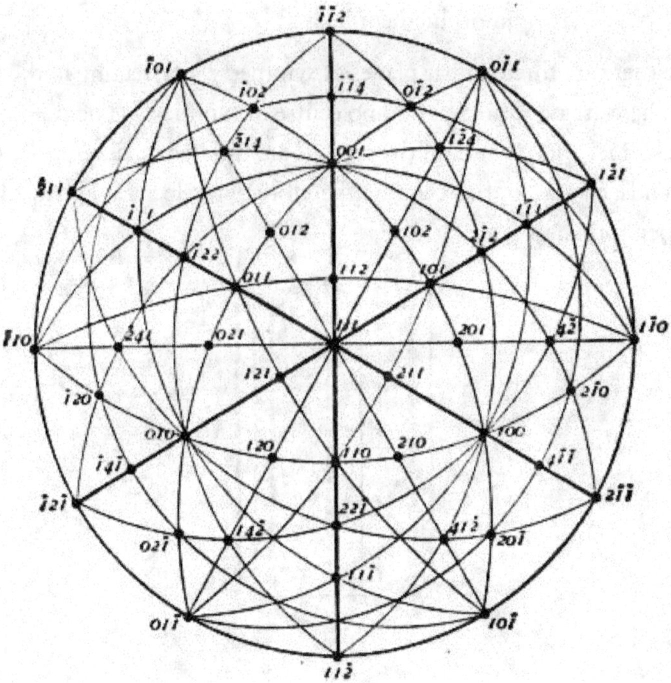

FIG. 72.—Stereographic Projection of a Holosymmetric Rhombohedral Crystal.

Fig. 71 shows a combination of a hexagonal prism (m) with the basal pinacoid (c). For figures of other combinations see CALCITE and CORUNDUM. The relation between rhombohedral forms and their indices are best studied with the aid of a stereographic projection (fig. 72); in this figure the thicker lines are the projections of the three planes of symmetry, and on these lie the poles of the rhombohedra (six of which are indicated).

Numerous substances, both natural and artificial, crystallize in this class; for example, calcite, chalybite, calamine, corundum (ruby and sapphire), haematite, chabazite; the elements arsenic, antimony, bismuth, selenium, tellurium and perhaps graphite; also ice, sodium nitrate, thymol, &c.

Ditrigonal Pyramidal Class

(Hemimorphic-hemihedral).

Here there are three similar planes of symmetry intersecting in the triad axis; there are no dyad axes and no centre of symmetry. The triad axis is uniterminal and polar, and the crystals are differently developed at the two ends; crystals of this class are therefore pyro-electric. The forms are all open forms:—

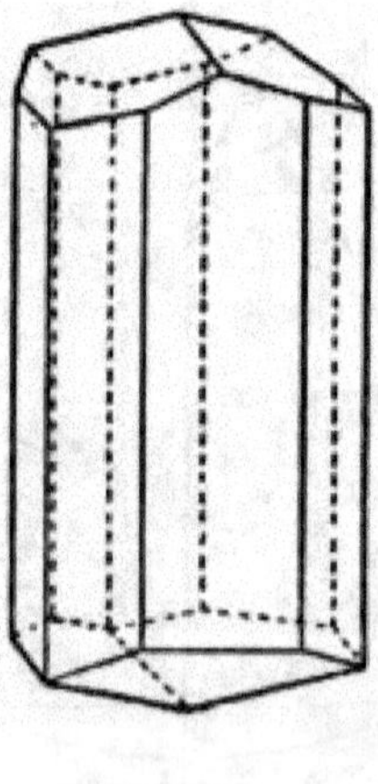

FIG. 73.—Crystal of Tourmaline.

Trigonal pyramid $\{hkk\}$, consisting of the three faces which correspond to the three upper or the three lower faces of a rhombohedron of the holosymmetric class.

Ditrigonal pyramid {*hkl*}, of six faces, corresponding to the six upper or lower faces of the scalenohedron.

Hexagonal pyramid (*hkl*) where ($h - 2k + l = 0$), of six faces, corresponding to the six upper or lower faces of the hexagonal bipyramid.

Trigonal prism {211} or {211}, two forms each consisting of three faces parallel to principal axis and perpendicular to the planes of symmetry.

Hexagonal prism {101}, which is geometrically the same as in the last class.

Ditrigonal prism {*hkl*} (where $h + k + l = 0$), of six faces parallel to the principal axis, and with two sets of angles between them.

Basal pedion (111) or (111), each consisting of a single plane perpendicular to the principal axis.

Fig. 73 represents a crystal of tourmaline with the trigonal prism (211), hexagonal prism (101), and a trigonal pyramid at each end. Other substances crystallizing in this class are pyrargyrite, proustite, iodyrite (AgI), greenockite, zincite, spangolite, sodium lithium sulphate, tolylphenylketone.

Trapezohedral Class

(Trapezohedral-hemihedral).

Here there are three similar dyad axes inclined to one another at 60° and perpendicular to the triad axis. There are no planes or centre of symmetry. The dyad axes are uniterminal, and are pyro-electric axes. Crystals of most substances of this class rotate the plane of polarization of a beam of light.

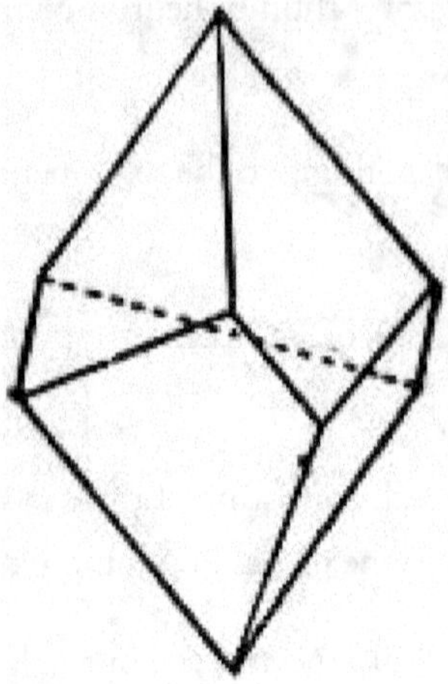

FIG. 74.—Trigonal Trapezohedron.

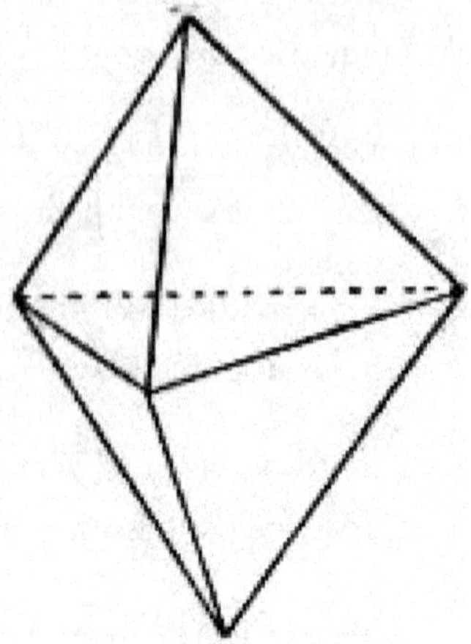

FIG. 75.—Trigonal Bipyramid.

In this class the rhombohedra {*hkk*}, the hexagonal prism {211}, and the basal pinacoid {111} are geometrically the same as in the holosymmetric class; the trigonal prism {101} and the ditrigonal prisms are as in the ditrigonal pyramidal class. The remaining simple forms are:—

Trigonal trapezohedron (fig. 74), bounded by six trapezoidal faces. There are two complementary and enantiomorphous trapezohedra, {*hkl*} and {*hlk*}, derivable from the scalenohedron.

Trigonal bipyramid (fig. 75), bounded by six isosceles triangles; the indices are {*hkl*}, where $h - 2k + l = 0$, as in the hexagonal bipyramid.

The only minerals crystallizing in this class are quartz (*q.v.*) and cinnabar, both of which rotate the plane of a beam of polarized light transmitted along the triad axis. Other examples are dithionates of lead ($PbS_2O_6{\cdot}4H_2O$), calcium and strontium, and of potassium ($K_2S_2O_6$), benzil, matico-stearoptene.

Rhombohedral Class

(Parallel-faced hemihedral).

The only elements of symmetry are the triad axis and a centre of symmetry. The general form {*hkl*} is a rhombohedron, and is a hemihedral form, with parallel faces, of the scalenohedron. The form {*hkl*}, where $h - 2k + l = 0$, is also a rhombohedron, being the hemihedral form of the hexagonal bipyramid. The dihexagonal prism {*hkl*} of the holosymmetric class becomes here a hexagonal prism. The rhombohedra (hkk), hexagonal prisms {211} and {101}, and the basal pinacoid {111} are geometrically the same in this class as in the holosymmetric class.

Fig. 76 represents a crystal of dioptase with the fundamental rhombohedron *r* {100} and the hexagonal prism of the second order *m* {101} combined with the rhombohedron *s* {031}.

Examples of minerals which crystallize in this class are phenacite, dioptase, willemite, dolomite, ilmenite and pyrophanite: amongst artificial substances is ammonium periodate ($(NH_4)_4I_2O_9{\cdot}3H_2O$).

Trigonal Pyramidal Class

(Hemimorphic-tetartohedral).

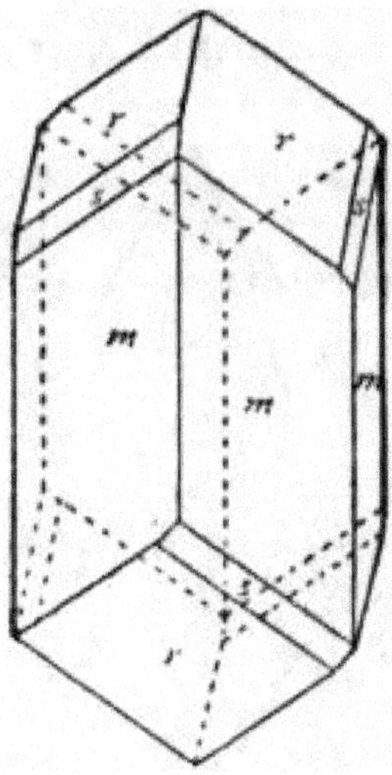

Fig. 76.—Crystal of Dioptase.

Here there is only the triad axis of symmetry, which is uniterminal. The general form {*hkl*} is a trigonal pyramid consisting of three faces at one end of the crystal. All other forms, in which the faces are neither parallel nor perpendicular to the triad axis, are trigonal pyramids. All the prisms are trigonal prisms; and perpendicular to these are two pedions.

The only substance known to crystallize in this class is sodium periodate ($NaIO_4 \cdot 3H_2O$), the crystals of which are circularly polarizing.

Trigonal Bipyramidal Class

Here there is a plane of symmetry perpendicular to the triad axis. The trigonal pyramids of the last class are here trigonal bipyramids (fig. 75); the prisms are all trigonal prisms, and parallel to the plane of symmetry is the basal pinacoid. No example is known for this class.

Ditrigonal Bipyramidal Class

Here there are three similar planes of symmetry intersecting in the triad axis, and perpendicular to them is a fourth plane of symmetry; at the intersection of the three vertical planes with the horizontal plane are three similar dyad axes; there is no centre of symmetry.

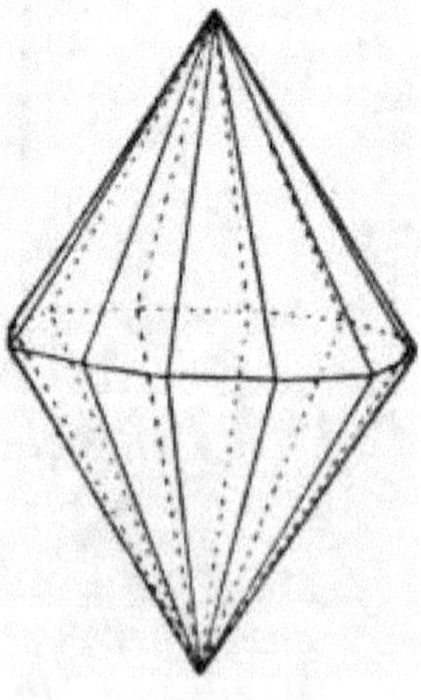

Fig. 77.—Dihexagonal Bipyramid.

The general form is bounded by twelve scalene triangles and is a ditrigonal bipyramid. Like the general form of the last class, this has two sets of indices {*hkl, pqr*}, (hkl) for faces above the equatorial plane of symmetry and (pqr) for faces below: with hexagonal axes there would be only one set of indices. The hexagonal bipyramids, the hexagonal prism {101} and the basal pinacoid {111} are geometrically the same in this class as in the holosymmetric class. The trigonal prism {211} and ditrigonal prisms {*hkl*} are the same as in the ditrigonal pyramidal class.

The only representative of this type of symmetry is the mineral benitoite (*q.v.*).

Hexagonal Division.

In crystals of this division of the hexagonal system the principal axis is a hexad axis of symmetry. Hexagonal axes of reference are used: if rhombohedral axes be used many of the simple forms will have two sets of indices.

Holosymmetric Class

(Holohedral; Dihexagonal bipyramidal).

Intersecting in the hexad axis are six planes of symmetry of two kinds, and perpendicular to them is an equatorial plane of symmetry. Perpendicular to the hexad axis are six dyad axes of two kinds and each perpendicular to a vertical plane of symmetry. The seven simple forms are:—

Dihexagonal bipyramid, bounded by twenty-four scalene triangles (fig. 77; *v* in fig. 80). The indices are {2131}, &c., or in general {*hikl*}. This form may be considered as a combination of two scalenohedra, a direct and an inverse.

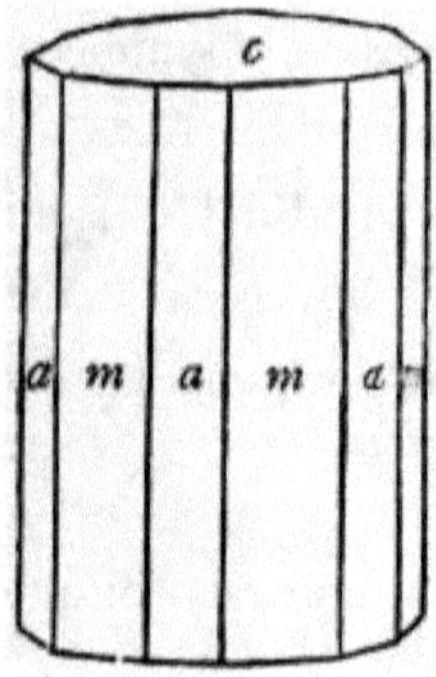

FIG. 79.

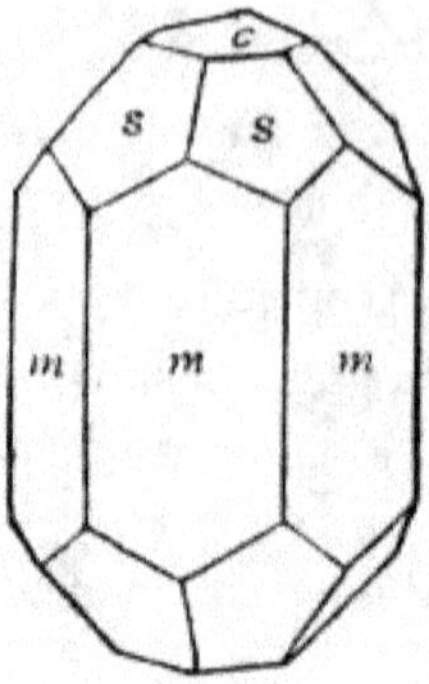

FIG. 80.

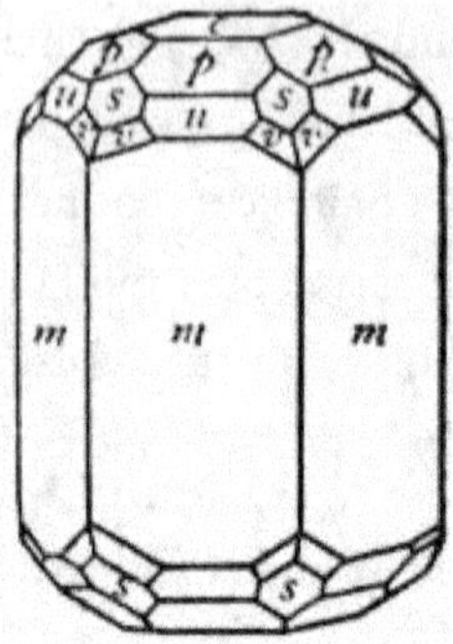

FIG. 81.

Combinations of Hexagonal forms.

Hexagonal bipyramid of the first order, bounded by twelve isosceles triangles (fig. 70; *p* and *u* in fig. 80); indices {1011}, {2021} ... (hohl). The hexagonal bipyramid so common in quartz is geometrically similar to this form, but it really is a combination of two rhombohedra, a direct and an inverse, the faces of which differ in surface characters and often also in size.

Hexagonal bipyramid of the second order, bounded by twelve faces (*s* in figs. 79 and 80); indices {1121}, {1122} ... {*h.h.2h.l*}.

Dihexagonal prism, consisting of twelve faces parallel to the hexad axis and inclined to the vertical planes of symmetry; indices {*hiko*}.

Hexagonal prism of the first order {1010}, consisting of six faces parallel to the hexad axis and perpendicular to one set of three vertical planes of symmetry (*m* in figs. 71, 78-80).

Hexagonal prism of the second order {1120}, consisting of six faces also parallel to the hexad axis, but perpendicular to the other set of three vertical planes of symmetry (*a* in fig. 78).

Basal pinacoid {0001}, consisting of a pair of parallel planes perpendicular to the hexad axis (*c* in figs. 71, 78-80).

Beryl (emerald), connellite, zinc, magnesium and beryllium crystallize in this class.

Bipyramidal Class

(Parallel-faced hemihedral).

Here there is a plane of symmetry perpendicular to the hexad axis; there is also a centre of symmetry. All the closed forms are hexagonal bipyramids; the open forms are hexagonal prisms or the basal pinacoid. The general form {*hikl*} is hemihedral with parallel faces with respect to the general form of the holosymmetric class.

Apatite (*q.v.*), pyromorphite, mimetite and vanadinite possess this degree of symmetry.

Dihexagonal Pyramidal Class

(Hemimorphic-hemihedral).

Six planes of symmetry of two kinds intersect in the hexad axis. The hexad axis is uniterminal and all the forms are open forms. The general form {*hikl*} consists of twelve faces at one end of the crystal, and is a dihexagonal pyramid. The hexagonal pyramids {*hohl*} and (*h.h.2h.l*) each consist of six faces at one end of the crystal. The prisms are geometrically the same as in the holosymmetric class. Perpendicular to the hexad axis are the pedions (0001) and (0001).

Iodyrite (AgI), greenockite (CdS), wurtzite (ZnS) and zincite (ZnO) are often placed in this class, but they more probably belong to the hemimorphic-hemihedral class of the rhombohedral division of this system.

Trapezohedral Class

(Trapezohedral-hemihedral).

Six dyad axes of two kinds are perpendicular to the hexad axis. The general form {*hikl*} is the hexagonal trapezohedron bounded by twelve trapezoidal faces. The other simple forms are geometrically the same as in the holosymmetric class. Barium-anti-monyldextro-tartrate + potassium nitrate ($Ba(SbO)_2(C_4H_4O_6)_2 \cdot KNO_3$) and the corresponding lead salt crystallize in this class.

Hexagonal Pyramidal Class

(Hemimorphic-tetartohedral).

No other element is here associated with the hexad axis, which is uniterminal. The pyramids all consist of six faces at one end of the crystal, and prisms are all hexagonal prisms; perpendicular to the hexad axis are the pedions.

Lithium potassium sulphate, strontium-antimonyl dextro-tartrate, and lead-antimonyl dextro-tartrate are examples of this type of symmetry. The mineral nepheline is placed in this class because of the absence of symmetry in the etched figures on the prism faces (fig. 92).

(g) Regular Grouping of Crystals.

Crystals of the same kind when occurring together may sometimes be grouped in parallel position and so give rise to special structures, of which the dendritic (from δένδρον, a tree) or branch-like aggregations of native copper or of magnetite and the fibrous structures of many minerals furnish examples. Sometimes, owing to changes in the surrounding conditions, the crystal may continue its growth with a different external form or colour, *e.g.* sceptre-quartz.

Regular intergrowths of crystals of totally different substances such as staurolite with cyanite, rutile with haematite, blende with chalcopyrite, calcite with sodium nitrate, are not uncommon. In these cases certain planes and edges of the two crystals are parallel. (See O. Mügge, "Die regelmässigen Verwachsungen von Mineralien verschiedener Art," *Neues Jahrbuch für Mineralogie*, 1903, vol. xvi. pp. 335-475).

But by far the most important kind of regular conjunction of crystals is that known as "twinning." Here two crystals or individuals of the same kind have grown together in a certain symmetrical manner, such that one portion of the twin may be brought into the position of the other by reflection across a plane or by rotation about an axis. The plane of reflection is called the twin-plane, and is parallel to one of the faces, or to a possible face, of the crystal: the axis of rotation, called the twin-axis, is parallel to one of the edges or perpendicular to a face of the crystal.

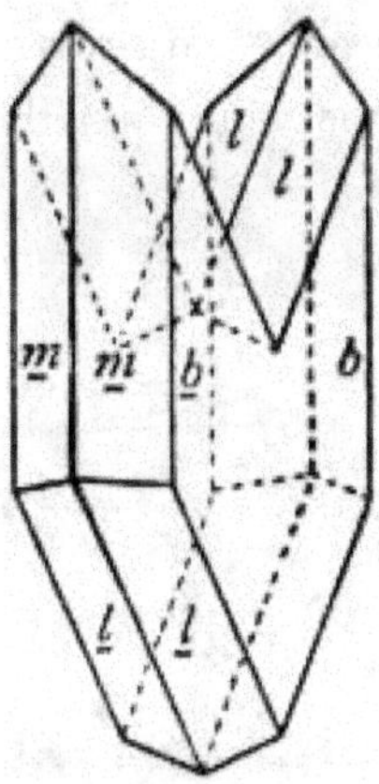

FIG. 81.—Twinned Crystal of Gypsum.

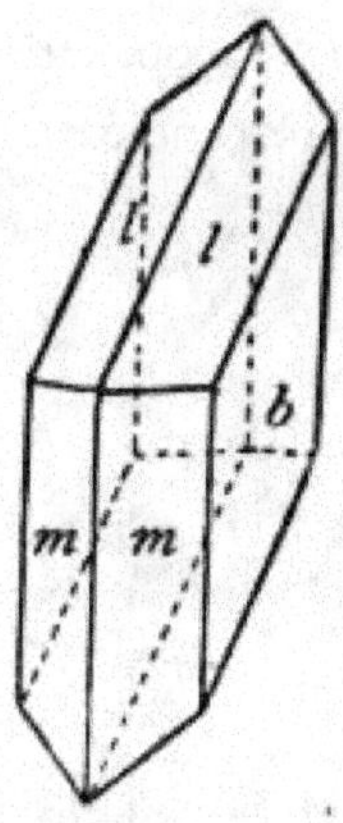

FIG. 82.—Simple Crystal of Gypsum.

In the twinned crystal of gypsum represented in fig. 81 the two portions are symmetrical with respect to a plane parallel to the ortho-pinacoid (100), *i.e.* a vertical plane perpendicular to the face *b*. Or we may consider the simple crystal (fig. 82) to be cut in half by this plane and one portion to be rotated through 180° about the normal to the same plane. Such a crystal (fig. 81) is therefore described as being twinned on the plane (100).

An octahedron (fig. 83) twinned on an octahedral face (111) has the two portions symmetrical with respect to a plane parallel to this face (the large triangular face in the figure); and either portion may be brought into the position of the other by a rotation through 180° about the triad axis of symmetry which is perpendicular to this face. This kind of twinning is especially frequent in crystals of spinel, and is consequently often referred to as the "spinel twin-law."

In these two examples the surface of the union, or composition-plane, of the two portions is a regular surface coinciding with the twin-plane; such twins are called "juxtaposition-twins." In other juxtaposed twins the plane of composition is, however, not necessarily the twin-plane. Another type of twin is the "interpenetration twin," an example of which is shown in fig. 84. Here one cube may be brought into the position of the other by a rotation of 180° about a triad axis, or by reflection across the octahedral plane which is perpendicular to this axis; the twin-plane is therefore (111).

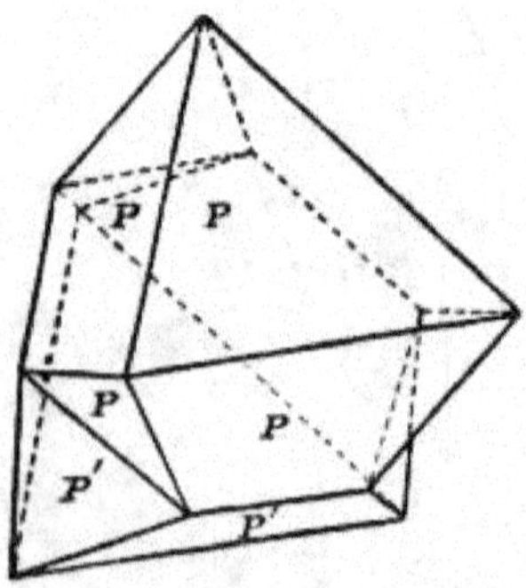

FIG. 83.—Spinel-twin.

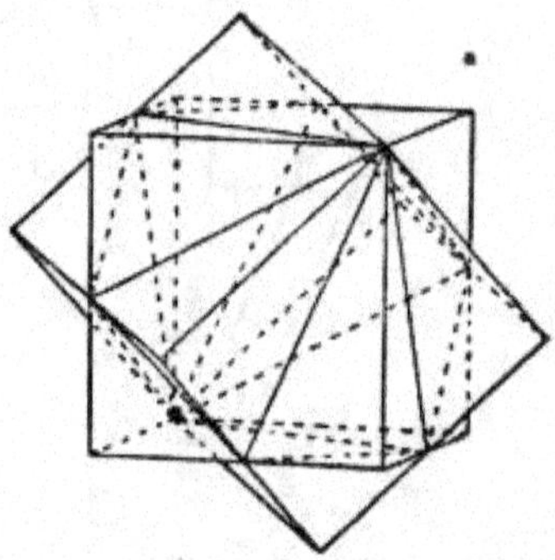

FIG. 84.—Interpenetrating Twinned Cubes.

Since in many cases twinned crystals may be explained by the rotation of one portion through two right angles, R. J. Haüy introduced the term "hemitrope" (from the Gr. ἡμι-, half, and τρόπος, a turn); the word "macle" had been earlier used by Romé d'Isle. There are, however, some rare types of twins which cannot be explained by rotation about an axis, but only by reflection across a plane; these are known as "symmetric twins," a good example of which is furnished by one of the twin-laws of chalcopyrite.

Twinned crystals may often be recognized by the presence of re-entrant angles between the faces of the two portions, as may be seen from the above figures. In some twinned crystals (*e.g.* quartz) there are, however, no re-entrant angles. On the other hand, two crystals accidentally grown together without any symmetrical relation between them will usually show some re-entrant angles, but this must not be taken to indicate the presence of twinning.

Twinning may be several times repeated on the same plane or on other similar planes of the crystal, giving rise to triplets, quartets and other complex groupings. When often repeated on the same plane, the twinning is said to be "polysynthetic," and gives rise to a laminated structure in the crystal. Sometimes such a crystal (*e.g.* of corundum or pyroxene) may be readily broken in this direction, which is thus a "plane of parting," often closely resembling a true cleavage in character. In calcite and some other substances this lamellar twinning may be produced artificially by pressure (see below, Sect. II. (*a*), *Glide-plane*).

Another curious result of twinning is the production of forms which apparently display a higher degree of symmetry than that actually possessed by the substance. Twins of this kind are known as "mimetic-twins or pseudo-symmetric twins." Two hemihedral or hemimorphic crystals (*e.g.* of diamond or of hemimorphite) are often united in twinned position to produce a group with apparently the same degree of symmetry as the holosymmetric class of the same system. Or again, a substance crystallizing in, say, the orthorhombic system (*e.g.* aragonite) may, by twinning, give rise to pseudo-hexagonal forms: and pseudo-cubic forms often result by the complex twinning of crystals (*e.g.* stannite, phillipsite, &c.) belonging to other systems. Many of the so-called "optical anomalies" of crystals may be explained by this pseudo-symmetric twinning.

(h) Irregularities of Growth of Crystals; Character of Faces.

Only rarely do actual crystals present the symmetrical appearance shown in the figures given above, in which similar faces are all represented as of equal size. It frequently happens that the crystal is so placed with respect to the liquid in which it grows that there will be a more rapid deposition of material on one part than on another; for instance, if the crystal be attached to some other solid it cannot grow in that direction. Only when a crystal is freely suspended in the mother-liquid and material for growth is supplied at the same rate on all sides does an equably developed form result.

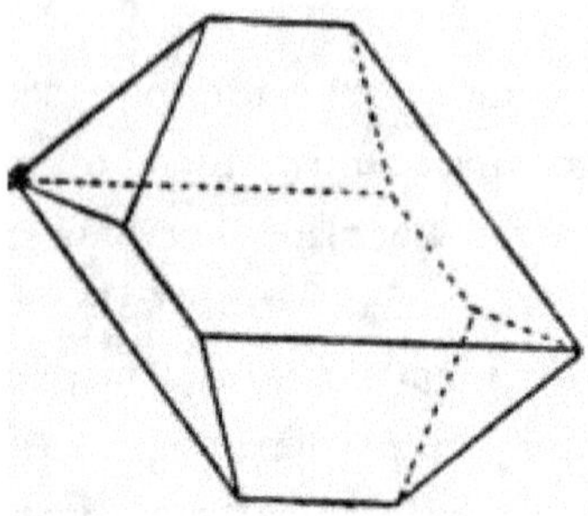

FIG. 85.

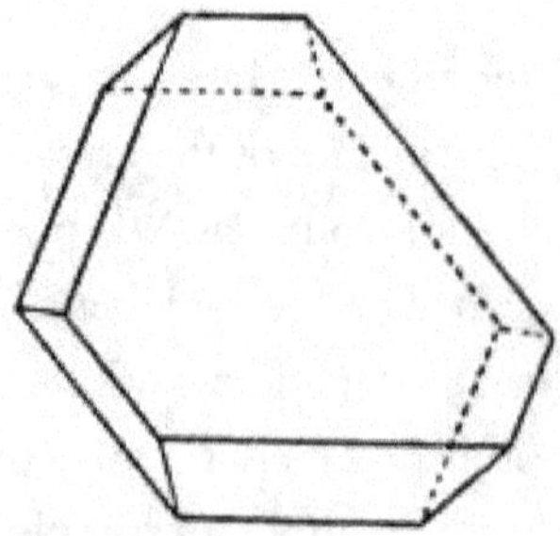

FIG. 86.

Misshapen Octahedra.

Two misshapen or distorted octahedra are represented in figs. 85 and 86; the former is elongated in the direction of one of the edges of the octahedron, and the latter is flattened parallel to one pair of faces. It will be noticed in these figures that the edges in which the faces intersect have the same directions as before, though here there are additional edges not present in fig. 3. The angles (70° 32′ or 109° 28′) between the faces also remain the same; and the faces have the same inclinations to the axes and planes of symmetry as in the equably developed form. Although from a geometrical point of view these figures are no longer symmetrical with respect to the axes and planes of symmetry, yet crystallographically they are just as symmetrical as the ideally developed form, and, however much their irregularity of development, they still are regular (cubic) octahedra of crystallography. A remarkable case of irregular development is presented by the mineral cuprite, which is often found as well-developed octahedra; but in the variety known as chalcotrichite it occurs as a matted aggregate of delicate hairs, each of which is an individual crystal enormously elongated in the direction of an edge or diagonal of the cube.

The symmetry of actual crystals is sometimes so obscured by irregularities of growth that it can only be determined by measurement of the angles. An extreme case, where several of the planes have not been developed at all, is illustrated in fig. 87, which shows the actual shape of a crystal of zircon from Ceylon; the ideally developed form (fig. 88) is placed at the side for comparison, and the parallelism of the edges between corresponding faces will be noticed. This crystal is a combination of five simple forms, viz. two tetragonal prisms (*a* and *m*,) two tetragonal bipyramids (*e* and *p*), and one ditetragonal bipyramid (*x*, with 16 faces).

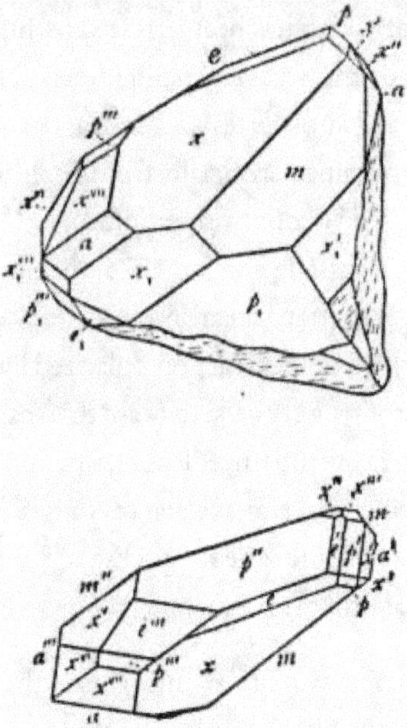

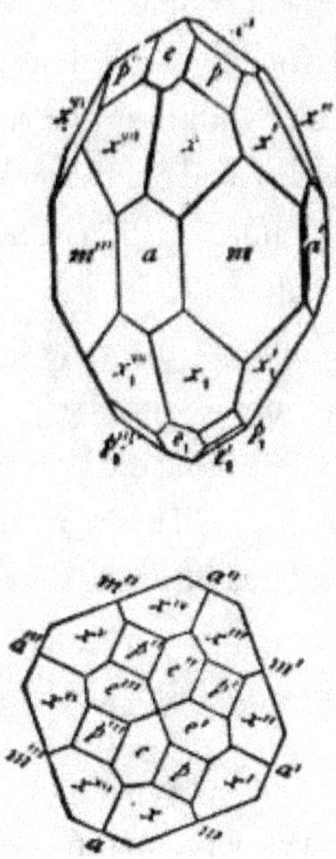

FIG. 87.—Actual Crystal. FIG. 88.—Ideal Development.

Crystal of Zircon (clinographic drawings and plans).

The actual form, or "habit," of crystals may vary widely in different crystals of the same substance, these differences depending largely on the conditions under which the growth has taken place. The material may have crystallized from a fused mass or from a solution; and in the latter case the solvent may be of different kinds and contain other substances in solution, or the temperature may vary. Calcite (*q.v.*) affords a good example of a substance crystallizing in widely different habits, but all crystals are referable to the same type of symmetry and may be reduced to the same fundamental form.

When crystals are aggregated together, and so interfere with each other's growth, special structures and external shapes often result, which are sometimes characteristic of certain substances, especially amongst minerals.

Incipient crystals, the development of which has been arrested owing to unfavourable conditions of growth, are known as crystallites (*q.v.*). They are met with in imperfectly crystallized substances and in glassy rocks (obsidian and pitchstone), or may be obtained artificially from a solution of sulphur in carbon disulphide rendered viscous by the addition of Canada-balsam. To the various forms H. Vogelsang gave, in 1875, the names "globulites," "margarites" (from μαργαρίτης, a pearl), "longulites," &c. At a more advanced stage of growth these bodies react on polarized light, thus possessing the internal structure of true crystals; they are then called "microlites." These have the form of minute rods, needles or hairs, and are aggregated into feathery and spherulitic forms or skeletal crystals. They are common constituents of microcrystalline igneous rocks, and often occur as inclusions in larger crystals of other substances.

Inclusions of foreign matter, accidentally caught up during growth, are frequently present in crystals. Inclusions of other minerals are specially frequent and conspicuous in crystals of quartz, and crystals of calcite may contain as much as 60% of included sand. Cavities, either with rounded boundaries or with the same shape ("negative crystals") as the surrounding crystal, are often to be seen; they may be empty or enclose a liquid with a movable bubble of gas.

The faces of crystals are rarely perfectly plane and smooth, but are usually striated, studded with small angular elevations, pitted or cavernous, and sometimes curved or twisted. These irregularities, however, conform with the symmetry of the crystal, and much may be learnt by their study. The parallel grooves or furrows, called "striae," are the result of oscillatory combination between adjacent faces, narrow strips of first one face and then another being alternately developed. Sometimes the striae on crystal-faces are due to repeated lamellar twinning, as in the plagioclase felspars. The directions of the striations are very characteristic features of many crystals: *e.g.* the faces of the hexagonal prism of quartz are always striated horizontally, whilst in beryl they are striated vertically. Cubes of pyrites (fig. 89) are striated parallel to one edge, the striae on adjacent faces being at right angles, and due to oscillatory combination of the cube and the pentagonal dodecahedron (compare fig. 36); whilst cubes of blende (fig. 90) are striated parallel to one diagonal of each face, *i.e.* parallel to the tetrahedron faces (compare fig. 31). These striated cubes thus possess different degrees of symmetry and belong to different symmetry-classes. Oscillatory combination of faces gives rise also to curved surfaces. Crystals with twisted surfaces (see DOLOMITE) are, however, built up of smaller crystals arranged in nearly parallel position. Sometimes a face is entirely replaced by small faces of other forms, giving rise to a drusy surface; an example of this is shown by some octahedral crystals of fluorspar (fig. 2) which are built up of minute cubes.

FIG. 89.—Striated Cube of Pyrites.

FIG. 90.—Striated Cube of Blende.

The faces of crystals are sometimes partly or completely replaced by smooth bright surfaces inclined at only a few minutes of arc from the true position of the face; such surfaces are called "vicinal faces," and their indices can be expressed only by very high numbers. In apparently perfectly developed crystals of alum the octahedral face, with the simple indices (111), is usually replaced by faces of very low triakis-octahedra, with indices such as (251·251·250); the angles measured on such crystals will therefore deviate slightly from the true octahedral angle. Vicinal faces of this character are formed during the growth of crystals, and have been studied by H. A. Miers (*Phil. Trans.*, 1903, Ser. A. vol. 202). Other faces with high indices, viz. "prerosion faces" and the minute faces forming the sides of etched figures (see below), as well as rounded edges and other surface irregularities, may, however, result from the corrosion of a crystal subsequent to its growth. The pitted and cavernous faces of artificially grown crystals of sodium chloride and of bismuth are, on the other hand, a result of rapid growth, more material being supplied at the edges and corners of the crystal than at the centres of the faces.

(i) Theories of Crystal Structure.

The ultimate aim of crystallographic research is to determine the internal structure of crystals from both physical and chemical data. The problem is essentially twofold: in the first place it is necessary to formulate a theory as to the disposition of the molecules, which conforms with the observed types of symmetry—this is really a mathematical problem; in the second place, it is necessary to determine the orientation of the atoms (or groups of atoms) composing the molecules with regard to the crystal axes—this involves a knowledge of the atomic structure of the molecule. As appendages to the second part of our problem, there have to be considered: (1) the possibility of the existence of the same substance in two or more distinct crystalline forms—polymorphism, and (2) the relations between the chemical structure of compounds which affect nearly identical or related crystal habits—isomorphism and morphotropy. Here we shall discuss the modern theory of crystal structure; the relations between chemical composition and crystallographical form are discussed in Part III. of this article; reference should also be made to the article CHEMISTRY: *Physical*.

The earliest theory of crystal structure of any moment is that of Haüy, in which, as explained above, he conceived a crystal as composed of elements bounded by the cleavage planes of the crystal, the elements being arranged ***Haüy.*** contiguously and along parallel lines. There is, however, no reason to suppose that matter is continuous throughout a crystalline body; in fact, it has been shown that space does separate the molecules, and we may therefore replace the contiguous elements of Haüy by particles equidistantly distributed along parallel lines; by this artifice we retain the reticulated or net-like structure, but avoid the continuity of matter which characterizes Haüy's theory; the permanence of crystal form being due to equilibrium between the intermolecular (and interatomic) forces. The crystal is thus conjectured as a "space-lattice," composed of three sets of parallel planes which enclose parallelopipeda, at the corners of which are placed the constituent molecules (or groups of molecules) of the crystal.

The geometrical theory of crystal structure (*i.e.* the determination of the varieties of crystal symmetry) is thus reduced to the mathematical problem: "in how many ways can space be partitioned?" M. L. Frankenheim, in 1835, ***Frankenheim; Bravais.*** determined this number as fifteen, but A. Bravais, in 1850, proved the identity of two of Frankenheim's forms, and showed how the remaining fourteen coalesced by pairs, so that really these forms only corresponded to seven distinct systems and fourteen classes of crystal symmetry. These systems, however, only represented holohedral forms, leaving the hemihedral and tetartohedral classes to be explained. Bravais attempted an explanation by attributing differences in the symmetry of the crystal elements, or, what comes to the same thing, he assumed the crystals to exhibit polar differences along any member of the lattice; for instance, assume the particles to be (say) pear-shaped, then the sharp ends point in one direction, the blunt ends in the opposite direction.

A different view was adopted by L. Sohncke in 1879, who, by developing certain considerations published by Camille Jordan in 1869 on the possible types of regular repetition in space of identical parts, showed that the ***Sohncke.*** lattice-structure of Bravais was unnecessary, it being sufficient that each molecule of an indefinitely extended crystal, represented by its "point" (or centre of gravity), was identically situated with respect to the molecules surrounding it. The problem then resolves itself into the determination of the number of "point-systems" possible; Sohncke derived sixty-five such arrangements, which may also be obtained from the fourteen space-lattices of Bravais, by interpenetrating any one space-lattice with one or more identical lattices, with the condition that the resulting structure should conform with the homogeneity characteristic of crystals. But the sixty-five arrangements derived by Sohncke, of which Bravais' lattices are particular cases, did not complete the solution, for certain of the known types of crystal symmetry still remained unrepresented. These missing forms are characterized as being enantiomorphs consequently, with the introduction of this principle of repetition over a plane, *i.e.* mirror images. E. S. Fedorov (1890), A. Schoenflies (1891), and W. Barlow (1894), independently and by different methods, showed how Sohncke's theory of regular point-systems explained the whole thirty-two classes of crystal symmetry, 230 distinct types of crystal structure falling into these classes.

By considering the atoms instead of the centres of gravity of the molecules, Sohncke (*Zeits. Kryst. Min.*, 1888, 14, p. 431) has generalized his theory, and propounded the structure of a crystal in the following terms: "A crystal consists of a finite number of interpenetrating regular point-systems, which all possess like and like-directed coincidence movements. Each separate point-system is occupied by similar material particles, but these may be different for the different interpenetrating partial systems which form the complex system." Or we may quote the words of P. von Groth (*British Assoc. Rep.*, 1904): "A crystal—considered as indefinitely extended—consists of n interpenetrating regular point-systems, each of which is formed of similar atoms; each of these point-systems is built up from a number of interpenetrating space-lattices, each of the latter being formed from similar atoms occupying parallel positions. All the space-lattices of the combined system are geometrically identical, or are characterized by the same elementary parallelopipedon."

A complete résumé, with references to the literature, will be found in "Report on the Development of the Geometrical Theories of Crystal Structure, 1666-1901" (*British Assoc. Rep.*, 1901).

II. PHYSICAL PROPERTIES OF CRYSTALS.

Many of the physical properties of crystals vary with the direction in the material, but are the same in certain directions; these directions obeying the same laws of symmetry as do the faces on the exterior of the crystal. The symmetry of the internal structure of crystals is thus the same as the symmetry of their external form.

(a) Elasticity and Cohesion.

The elastic constants of crystals are determined by similar methods to those employed with amorphous substances, only the bars and plates experimented upon must be cut from the crystal with known orientations. The "elasticity surface" expressing the coefficients in various directions within the crystal has a configuration symmetrical with respect to the same planes and axes of symmetry as the crystal itself. In calcite, for instance, the figure has roughly the shape of a rounded rhombohedron with depressed faces and is symmetrical about three vertical planes. In the case of homogeneous elastic deformation, produced by pressure on all sides, the effect on the crystal is the same as that due to changes of temperature; and the surfaces expressing the compression coefficients in different directions have the same higher degree of symmetry, being either a sphere, spheroid or ellipsoid. When strained beyond the limits of elasticity, crystalline matter may suffer permanent deformation in one or other of two ways, or may be broken along cleavage surfaces or with an irregular fracture. In the case of plastic deformation, *e.g.* in a crystal of ice, the crystalline particles are displaced but without any change in their orientation. Crystals of some substances (*e.g.* para-azoxyanisol) have such a high degree of plasticity that they are deformed even by their surface tension, and the crystals take the form of drops of doubly refracting liquid which are known as "liquid crystals." (See O. Lehmann, *Flüssige Kristalle*,

Leipzig, 1904; F. R. Schenck, *Kristallinische Flüssigkeiten und flüssige Krystalle*, Leipzig, 1905.)

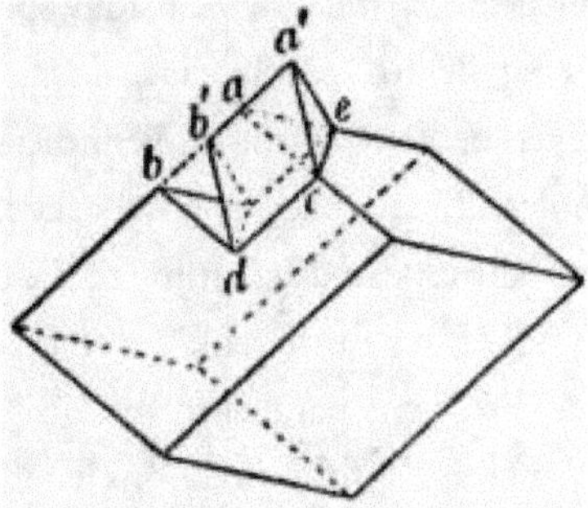

Fig. 91.—Glide-plane of Calcite.

In the second, and more usual kind of permanent deformation without fracture, the particles glide along certain planes into a new (twinned) position of equilibrium. If a knife blade be pressed into the edge of a cleavage rhombohedron of calcite (at *b*, fig. 91) the portion *abcde* of the crystal will take up the position *a'b'cde*. The obtuse solid angle at *a* becomes acute (*a'*), whilst the acute angle at *b* becomes obtuse (*b'*); and the new surface *a'ce* is as bright and smooth as before. This result has been effected by the particles in successive layers gliding or rotating over each other, without separation, along planes parallel to *cde*. This plane, which truncates the edge of the rhombohedron and has the indices (110), is called a "glide-plane." The new portion is in twinned position with respect to the rest of the crystal, being a reflection of it across the plane *cde*, which is therefore a plane of twinning. This secondary twinning is often to be observed as a repeated lamination in the grains of calcite composing a crystalline limestone, or marble, which has been subjected to earth movements. Planes of gliding have been observed in many minerals (pyroxene, corundum, &c.) and their crystals may often be readily broken along these directions, which are thus "planes of parting" or "pseudo-cleavage." The characteristic transverse striae, invariably present on the cleavage surfaces of stibnite and cyanite are due to secondary twinning along glide-planes, and have resulted from the bending of the crystals.

One of the most important characters of crystals is that of "cleavage"; there being certain plane directions across which the cohesion is a minimum, and along which the crystal may be readily split or cleaved. These directions are always parallel to a possible face on the crystal and usually one prominently developed and with simple indices, it being a face in which the crystal molecules are most closely packed. The directions of cleavage are symmetrically repeated according to the degree of symmetry possessed by the crystal. Thus in the cubic system, crystals of salt and galena cleave in three directions parallel to the faces of the cube {100}, diamond and fluorspar cleave in four directions parallel to the octahedral faces {111}, and blende in six directions parallel to the faces of the rhombic dodecahedron {110}. In crystals of other systems there will be only a single direction of cleavage if this is parallel to the faces of a pinacoid; *e.g.* the basal pinacoid in tetragonal (as in apophyllite) and hexagonal crystals; or parallel (as in gypsum) or perpendicular (as in mica and cane-sugar) to the plane of symmetry in monoclinic crystals. Calcite cleaves in three directions parallel to the faces of the primitive rhombohedron. Barytes, which crystallizes in the orthorhombic system, has two sets of cleavages, viz. a single cleavage parallel to the basal pinacoid {001} and also two directions parallel to the faces of the prism {110}. In all of the examples just quoted the cleavage is described as perfect, since cleavage flakes with very smooth and bright surfaces may be readily detached from the crystals. Different substances, however, vary widely in their character of cleavage; in some it can only be described as good or distinct, whilst in others, *e.g.* quartz and alum, there is little or no tendency to split along certain directions and the surfaces of fracture are very uneven. Cleavage is therefore a character of considerable determinative value, especially for the purpose of distinguishing different minerals.

Another result of the presence in crystals of directions of minimum cohesion are the "percussion figures," which are produced on a crystal-face when this is struck with a sharp point. A percussion figure consists of linear cracks radiating from the point of impact, which in their number and orientation agree with the symmetry of the face. Thus on a cube face of a crystal of salt the rays of the percussion figure are parallel to the diagonals of the face, whilst on an octahedral face a three-rayed star is developed. By pressing a blunt point into a crystal face a somewhat similar figure, known as a "pressure figure," is produced. Percussion and pressure figures are readily developed in cleavage sheets of mica (*q.v.*).

Closely allied to cohesion is the character of "hardness," which is often defined, and measured by, the resistance which a crystal face offers to scratching. That hardness is a character depending largely on crystalline structure is well illustrated by the two crystalline modifications of carbon: graphite is one of the softest of minerals, whilst diamond is the hardest of all. The hardness of crystals of different substances thus varies widely, and with minerals it is a character of considerable determinative value; for this purpose a scale of hardness is employed (see MINERALOGY). Various attempts have been made with the view of obtaining accurate determinations of degrees of hardness, but with varying results; an instrument used for this purpose is called a sclerometer (from σκληρός, hard). It may, however, be readily demonstrated that the degree of hardness on a crystal face varies with the direction, and that a curve expressing these relations possesses the same geometrical symmetry as the face itself. The mineral cyanite is remarkable in having widely different degrees of hardness on different faces of its crystals and in different directions on the same face.

Another result of the differences of cohesion in different directions is that crystals are corroded, or acted upon by chemical solvents, at different rates in different directions. This is strikingly shown when a sphere cut from a crystal, say of calcite or quartz, is immersed in acid; after some time the resulting form is bounded by surfaces approximating to crystal faces, and has the same symmetry as that of the crystal from which the sphere was cut. When a crystal bounded by faces is immersed in a solvent the edges and corners become rounded and "prerosion faces" developed in their place; the faces become marked all over with minute pits or shallow depressions, and as these are extended by further solution they give place to small elevations on the corroded face. The sides of the pits and elevations are bounded by small faces which have the character of vicinal faces. These markings are known as "etched figures" or "corrosion figures," and they are extremely important aids in determining the symmetry of crystals. Etched figures are sometimes beautifully developed on the faces of natural crystals, *e.g.* of diamond, and they may be readily produced artificially with suitable solvents.

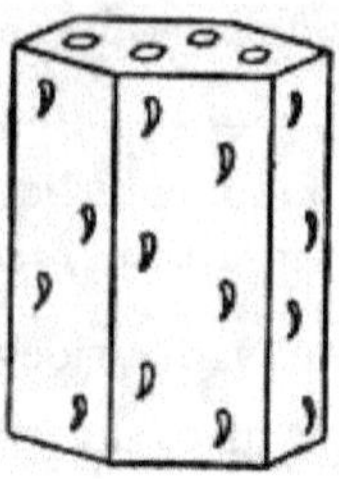

Fig. 92.—Nepheline.

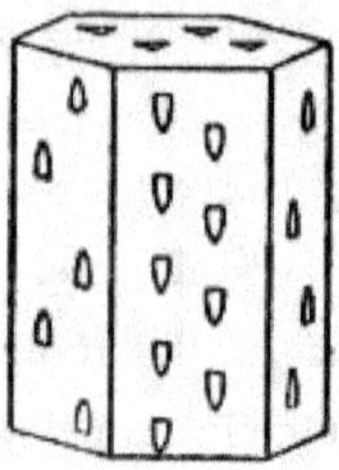

Fig. 93.—Calcite.

Fig. 94.—Beryl.

Etched Figures on Hexagonal Prisms.

As an example, the etched figures on the faces of a hexagonal prism and the basal plane are illustrated in figs. 92-94 for three of the several symmetry-classes of the hexagonal system. The classes chosen are those in which nepheline, calcite and beryl (emerald) crystallize, and these minerals often have the simple form of crystal represented in the figures. In nepheline (fig. 92) the only element of symmetry is a hexad axis; the etched figures on the prism are therefore unsymmetrical, though similar on all the faces; the hexagonal markings on the basal plane have none of their edges parallel to the edges of the face; further the crystals being hemimorphic, the etched figures on the basal planes at the two ends will be different in character. The facial development of crystals of nepheline give no indication of this type of symmetry, and the mineral has been referred to this class solely on the evidence afforded by the etched figures. In calcite there is a triad axis of symmetry parallel to the prism edges, three dyad axes each perpendicular to a pair of prism edges and three planes of symmetry perpendicular to the prism faces; the etched figures shown in fig. 93 will be seen to conform to all these elements of symmetry. There being in calcite also a centre of symmetry, the equilateral triangles on the basal plane at the lower end of the crystal will be the same in form as those at the top, but they will occupy a reversed position. In beryl, which crystallizes in the holosymmetric class of the hexagonal system, the etched figures (fig. 94) display the fullest possible degree of symmetry; those on the prism faces are all similar and are each symmetrical with respect to two lines, and the hexagonal markings on the basal planes at both ends of the crystal are symmetrically placed with respect to six lines. A detailed account of the etched figures of crystals is given by H. Baumhauer, *Die Resultate der Ätzmethode in der krystallographischen Forschung* (Leipzig, 1894).

(b) Optical Properties.

The complex optical characters of crystals are not only of considerable interest theoretically, but are of the greatest practical importance. In the absence of external crystalline form, as with a faceted gem-stone, or with the minerals constituting a rock (thin, transparent sections of which are examined in the polarizing microscope), the mineral species may often be readily identified by the determination of some of the optical characters.

According to their action on transmitted plane-polarized light (see Polarization of Light) all crystals may be referred to one or other of the five groups enumerated below. These groups correspond with the six systems of crystallization (in the second group two systems being included together). The several symmetry-classes of each system are optically the same, except in the rare cases of substances which are circularly polarizing.

(1) Optically isotropic crystals—corresponding with the cubic system.

(2) Optically uniaxial crystals—corresponding with the tetragonal and hexagonal systems.

(3) Optically biaxial crystals in which the three principal optical directions

coincide with the three crystallographic axes—corresponding with the orthorhombic system.

(4) Optically biaxial crystals in which only one of the three principal optical directions coincides with a crystallographic axis—corresponding with the monoclinic system.

(5) Optically biaxial crystals in which there is no fixed and definite relation between the optical and crystallographic directions—corresponding with the anorthic system.

Optically Isotropic Crystals.—These belong to the cubic system, and like all other optically isotropic (from ἴσος, like, and τρόπος, character) bodies have only one index of refraction for light of each colour. They have no action on polarized light (except in crystals which are circularly polarizing); and when examined in the polariscope or polarizing microscope they remain dark between crossed nicols, and cannot therefore be distinguished optically from amorphous substances, such as glass and opal.

Optically Uniaxial Crystals.—These belong to the tetragonal and hexagonal (including rhombohedral) systems, and between crystals of these systems there is no optical distinction. Such crystals are anisotropic or doubly refracting (see Refraction: *Double*); but for light travelling through them in a certain, single direction they are singly refracting. This direction, which is called the optic axis, is the same for light of all colours and at all temperatures; it coincides in direction with the principal crystallographic axis, which in tetragonal crystals is a tetrad (or dyad) axis of symmetry, and in the hexagonal system a triad or hexad axis.

For light of each colour there are two indices of refraction; namely, the ordinary index (ω) corresponding with the ordinary ray, which vibrates perpendicular to the optic axis; and the extraordinary index (ε) corresponding with the extraordinary ray, which vibrates parallel to the optic axis. If the ordinary index of refraction be greater than the extraordinary index, the crystal is said to be optically negative, whilst if less the crystal is optically positive. The difference between the two indices is a measure of the strength of the double refraction or birefringence. Thus in calcite, for sodium (D) light, $\omega = 1.6585$ and $\varepsilon = 1.4863$; hence this substance is optically negative with a relatively high double refraction of $\omega - \varepsilon = 0.1722$. In quartz $\omega = 1.5442$, $\varepsilon = 1.5533$ and $\varepsilon - \omega = 0.0091$; this mineral is therefore optically positive with low double refraction. The indices of refraction vary, not only for light of different colours, but also slightly with the temperature.

The optical characters of uniaxial crystals are symmetrical not only with respect to the full number of planes and axes of symmetry of tetragonal and hexagonal crystals, but also with respect to all vertical planes, *i.e.* all planes containing the optic axis. A surface expressing the optical relations of such crystals is thus an ellipsoid of revolution about the optic axis. (In cubic crystals the corresponding surface is a sphere.) In the "optical indicatrix" (L. Fletcher, *The Optical Indicatrix and the Transmission of Light in Crystals*, London, 1892), the length of the principal axis, or axis of rotation, is proportional to the index of refraction, (*i.e.* inversely proportional to the velocity) of the extraordinary rays, which vibrate along this axis and are transmitted in directions perpendicular thereto; the equatorial diameters are proportional to the index of refraction of the ordinary rays, which vibrate perpendicular to the optic axis. For positive uniaxial crystals the indicatrix is thus a prolate spheroid (egg-shaped), and for negative crystals an oblate spheroid (orange-shaped).

In "Fresnel's ellipsoid" the axis of rotation is proportional to the velocity of the extraordinary ray, and the equatorial diameters proportional to the velocity of the ordinary ray; it is therefore an oblate spheroid for positive crystals, and a prolate spheroid for negative crystals. The "ray-surface," or "wave-surface," which represents the distances traversed by the rays during a given interval of time in various directions from a point of origin within the crystal, consists in uniaxial crystals of two sheets; namely, a sphere, corresponding to the ordinary rays, and an ellipsoid of revolution, corresponding to the extraordinary rays. The difference in form of the ray-surface for positive and negative crystals is shown in figs. 95 and 96.

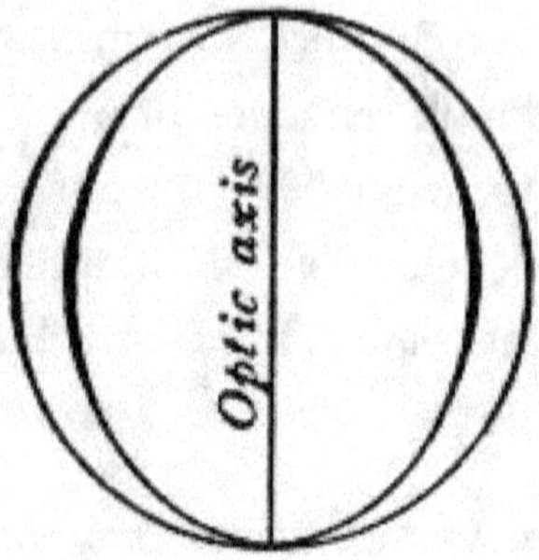

Fig. 95.—Section of the Ray-Surface of a Positive Uniaxial Crystal.

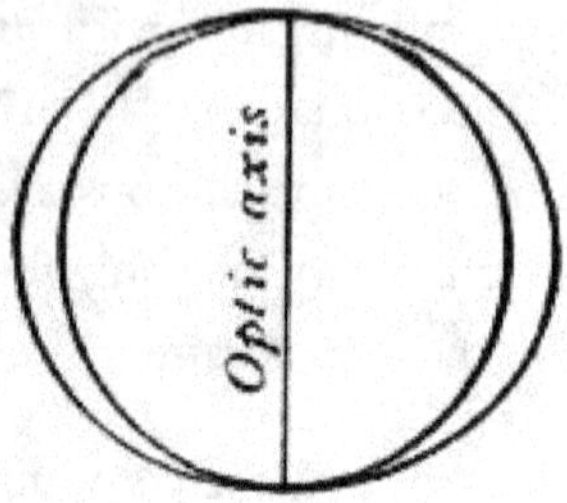

Fig. 96.—Section of the Ray-Surface of a Negative Uniaxial Crystal.

When a uniaxial crystal is examined in a polariscope or polarizing microscope between crossed nicols (*i.e.* with the principal planes of the polarizer or analyser at right angles, and so producing a dark field of view) its behaviour differs according to the direction in which the light travels through the crystal, to the position of the crystal with respect to the principal planes of the nicols, and further, whether convergent or parallel polarized light be employed. A tetragonal or hexagonal crystal viewed, in parallel light, through the basal plane, *i.e.* along the principal axis, will remain dark as it is rotated between crossed nicols, and will thus not differ in its behaviour from a cubic crystal or other isotropic body. If, however, the crystal be viewed in any other direction, for example, through a prism face, it will, except in certain positions, have an action on the polarized light. A plane-polarized ray entering the crystal will be resolved into two polarized rays with the directions of vibration parallel to the vibration-directions in the crystal. These two rays on leaving the crystal will be combined again in the analyser, and a portion of the light transmitted through the instrument; the crystal will then show up brightly against the dark field. Further, owing to interference of these two rays in the analyser, the light will be brilliantly coloured, especially if the crystal be thin, or if a thin section of a crystal be examined. The particular colour seen will depend on the strength of the double refraction, the orientation of the crystal or section, and upon its thickness. If now, the crystal be rotated with the stage of the microscope, the nicols remaining fixed in position, the light transmitted through the instrument will vary in intensity, and in certain positions will be cut out altogether. The latter happens when the vibration-directions of the crystal are parallel to the vibration-directions of the nicols (these being indicated by cross-wires in the microscope). The crystal, now being dark, is said to be in position of extinction; and as it is turned through a complete rotation of 360° it will extinguish four times. If a prism face be viewed through, it will be seen that, when the crystal is in a position of extinction, the cross-wires of the microscope are parallel to the edges of the prism: the crystal is then said to give "straight extinction."

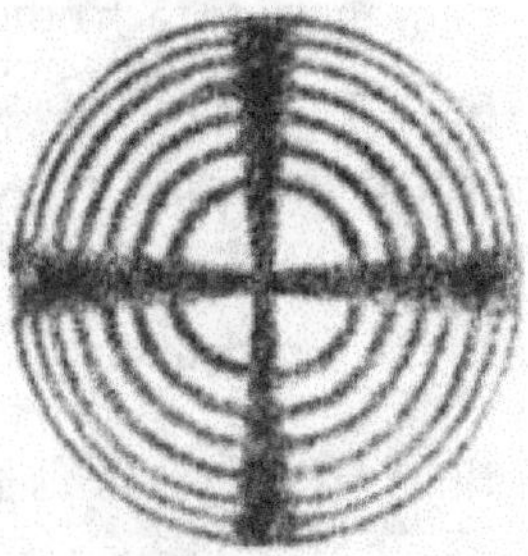

FIG. 97.—Interference Figure of a Uniaxial Crystal.

In convergent light, between crossed nicols, a very different phenomenon is to be observed when a uniaxial crystal, or section of such a crystal, is placed with its optic axis coincident with the axis of the microscope. The rays of light, being convergent, do not travel in the direction of the optic axis and are therefore doubly refracted in the crystal; in the analyser the vibrations will be reduced to the same plane and there will be interference of the two sets of rays. The result is an "interference figure" (fig. 97), which consists of a number of brilliantly coloured concentric rings, each showing the colours of the spectrum of white light; intersecting the rings is a black cross, the arms of which are parallel to the principal planes of the nicols. If monochromatic light be used instead of white light, the rings will be alternately light and dark. The number and distance apart of the rings depend on the strength of the double refraction and on the thickness of the crystal. By observing the effect produced on such a uniaxial interference figure when a "quarter undulation (or wave-length) mica-plate" is superposed on the crystal, it may be at once decided whether the crystal is optically positive or negative. Such a simple test may, for example, be applied for distinguishing certain faceted gem-stones: thus zircon and phenacite are optically positive, whilst corundum (ruby and sapphire) and beryl (emerald) are optically negative.

Optically Biaxial Crystals.—In these crystals there are three principal indices of refraction, denoted by α, β and γ; of these γ is the greatest and α the least ($\gamma > \beta > \alpha$). The three principal vibration-directions, corresponding to these indices, are at right angles to each other, and are the directions of the three rectangular axes of the optical indicatrix. The indicatrix (fig. 98) is an ellipsoid with the lengths of its axes proportional to the refractive indices; $OC = \gamma$, $OB = \beta$, $OA = \alpha$, where $OC > OB > OA$. The figure is symmetrical with respect to the principal planes *OAB*, *OAC*, *OBC*.

In Fresnel's ellipsoid the three rectangular axes are proportional to $1/\alpha$, $1/\beta$, and $1/\gamma$, and are usually denoted by **a**, **b** and **c** respectively, where **a** > **b** > **c:** these have often been called "axes of optical elasticity," a term now generally discarded.

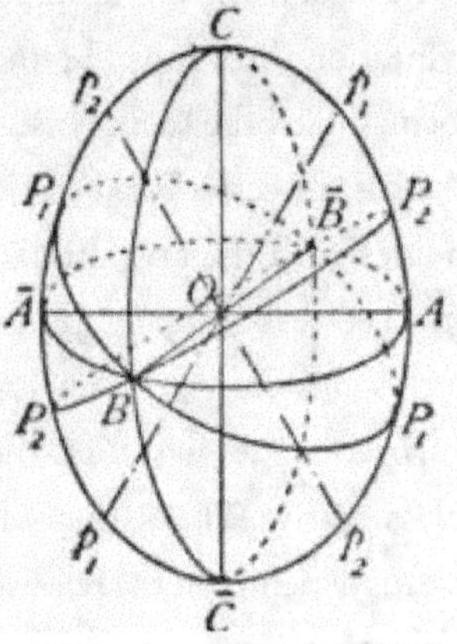

FIG. 98.—Optical Indicatrix of a Biaxial Crystal.

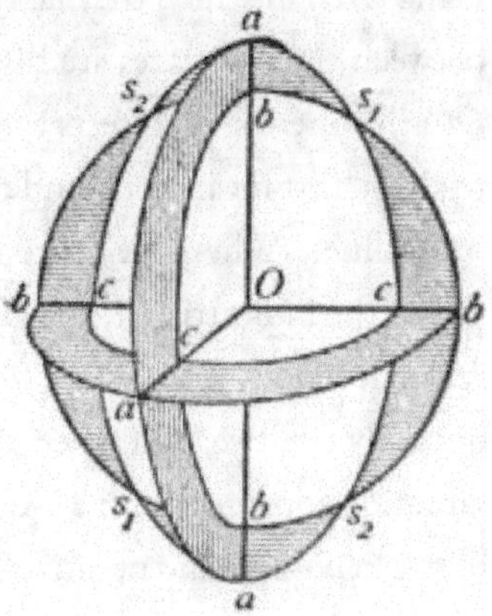

FIG. 99.—Ray-Surface of a Biaxial Crystal.

The ray-surface (represented in fig. 99 by its sections in the three principal planes) is derived from the indicatrix in the following manner. A ray of light entering the crystal and travelling in the direction *OA* is resolved into polarized rays vibrating parallel to *OB* and *OC*, and therefore propagated with the velocities $1/\beta$ and $1/\gamma$ respectively: distances *Ob* and *Oc* (fig. 99) proportional to these velocities are marked off in the direction *OA*. Similarly, rays travelling along *OC* have the velocities $1/\alpha$ and $1/\beta$, and those along *OB* the velocities $1/\alpha$ and $1/\gamma$. In the two directions Op_1 and Op_2 (fig. 98), perpendicular to the two circular sections P_1P_1 and P_2P_2 of the indicatrix, the two rays will be transmitted with the same velocity $1/\beta$. These two directions are called the optic axes ("primary optic axis"), though they have not all the properties which are associated with the optic axis of a uniaxial crystal. They have very nearly the same direction as the lines Os_1 and Os_2 in fig. 99, which are distinguished as the "secondary optic axes." In most crystals the primary and secondary optic axes are inclined to each other at not more than a few minutes, so that for practical purposes there is no distinction between them.

The angle between Op_1 and Op_2 is called the "optic axial angle"; and the plane *OAC* in which they lie is called the "optic axial plane." The angles between the optic axes are bisected by the vibration-directions *OA* and *OC*; the one which bisects the acute angle being called the "acute bisectrix" or "first mean line," and the other the "obtuse bisectrix" or "second mean line." When the acute bisectrix coincides with the greatest axis *OC* of the indicatrix, *i.e.* the vibration-direction corresponding with the refractive index γ (as in figs. 98 and 99), the crystal is described as being optically positive; and when the acute bisectrix coincides with *OA*, the vibration-direction for the index α, the crystal is negative. The distinction between positive and negative biaxial crystals thus depends on the relative magnitude of the three principal indices of refraction; in positive crystals β is nearer to α than to γ, whilst in negative crystals the reverse is the case. Thus in topaz, which is optically positive, the refractive indices for sodium light are $\alpha = 1.6120$, $\beta = 1.6150$, $\gamma = 1.6224$; and for orthoclase which is optically negative, $\alpha = 1.5190$, $\beta = 1.5237$, $\gamma = 1.5260$. The difference $\gamma - \alpha$ represents the strength of the double refraction.

Since the refractive indices vary both with the colour of the light and with the temperature, there will be for each colour and temperature slight differences in the form of both the indicatrix and the ray-surface: consequently there will be variations in the positions of the optic axes and in the size of the optic axial angle. This phenomenon is known as the "dispersion of the optic axes." When the axial angle is greater for red light than for blue the character of the dispersion is expressed by $\rho > \upsilon$, and when less by $\rho < \upsilon$. In some crystals, *e.g.* brookite, the optic axes for red light and for blue light may be, at certain temperatures, in planes at right angles.

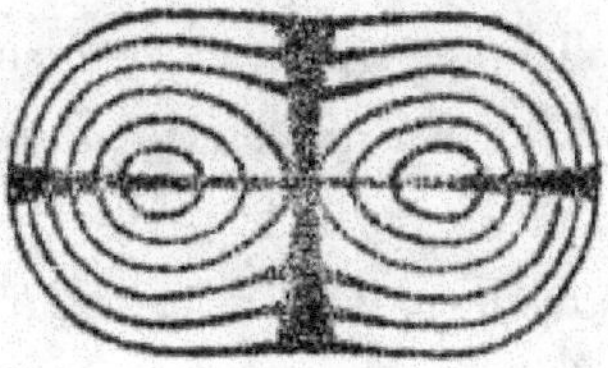

FIG. 100.

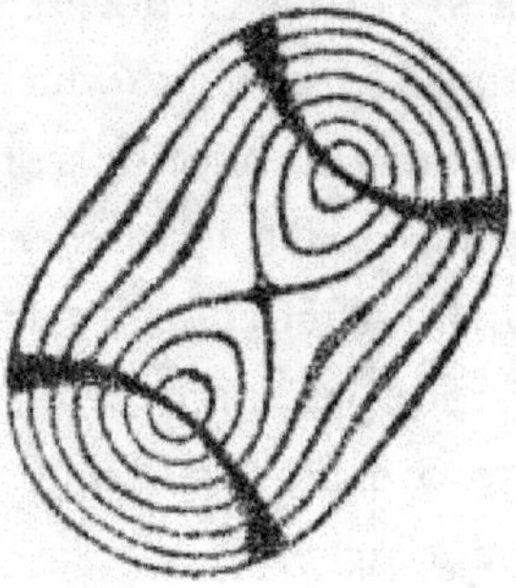

FIG. 101.

Interference Figures of a Biaxial Crystal.

The type of interference figure exhibited by a biaxial crystal in convergent polarized light between crossed nicols is represented in figs. 100 and 101. The crystal must be viewed along the acute bisectrix, and for this purpose it is often necessary to cut a plate from the crystal perpendicular to this direction: sometimes, however, as in mica and topaz, a cleavage flake will be perpendicular to the acute bisectrix. When seen in white light, there are around each optic axis a series of brilliantly coloured ovals, which at the centre join to form an 8-shaped loop, whilst further from the centre the curvature of the rings is approximately that of lemniscates. In the position shown in fig. 100 the vibration-directions in the crystal are parallel to those of the nicols, and the figure is intersected by two black bands or "brushes" forming a cross. When, however, the crystal is rotated with the stage of the microscope the cross breaks up into the two branches of a hyperbola, and when the vibration-directions of the crystal are inclined at 45° to those of the nicols the figure is that shown in fig. 101. The points of emergence of the optic axes are at the middle of the hyperbolic brushes when the crystal is in the diagonal position: the size of the optic axial angle can therefore be directly measured with considerable accuracy.

In orthorhombic crystals the three principal vibration-directions coincide with the three crystallographic axes, and have therefore fixed positions in the crystal, which are the same for light of all colours and at all temperatures. The optical orientation of an orthorhombic crystal is completely defined by stating to which crystallographic planes the optic axial plane and the acute bisectrix are respectively parallel and perpendicular. Examined in parallel light between crossed nicols, such a crystal extinguishes parallel to the crystallographic axes, which are often parallel to the edges of a face or section; there is thus usually "straight extinction." The interference figure seen in convergent polarized light is symmetrical about two lines at right angles.

In monoclinic crystals only one vibration-direction has a fixed position within the crystal, being parallel to the ortho-axis (*i.e.* perpendicular to the plane of symmetry or the plane (010)). The other two vibration-directions lie in the plane (010), but they may vary in position for light of different colours and at different temperatures. In addition to dispersion of the optic axes there may thus, in crystals of this system, be also "dispersion of the bisectrices." The latter may be of one or other of three kinds, according to which of the three vibration-directions coincides with the ortho-axis of the crystal. When the acute bisectrix is fixed in position, the optic axial planes for different colours may be crossed, and the interference figure will then be symmetrical with respect to a point only ("crossed dispersion"). When the obtuse bisectrix is fixed, the axial planes may be inclined to one another, and the interference figure is symmetrical only about a line which is perpendicular to the axial planes ("horizontal dispersion"). Finally, when the vibration-direction corresponding to the refractive index β, or the "third mean line," has a fixed position, the optic axial plane lies in the plane (010), but the acute bisectrix may vary in position in this plane; the interference figure will then be symmetrical only about a line joining the optic axes ("inclined dispersion"). Examples of substances exhibiting these three kinds of dispersion are borax, orthoclase and gypsum respectively. In orthoclase and gypsum, however, the optic axial angle gradually diminishes as the crystals are heated, and after passing through a uniaxial position they open out in a plane at right angles to the one they previously occupied; the character of the dispersion thus becomes reversed in the two examples quoted. When examined in parallel light between crossed nicols monoclinic crystals will give straight extinction only in faces and sections which are perpendicular to the plane of symmetry (or the plane (010)); in all other faces and sections the extinction-directions will be inclined to the edges of the crystal. The angles between these directions and edges are readily measured, and, being dependent on the optical orientation of the crystal, they are often characteristic constants of the substance (see, *e.g.*, PLAGIOCLASE).

In anorthic crystals there is no relation between the optical and crystallographic directions, and the exact determination of the optical orientation is often a matter of considerable difficulty. The character of the dispersion of the bisectrices and optic axes is still more complex than in monoclinic crystals, and the interference figures are devoid of symmetry.

Absorption of Light in Crystals: Pleochroism.—In crystals other than those of the cubic system, rays of light with different vibration-directions will, as a rule, be differently absorbed; and the polarized rays on emerging from the crystal may be of different intensities and (if the observation be made in white light and the crystal is coloured) differently coloured. Thus, in tourmaline the ordinary ray, which vibrates perpendicular to the principal axis, is almost completely absorbed, whilst the extraordinary ray is allowed to pass through the crystal. A plate of tourmaline cut parallel to the principal axis may therefore be used for producing a beam of polarized light, and two such plates placed in crossed position form the polarizer or analyser of "tourmaline tongs," with the aid of which the interference figures of crystals may be simply shown. Uniaxial (tetragonal and hexagonal) crystals when showing perceptible differences in colour for the ordinary and extraordinary rays are said to be "dichroic." In biaxial (orthorhombic, monoclinic and anorthic) crystals, rays vibrating along each of the three principal vibration-directions may be differently absorbed, and, in coloured crystals, differently coloured; such crystals are therefore said to be "trichroic" or in general "pleochroic" (from πλέων, more, and χρόα, colour). The directions of maximum absorption in biaxial crystals have, however, no necessary relation with the axes of the indicatrix, unless these have fixed crystallographic directions, as in the orthorhombic system and the ortho-axis in the monoclinic. In epidote it has been shown that the two directions of maximum absorption which lie in the plane of symmetry are not even at right angles.

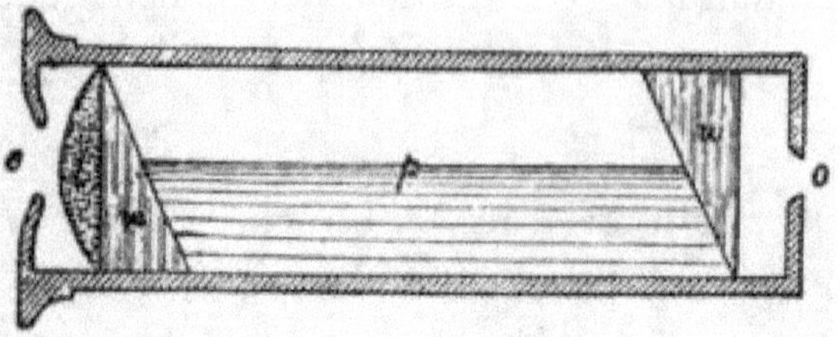

FIG. 102.—Dichroscope.

The pleochroism of some crystals is so strong that when they are viewed through in different directions they exhibit marked differences in colour. Thus a crystal of the mineral iolite (called also dichroite because of its strong pleochroism) will be seen to be dark blue, pale blue or pale yellow according to which of three perpendicular directions it is viewed. The "face colours" seen directly in this way result, however, from the mixture of two "axial colours" belonging to rays vibrating in two directions. In order to see the axial colours separately the crystal must be examined with a dichroscope, or in a polarizing microscope from which the analyser has been removed. The dichroscope, or dichroiscope (fig. 102), consists of a cleavage rhombohedron of calcite (Iceland-spar) *p*, on the ends of which glass prisms *w* are cemented: the lens *l* is focused on a small square aperture *o* in the tube of the instrument. The eye of the observer placed at *e* will see two images of the square aperture, and if a pleochroic crystal be placed in front of this aperture the two images will be differently coloured. On rotating this crystal with respect to the instrument the maximum difference in the colours will be obtained when the vibration-directions in the crystal coincide with those in the calcite. Such a simple instrument is especially useful for the examination of faceted gem-stones, even when they are mounted in their settings. A single glance suffices to distinguish between a ruby and a "spinel-ruby," since the former is dichroic and the latter isotropic and therefore not dichroic.

The characteristic absorption bands in the spectrum of white light which has been transmitted through certain crystals, particularly those of salts of the cerium metals, will, of course, be different according to the direction of vibration of the rays.

Circular Polarization in Crystals.—Like the solutions of certain optically active organic substances, such as sugar and tartaric acid, some optically isotropic and uniaxial crystals possess the property of rotating the plane of polarization of a beam of light. In uniaxial (tetragonal and hexagonal) crystals it is only for light transmitted in the direction of the optic axis that there is rotatory action, but in isotropic (cubic) crystals all directions are the same in this respect. Examples of circularly polarizing cubic crystals are sodium chlorate, sodium bromate, and sodium uranyl acetate; amongst tetragonal crystals are strychnine sulphate and guanidine carbonate; amongst rhombohedral are quartz (*q.v.*) and cinnabar (*q.v.*) (these being the only two mineral substances in which the phenomenon has been observed), dithionates of potassium, lead, calcium and strontium, and sodium periodate; and amongst hexagonal crystals is potassium lithium sulphate. Crystals of all these substances belong to one or other of the several symmetry-classes in which there are neither planes nor centre of symmetry, but only axes of symmetry. They crystallize in two complementary hemihedral forms, which are respectively right-handed and left-handed, *i.e.* enantiomorphous forms. Some other substances which crystallize in enantiomorphous forms are, however, only "optically active" when in solution (*e.g.* sugar and tartaric acid); and there are many other substances presenting this peculiarity of crystalline form which are not circularly polarizing either when crystallized or when in solution. Further, in the examples quoted above, the rotatory power is lost when the crystals are dissolved (except in the case of strychnine sulphate, which is only feebly active in solution). The rotatory power is thus due to different causes in the two cases, in the one depending on a spiral arrangement of the crystal particles, and in the other on the structure of the molecules themselves.

The circular polarization of crystals may be imitated by a pile of mica plates, each plate being turned through a small angle on the one below, thus giving a spiral arrangement to the pile.

"Optical Anomalies" of Crystals.—When, in 1818, Sir David Brewster established the important relations existing between the optical properties of crystals and their external form, he at the same time noticed many apparent exceptions. For example, he observed that crystals of leucite and boracite, which are cubic in external form, are always doubly refracting and optically biaxial, but with a complex internal structure; and that cubic crystals of garnet and analcite sometimes exhibit the same phenomena. Also some tetragonal and hexagonal crystals, *e.g.* apophyllite, vesuvianite, beryl, &c., which should normally be optically uniaxial, sometimes consist of several biaxial portions arranged in sectors or in a quite irregular manner. Such exceptions to the general rule have given rise to much discussion. They have often been considered to be due to internal strains in the crystals, set up as a result of cooling or by earth pressures, since similar phenomena are observed in chilled and compressed glasses and in dried gelatine. In many cases, however, as shown by E. Mallard, in 1876, the higher degree of symmetry exhibited by the external form of the crystals is the result of mimetic twinning, as in the pseudo-cubic crystals of leucite (*q.v.*) and boracite (*q.v.*). In other instances, substances not usually regarded as cubic, *e.g.* the monoclinic phillipsite (*q.v.*), may by repeated twinning give rise to pseudo-cubic forms. In some cases it is probable that the substance originally crystallized in one modification at a higher temperature, and when the temperature fell it became transformed into a dimorphous modification, though still preserving the external form of the original crystal (see Boracite). A summary of the literature is given by R. Brauns, *Die optischen Anomalien der Krystalle* (Leipzig, 1891).

(c) Thermal Properties.

FIG. 103.—Conductivity of Heat in Quartz.

The thermal properties of crystals present certain points in common with the optical properties. Heat rays are transmitted and doubly refracted like light rays; and surfaces expressing the conductivity and dilatation in different directions possess the same degree of symmetry and are related in the same way to the crystallographic axes as the ellipsoids expressing the optical relations. That crystals conduct heat at different rates in different directions is well illustrated by the following experiment. Two plates (fig. 103) cut from a crystal of quartz, one parallel to the principal axis and the other perpendicular to it, are coated with a thin layer of wax, and a hot wire is applied to a point on the surface. On the transverse section the wax will be melted in a circle, and on the longitudinal section (or on the natural prism faces) in an ellipse. The isothermal surface in a uniaxial crystal is therefore a spheroid; in cubic crystals it is a sphere; and in biaxial crystals an ellipsoid, the three axes of which coincide, in orthorhombic crystals, with the crystallographic axes.

With change of temperature cubic crystals expand equally in all directions, and the angles between the faces are the same at all temperatures. In uniaxial crystals there are two principal coefficients of expansion; the one measured in the direction of the principal axis may be either greater or less than that measured in directions perpendicular to this axis. A sphere cut from a uniaxial crystal at one temperature will be a spheroid at another temperature. In biaxial crystals there are different coefficients of expansion along three rectangular axes, and a sphere at one temperature will be an ellipsoid at another. A result of this is that for all crystals, except those belonging to the cubic system, the angles between the faces will vary, though only slightly, with changes of temperature. E. Mitscherlich found that the rhombohedral angle of calcite decreases 8′ 37″ as the crystal is raised in temperature from 0° to 100° C.

As already mentioned, the optical properties of crystals vary considerably with the temperature. Such characters as specific heat and melting-point, which do not vary with the direction, are the same in crystals as in amorphous substances.

(d) Magnetic and Electrical Properties.

Crystals, like other bodies, are either paramagnetic or diamagnetic, *i.e.* they are either attracted or repelled by the pole of a magnet. In crystals other than those belonging to the cubic system, however, the relative strength of the induced magnetization is different in different directions within the mass. A sphere cut from a tetragonal or hexagonal (uniaxial) crystal will if freely suspended in a magnetic field (between the poles of a strong electro-magnet) take up a position such that the principal axis of the crystal is either parallel or perpendicular to the lines of force, or to a line joining the two poles of the magnet. Which of these two directions is taken by the axis depends on whether the crystal is paramagnetic or diamagnetic, and on whether the principal axis is the direction of maximum or minimum magnetization. The surface expressing the magnetic character in different directions is in uniaxial crystals a spheroid; in cubic crystals it is a sphere. In orthorhombic, monoclinic and anorthic crystals there are three principal axes of magnetic induction, and the surface is an ellipsoid, which is related to the symmetry of the crystal in the same way as the ellipsoids expressing the thermal and optical properties.

Similarly, the dielectric constants of a non-conducting crystal may be expressed by a sphere, spheroid or ellipsoid. A sphere cut from a crystal will when suspended in an electro-magnetic field set itself so that the axis of maximum induction is parallel to the lines of force.

The electrical conductivity of crystals also varies with the direction, and bears the same relation to the symmetry as the thermal conductivity. In a rhombohedral crystal of haematite the electrical conductivity along the principal axis is only half as great as in directions perpendicular to this axis; whilst in a crystal of bismuth, which is also rhombohedral, the conductivities along and perpendicular to the axis are as 1.6 : 1.

Conducting crystals are thermo-electric: when placed against another conducting substance and the contact heated there will be a flow of electricity from one body to the other if the circuit be closed. The thermo-electric force depends not only on the nature of the substance, but also on the direction within the crystal, and may in general be expressed by an ellipsoid. A remarkable case is, however, presented by minerals of the pyrites group: some crystals of pyrites are more strongly thermo-electrically positive than antimony, and others more negative than bismuth, so that the two when placed together give a stronger thermo-electric couple than do antimony and bismuth. In the thermo-electrically positive crystals of pyrites the faces of the pentagonal dodecahedron are striated parallel to the cubic edges, whilst in the rarer negative crystals the faces are striated perpendicular to these edges. Sometimes both sets of striae are present on the same face, and the corresponding areas are then thermo-electrically positive and negative.

The most interesting relation between the symmetry of crystals and their electrical properties is that presented by the pyro-electrical phenomena of certain crystals. This is a phenomenon which may be readily observed, and one which often aids in the determination of the symmetry of crystals. It is exhibited by crystals in which there is no centre of symmetry, and the axes of symmetry are uniterminal or polar in character, being associated with different faces on the crystal at their two ends. When a non-conducting crystal possessing this hemimorphic type of symmetry is subjected to changes of temperature a charge of positive electricity will be developed on the faces in the region of one end of the uniterminal axis, whilst the faces at the opposite end will be negatively charged. With rising temperature the pole which becomes positively charged is called the "analogous pole," and that negatively charged the "antilogous pole": with falling temperature the charges are reversed. The phenomenon was first observed in crystals of tourmaline, the principal axis of which is a uniterminal triad axis of symmetry. In crystals of quartz there are three uniterminal dyad axes of symmetry perpendicular to the principal triad axis (which is here similar at its two ends): the dyad axes emerge at the edges of the hexagonal prism, alternate edges of which become positively and negatively charged on change of temperature. In boracite there are four uniterminal triad axes, and the faces of the two tetrahedra perpendicular to them will bear opposite charges. Other examples of pyro-electric crystals are the orthorhombic mineral hemimorphite (called also, for this reason, "electric calamine") and the monoclinic tartaric acid and cane-sugar, each of which possesses a uniterminal dyad axis of symmetry. In some exceptional cases, *e.g.* axinite, prehnite, &c., there is no apparent relation between the distribution of the pyro-electric charges and the symmetry of the crystals.

The distribution of the electric charges may be made visible by the following simple method, which may be applied even with minute crystals observed under the microscope. A finely powdered mixture of red-lead and sulphur is dusted through a sieve over the cooling crystal. In passing through the sieve the particles of red-lead and sulphur become electrified by mutual friction, the former positively and the latter negatively. The red-lead is therefore attracted to the negatively charged parts of the crystal and the sulphur to those positively charged, and the distribution of the charges over the whole crystal becomes mapped out in the two colours red and yellow.

Since, when a crystal changes in temperature, it also expands or contracts, a similar distribution of "piezo-electric" (from πιέζειν, to press) charges are developed when a crystal is subjected to changes of pressure in the direction of a uniterminal axis of symmetry. Thus increasing pressure along the principal axis of a tourmaline crystal produces the same electric charges as decreasing temperature.

III. RELATIONS BETWEEN CRYSTALLINE FORM AND CHEMICAL COMPOSITION.

That the general and physical characters of a chemical substance are profoundly modified by crystalline structure is strikingly illustrated by the two crystalline modifications of the element carbon—namely, diamond and graphite. The former crystallizes in the cubic system, possesses four directions of perfect cleavage, is extremely hard and transparent, is a non-conductor of heat and electricity, and has a specific gravity of 3.5; whilst graphite crystallizes in the hexagonal system, cleaves in a single direction, is very soft and opaque, is a good conductor of heat and electricity, and has a specific gravity of 2.2. Such substances, which are identical in chemical composition, but different in crystalline form and consequently in their physical properties, are said to be "dimorphous." Numerous examples of dimorphous substances are known; for instance, calcium carbonate occurs in nature either as calcite or as aragonite, the former being rhombohedral and the latter orthorhombic; mercuric iodide crystallizes from solution as red tetragonal crystals, and by sublimation as yellow orthorhombic crystals. Some substances crystallize in three different modifications, and these are said to be "trimorphous"; for example, titanium dioxide is met with as the minerals rutile, anatase and brookite (*q.v.*). In general, or in cases where more than three crystalline modifications are known (*e.g.* in sulphur no less than six have been described), the term "polymorphism" is applied.

On the other hand, substances which are chemically quite distinct may exhibit similarity of crystalline form. For example, the minerals iodyrite (AgI), greenockite (CdS), and zincite (ZnO) are practically identical in crystalline form; calcite ($CaCO_3$) and sodium nitrate ($NaNO_3$); celestite $(SrSO)_4$ and marcasite (FeS_2); epidote and azurite; and many others, some of which are no doubt only accidental coincidences. Such substances are said to be "homoeomorphous" (Gr. ὅμοιος, like, and μορφή, form).

Similarity of crystalline form in substances which are chemically related is frequently met with and is a relation of much importance: such substances are described as being "isomorphous." Amongst minerals there are many examples of isomorphous groups, *e.g.* the rhombohedral carbonates, garnet (*q.v.*), plagioclase (*q.v.*); and amongst crystals of artificially prepared salts isomorphism is equally common, *e.g.* the sulphates and selenates of potassium, rubidium and caesium. The rhombohedral carbonates have the general formula $R''CO_3$, where R″ represents calcium, magnesium, iron, manganese, zinc, cobalt or lead, and the different minerals (calcite, ankerite, magnesite, chalybite, rhodochrosite and calamine (*q.v.*)) of the group are not only similar in crystalline form, cleavage, optical and other characters, but the angles between corresponding faces do not differ by more than 1° or 2°. Further, equivalent amounts of the different chemical elements represented by R" are mutually replaceable, and two or more of these elements may be present together in the same crystal, which is then spoken of as a "mixed crystal" or isomorphous mixture.

In another isomorphous series of carbonates with the same general formula $R''CO_3$, where R″ represents calcium, strontium, barium, lead or zinc, the crystals are orthorhombic in form, and are thus dimorphous with those of the previous group (*e.g.* calcite and aragonite, the other members being only represented by isomorphous replacements). Such a relation is known as "isodimorphism." An even better example of this is presented by the arsenic and antimony trioxides, each of which occurs as two distinct minerals:—

As_2O_3, Arsenolite (cubic);
Claudetite (monoclinic).
Sb_2O_3, Senarmontite (cubic);
Valentinite (orthorhombic).

Claudetite and valentinite though crystallizing in different systems have the same cleavages and very nearly the same angles, and are strictly isomorphous.

Substances which form isodimorphous groups also frequently crystallize as double salts. For instance, amongst the carbonates quoted above are the minerals dolomite ($CaMg(CO_3)_2$) and barytocalcite ($CaBa(CO_3)_2$). Crystals of barytocalcite (*q.v.*) are monoclinic; and those of dolomite (*q.v.*), though closely related to calcite in angles and cleavage, possess a different degree of symmetry, and the specific gravity is not such as would result by a simple isomorphous mixture of the two carbonates. A similar case is presented by artificial crystals of silver nitrate and potassium nitrate. Somewhat analogous to double salts are the molecular compounds formed by the introduction of "water of crystallization," "alcohol of crystallization," &c. Thus sodium sulphate may crystallize alone or with either seven or ten molecules of water, giving rise to three crystallographically distinct substances.

A relation of another kind is the alteration in crystalline form resulting from the replacement in the chemical molecule of one or more atoms by atoms or radicles of a different kind. This is known as a "morphotropic" relation (Gr. μορφή, form, τρόπος, habit). Thus when some of the hydrogen atoms of benzene are replaced by (OH) and (NO_2) groups the orthorhombic system of crystallization remains the same as before, and the crystallographic axis a is not much affected, but the axis c varies considerably:—

	a	: b	: c
Benzene, C6H6	0.891	: 1	: 0.799
Resorcin, C6H4(OH)2	0.910	: 1	: 0.540
Picric acid, C6H2(OH)(NO2)3	0.937	: 1	: 0.974

A striking example of morphotropy is shown by the humite (*q.v.*) group of minerals: successive additions of the group Mg_2SiO_4 to the molecule produce successive increases in the length of the vertical crystallographic axis.

In some instances the replacement of one atom by another produces little or no influence on the crystalline form; this happens in complex molecules of high molecular weight, the "mass effect" of which has a controlling influence on the isomorphism. An example of this is seen in the replacement of sodium or potassium by lead in the alunite (*q.v.*) group of minerals, or again in such a complex mineral as tourmaline, which, though varying widely in chemical composition, exhibits no variation in crystalline form.

For the purpose of comparing the crystalline forms of isomorphous and morphotropic substances it is usual to quote the angles or the axial ratios of the crystal, as in the table of benzene derivatives quoted above. A more accurate comparison is, however, given by the "topic axes," which are calculated from the axial ratios and the molecular volume; they express the relative distances apart of the crystal molecules in the axial directions.

The two isomerides of substances, such as tartaric acid, which in solution rotate the plane of polarized light either to the right or to the left, crystallize in related but enantiomorphous forms.

References.—An introduction to crystallography is given in most text-books of mineralogy, *e.g.* those of H. A. Miers and of E. S. Dana (see Mineralogy). The standard work treating of the subject generally is that of P. Groth, *Physikalische Kristallographie* (4th ed., Leipzig, 1905). A condensed summary is given by A. J. Moses, *The Characters of Crystals* (New York, 1899).

For geometrical crystallography, dealing exclusively with the external form of crystals, reference may be made to N. Story-Maskelyne, *Crystallography, a Treatise on the Morphology of Crystals* (Oxford, 1895) and W. J. Lewis, *A Treatise on Crystallography* (Cambridge, 1899). Theories of crystal structure are discussed by L. Sohncke, *Entwickelung einer Theorie der Krystallstruktur* (Leipzig, 1879); A. Schoenflies, *Krystallsysteme und Krystallstructur* (Leipzig, 1891); and H. Hilton, *Mathematical Crystallography and the Theory of Groups of Movements* (Oxford, 1903).

The physical properties of crystals are treated by T. Liebisch, *Physikalische Krystallographie* (Leipzig, 1891), and in a more elementary form in his *Grundriss der physikalischen Krystallographie* (Leipzig, 1896); E. Mallard, *Traité de cristallographie, Cristallographie physique* (Paris, 1884); C. Soret, *Éléments de cristallographie physique* (Geneva and Paris, 1893).

For an account of the relations between crystalline form and chemical composition, see A. Arzruni, *Physikalische Chemie der Krystalle* (Braunschweig, 1893); A. Fock, *An Introduction to Chemical Crystallography*, translated by W. J. Pope (Oxford, 1895); P. Groth, *An Introduction to Chemical Crystallography*, translated by H. Marshall (London, 1906); A. E. H. Tutton, *Crystalline Structure and Chemical Constitution*, 1910. Descriptive works giving the crystallographic constants of different substances are C. F. Rammelsberg, *Handbuch der krystallographisch-physikalischen Chemie* (Leipzig, 1881-1882); P. Groth, *Chemische Krystallographie* (Leipzig, 1906); and of minerals the treatises of J. D. Dana and C. Hintze.

(L. J. S.)

---

1 From the Greek letter δ, Δ; in general, a triangular-shaped object; also an alternative name for a trapezoid.
2 Named after pyrites, which crystallizes in a typical form of this class.
3 From πλάγιος, placed sideways, referring to the absence of planes and centre of symmetry.
4 From γῦρος, a ring or spiral, and εἶδος, form.
5 From μόνος, single, and κλίειν, to incline, since one axis is inclined to the plane of the other two axes, which are at right angles.

CRYSTAL PALACE, THE

A well-known English resort, standing high up in grounds just outside the southern boundary of the county of London, in the neighbourhood of Sydenham. The building, chiefly of iron and glass, is flanked by two towers and is visible from far over the metropolis. It measures 1608 ft. in length by 384 ft. across the transepts, and was opened in its present site in 1854. The materials, however, were mainly those of the hall set up in Hyde Park for the Great Exhibition of 1851. The designer was Sir Joseph Paxton. In the palace there are various permanent exhibitions, while special exhibitions are held from time to time, also concerts, winter pantomimes and other entertainments. In the extensive grounds there is accommodation for all kinds of games: the final tie of the Association Football Cup and other important football matches are played here, and there are also displays of fireworks and other attractions.

CSENGERY, ANTON

(1822-1880), Hungarian publicist, and a historical writer of great influence on his time, was born at Nagyvárad on the 2nd of June 1822. He took, at an early date, a very active part in the literary and political movements immediately preceding the Hungarian Revolution of 1848. He and Baron Sigismund Kemény may be considered as the two founders of high-class Magyar journalism. After 1867 the greatest of modern Hungarian statesmen, Francis Deák, attached Csengery to his personal service, and many of the momentous state documents inspired or suggested by Deák were drawn up by Csengery. In that manner his influence, as represented by the text of many a statute regulating the relations between Austria and Hungary, is one of an abiding character. As a historical writer he excelled chiefly in brilliant and thoughtful essays on the leading political personalities of his time, such as Paul Nagy, Bertalan, Szemere and others. He also commenced a translation of Macaulay's *History*. He died at Budapest on the 13th of July 1880.

CSIKY, GREGOR

(1842-1891), Hungarian dramatist, was born on the 8th of December 1842 at Pankota, in the county of Arad. He studied Roman Catholic theology at Pest and Vienna, and was professor in the Priests' College at Temesvár from 1870 to 1878. In the latter year, however, he joined the Evangelical Church, and took up literature. Beginning with novels and works on ecclesiastical history, which met with some recognition, he ultimately devoted himself to writing for the stage. Here his success was immediate. In his *Az ellenállhatatlan* ("L'Irrésistible"), which obtained a prize from the Hungarian Academy, he showed the distinctive features of his talent—directness, freshness, realistic vigour, and highly individual style. In rapid succession he enriched Magyar literature with realistic *genre*-pictures, such as *A Proletárok* ("Proletariate"), *Buborckok* ("Bubbles"), *Két szerelem* ("Two Loves"), *A szégyenlös* ("The Bashful"), *Athalia*, &c., in all of which he seized on one or another feature or type of modern life, dramatizing it with unusual intensity, qualified by chaste and well-balanced diction. Of the latter, his classical studies may, no doubt, be taken as the inspiration, and his translation of Sophocles and Plautus will long rank with the most successful of Magyar translations of the ancient classics. Among the best known of his novels are *Arnold*, *Az Atlasz család* ("The Atlas Family"). He died at Budapest on the 19th of November 1891.

CSOKONAI, MIHALY VITEZ

(1773-1805), Hungarian poet, was born at Debreczen in 1773. Having been educated in his native town, he was appointed while still very young to the professorship of poetry there; but soon after he was deprived of the post on account of the immorality of his conduct. The remaining twelve years of his short life were passed in almost constant wretchedness, and he died in his native town, and in his mother's house, when only thirty-one years of age. Csokonai was a genial and original poet with something of the lyrical fire of Petöfi, and wrote a mock-heroic poem called *Dorottya or the Triumph of the Ladies at the Carnival*, two or three comedies or farces, and a number of love-poems. Most of his works have been published, with a life, by Schedel (1844-1847).

CSOMA DE KÖRÖS, ALEXANDER

(*c.* 1790-1842), or, as the name is written in Hungarian, KÖRÖSI CSOMA SÁNDOR, Hungarian traveller and philologist, born about 1790 at Körös in Transylvania, belonged to a noble family which had sunk into poverty. He was educated at Nagy-Enyed and at Göttingen; and, in order to carry out the dream of his youth and discover the origin of his countrymen, he divided his attention between medicine and the Oriental languages. In 1820, having received from a friend the promise of an annuity of 100 florins (about £10) to support him during his travels, he set out for the East. He visited Egypt, and made his way to Tibet, where he spent four years in a Buddhist monastery studying the language and the Buddhist literature. To his intense disappointment he soon discovered that he could not thus obtain any assistance in his great object; but, having visited Bengal, his knowledge of Tibetan obtained him employment in the library of the Asiatic Society there, which possessed more than 1000 volumes in that language; and he was afterwards supported by the government while he published a Tibetan-English dictionary and grammar (both of which appeared at Calcutta in 1834). He also contributed several articles on the Tibetan language and literature to the *Journal of the Asiatic Society of Bengal*, and he published an analysis of the *Kah-Gyur*, the most important of the Buddhist sacred books. Meanwhile his fame had reached his native country, and procured him a pension from the government, which, with characteristic devotion to learning, he devoted to the purchase of books for Indian libraries. He spent some time in Calcutta, studying Sanskrit and several other languages; but, early in 1842, he commenced his second attempt to discover the origin of the Hungarians, but he died at Darjiling on the 11th of April 1842. An oration was delivered in his honour before the Hungarian Academy by Eötvös, the novelist.

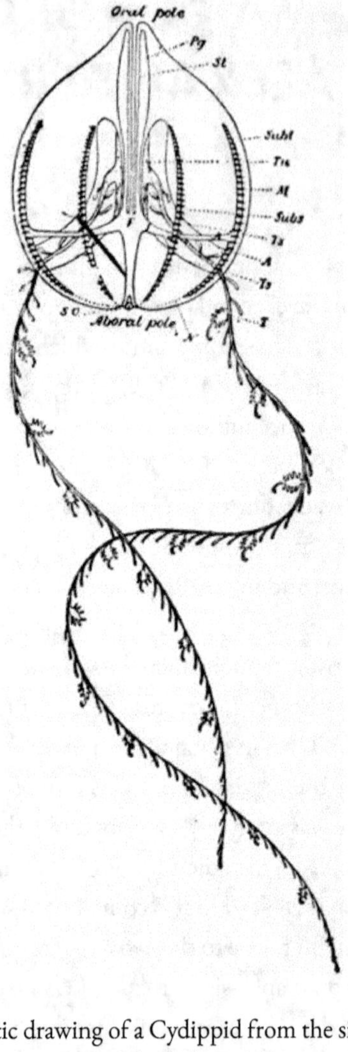

FIG. 1.—Schematic drawing of a Cydippid from the side. (After Chun.)

A, Adradial canals.

F, Infundibulum.

I, Interradial canal.

M, Meridianal canal lying under a costa.

N, Ciliated furrow from sense pole to costa.

Pg, Paragastric canal.

SO, Sense-organ.

St, Stomodaeum.

Subs, Subsagittal costa.
Subt, Subtentacular costa.
T, Tentacle.
Ts, Boundaries of tentacle-sheath.

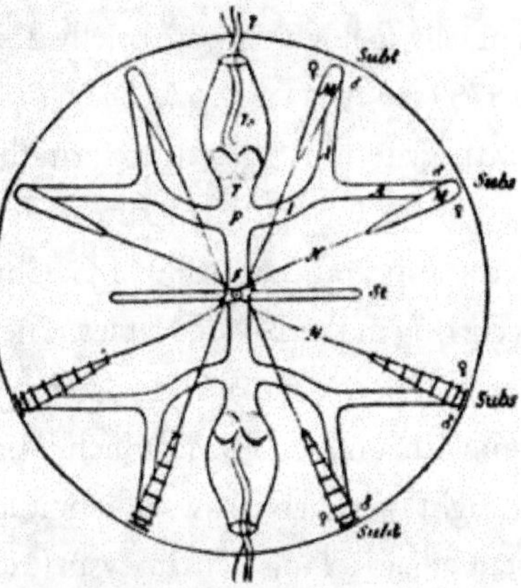

FIG. 2.—Schematic drawing of a Cydippid from the aboral pole. (After Chun.)
T (centrally), Tentacular canal, and (distally) tentacle.
□, Position of testes.
□, Position of ovaries; other letters in fig. 1. The stomodaeum lies in the sagittal plane, the funnel and tentacles in the transverse or tentacular plane.

CTENOPHORA,

In zoology, a class of jelly-fish which were briefly described by Professor T. H. Huxley in 1875 (see ACTINOZOA, *Ency. Brit.* 9th ed. vol. i.) as united with what we now term Anthozoa to form the group Actinozoa; but little was known of the intimate structure of those remarkable and beautiful forms till the appearance in 1880 of C. Chun's Monograph of the Ctenophora occurring in the Bay of Naples. They may be defined as Coelentera which exhibit both a radial and bilateral symmetry of organs; with a stomodaeum; with a mesenchyma which is partly gelatinous but partly cellular; with eight meridianal rows of vibratile paddles formed of long fused or matted cilia; lacking nematocysts (except in one genus). An example common on the British coasts is furnished by *Hormiphora* (*Cydippe*). In outward form this is an egg-shaped ball of clear jelly, having a mouth at the pointed (oral) pole, and a sense-organ at the broader (aboral) pole. It possesses eight meridians (costae) of iridescent paddles in constant vibration, which run from near one pole towards the other; it has also two pendent feathery tentacles of considerable length, which can be retracted into pouches. The mouth leads into an ectodermal stomodaeum ("stomach"), and the latter into an endodermal funnel (infundibulum); these two are compressed in planes at right angles to one another, the sectional long axis of the stomodaeum lying in the so-called sagittal (stomodaeal or gastric) plane, that of the funnel in the transverse (tentacular or funnel) plane. From the funnel, canals are given off in three directions; (*a*) a pair of paragastric (stomachal, or stomodaeal) canals run orally, parallel to the stomodaeum, and end blindly near the mouth; (*b*) a pair of perradial canals run in the transverse plane towards the equator of the animal; each of these becomes divided into two short canals at the base of the tentacle sheath which they supply, but has previously given off a pair of short interradial canals, which again bifurcate into two adradial canals; all these branches lie in the equatorial plane of the animal, but the eight adradial canals then open into eight meridianal canals which run orally and aborally under the costae; (*c*) a pair of aboral vessels which run towards the sense-organ, each of which bifurcates; of the four vessels thus formed, two only open at the sides of the sense-organ, forming the so-called excretory apertures. These three sets of structures, with the funnel from which they rise, make up the endodermal coelenteron, or gastro-vascular system. The generative organs are endodermal by origin, borne at the sides of the meridianal canals as indicated by the signs □ □. There exists a subepithelial plexus with nerve cells and fibres, similar to that of jelly-fishes. The sense-organ of the aboral pole is complex, and lies under a dome of fused cilia shaped like an inverted bell-jar; it consists of an otolith, formed of numerous calcareous spheroids, which is supported on four plates of fused cilia termed balancers, but is otherwise free. The ciliated ectoderm below the organ is markedly thickened, and perhaps functionally represents a nerve-ganglion: from it eight ciliated furrows radiate outwards, two passing under each balancer as through an archway, and diverge each to the head of a meridianal costa. These ciliated furrows stain deeply with osmic acid, and nervous impulses are certainly transmitted along them. Locomotion is effected by strokes of the paddles in an aboral direction, driving the animal mouth forwards through the water: each paddle or comb (Gr. κτείς; hence Ctenophora) consists of a plate of fused or matted cilia set transversely to the costa. The myoepithelial cells (formerly termed neuro-muscular cells), characteristic of other Coelentera, are not to be found in this group. On the other hand there are well-marked muscle fibres in definite layers, derived from special mesoblastic cells in the embryo, which are embedded in a jelly; these in their origin and arrangement are quite comparable to the mesoderm of Triploblastica, and, although the muscle-cells of some jelly-fish exhibit a somewhat similar condition, nothing so highly specialized as the mesenchyme of Ctenophora occurs in any other Coelenterate. The nematocysts being nearly absent from their group, their chief function is carried out by adhesive lasso-cells.

The Ctenophora are classified as follows:—

Sub-class i. Tentaculata,	Order	1. Cydippidea,	Hormiphora.
	"	2. Lobata,	Deiopea.
	"	3. Cestoidea,	Cestus.
" ii. Nuda,	"		Beroë.

The **Tentaculata**, as the name implies, may be recognized by the presence of tentacles of some sort. The CYDIPPIDEA are generally spherical or ovoid, with two long retrusible pinnate tentacles: the meridianal and paragastric canals end blindly. An example of these has already been briefly described. The LOBATA are of the same general type as the first Order, except for the presence of four circumoral auricles (processes of the subtransverse costae) and of a pair of sagittal outgrowths or lobes, on to which the subsagittal costae are continued. Small accessory tentacles lie in grooves, but there is no tentacular pouch; the meridianal vessels anastomose in the lobes. In the CESTOIDEA the body is compressed in the transverse plane, elongated in the sagittal plane, so as to become riband-like: the subtransverse costae are greatly reduced, the subsagittal costae extend along the aboral edge of the riband. The subsagittal canals lie immediately below their costae aborally, but continuations of the subtransverse canals round down the middle of the riband, and at its end unite, not only with the subsagittal but also with the paragastric canals which run along the oral edge of the riband. The tentacular bases and pouches are present, but there is no main tentacle as in Cydippidea; fine accessory tentacles lie in four grooves along the oral edge. The sub-class **Nuda** have no tentacles of any kind; they are conical or ovoid, with a capacious stomodaeum like the cavity of a thimble. There is a coelenteric network formed by anastomoses of the meridianal and paragastric canals all over the body.

The embryology of *Callianira* has been worked out by E. Mechnikov. Segmentation is complete and unequal, producing macromeres and micromeres marked by differences in the size and in yolk-contents. The micromeres give rise to the ectoderm; each of the sixteen macromeres, after budding off a small mesoblast cell, passes on as endoderm. A gastrula is established by a mixed process of embole and epibole. The mesoblast cells travel to the aboral pole of the embryo, and there form a cross-shaped mass, the arms of which lie in the sagittal and transverse planes (perradii).

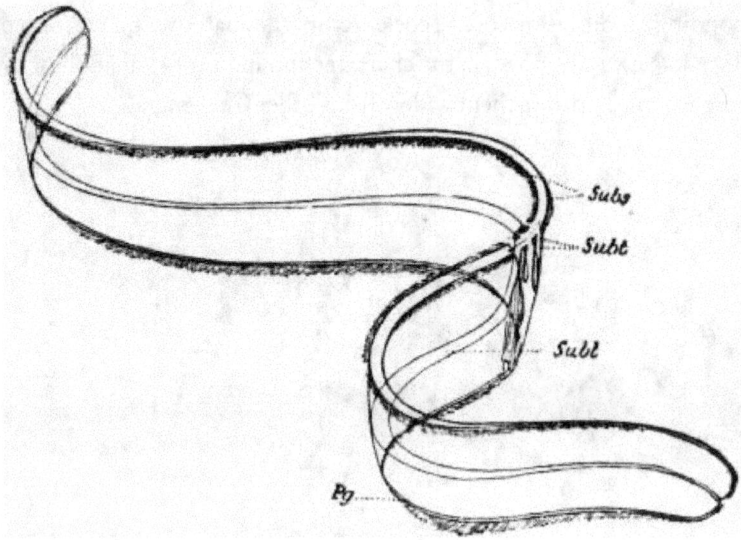

FIG. 3.—Schematic Drawing of Cestus. (After Chun.)

Subs, Subsagittal costae.
Subt, Much reduced subtentacular costae.
Subt, Branch of the subtentacular canal which runs along the centre of the riband.
Pg, Continuation of the paragastric canal at right angles to its original direction along the lower edge of the riband. At the right-hand end the last two are seen to unite with the subsagittal canal.

There can be but little question of the propriety of including Ctenophora among the Coelentera. The undivided coelenteron (gastro-vascular system) which constitutes the sole cavity of the body, the largely radial symmetry, the presence of endodermal generative organs on the coelenteric canals, the subepithelial nerve-plexus, the mesogloea-like matrix of the body—all these features indicate affinity to other Coelentera, but, as has been stated in the article under that title, the relation is by no means close. At what period the Ctenophora branched off from the line of descent, which culminated in the Hydromedusae and Scyphozoa of to-day, is not clear, but it is practically certain that they did so before the point of divergence of these two groups from one another. The peculiar sense-organ, the specialization of the cilia into paddles with the corresponding modifications of the coelenteron, the anatomy and position of the tentacles, and, above all, the character and mode of formation of the mesenchyme, separate them widely from other Coelentera.

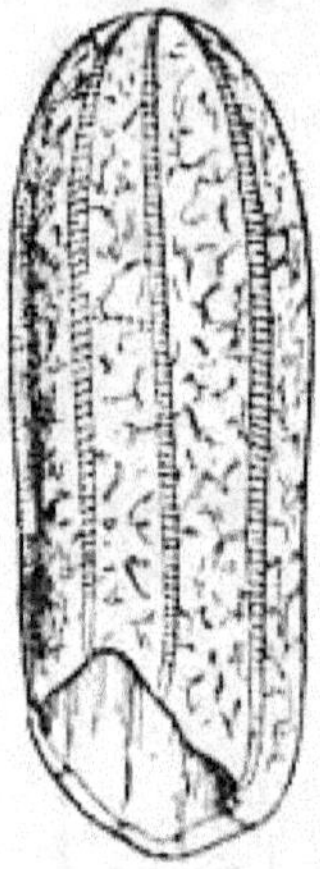

FIG. 4.—Schematic Drawing of *Beröe*. (After Chun.)

The last-named character, however, combined with the discovery of two remarkable organisms, *Coeloplana* and *Ctenoplana*, has suggested affinity to the flat-worms termed Turbellaria. *Ctenoplana*, the best known of these, has recently been redescribed by A. Willey (*Quart. Journ. Micr. Sci.* xxxix., 1896). It is flattened along the axis which unites sense-organ and mouth, so as to give it a dorsal (aboral) surface, and a ventral (oral) surface on which it frequently creeps. Its costae are very short, and retrusible; its two tentacles are pinnate and are also retrusible. Two crescentic rows of ciliated papillae lie in the transverse plane on each side of the sense-organ. The coelenteron exhibits six lobes, two of which Willey identifies with the stomodaeum of other Ctenophora; the other four give rise to a system of anastomosing canals such as are found in *Beroë* and Polyclad Turbellaria. An aboral vessel embraces the sense-organ, but has no external opening.

Ctenoplana is obviously a Ctenophoran flattened and of a creeping habit. *Coeloplana* is of similar form and habit, with two Ctenophoran tentacles: it has no costae, but is uniformly ciliated. These two forms at least indicate a possible stepping-stone from Ctenophora to Turbellaria, that is to say, from diploblastic to triploblastic Metazoa. By themselves they would present no very weighty argument for this line of descent from two-layered to three-layered forms, but the coincidences which occur in the development of Ctenophora and Turbellaria,—the methods of segmentation and gastrulation, of the separation of the mesoblast cells, and of mesenchyme formation,—together with the marked similarity of the adult mesenchyme in the two groups, have led many to accept this pedigree. In his Monograph on the Polyclad Turbellaria of the Bay of Naples, A. Lang regards a Turbellarian, so to say, as a Ctenophora, in which the sensory pole has rotated forwards in the sagittal plane through 90° as regards the original oral-aboral axis, a rotation which actually occurs in the development of *Thysanozoon* (Müller's larva); and he sees, in the eight lappets of the preoral ciliated ring of such a larva, the rudiments of the costal plates. According to his view, a simple early Turbellarian larva, such as that of *Stylochus*, most nearly represents for us to-day that ancestor from which Ctenophora and Turbellaria are alike derived. For details of this brilliant theory, the reader is referred to the original monograph.

Literature.—G. C. Bourne, "The Ctenophora," in Ray Lankester's *Treatise on Zoology* (1900), where a bibliography is given; G. Curreri, "Osservazioni sui ctenofori," *Boll. Soc. Zool. Ital.* (2), i. pp. 190-193 et ii. pp. 58-76; A. Garbe, "Untersuchungen über die Entstehung der Geschlechtsorgane bei den Ctenophoren.," *Zeitschr. Wiss. Zool.* lxix. pp. 472-491; K. C. Schneider, *Lehrbuch der vergleich. Histologie* (1902).

(G. H. Fo.)

CTESIAS

Of Cnidus in Caria, Greek physician and historian, flourished in the 5th century B.C. In early life he was physician to Artaxerxes Mnemon, whom he accompanied (401) on his expedition against his brother Cyrus the Younger. Ctesias was the author of treatises on rivers, and on the Persian revenues, of an account of India (which is of value as recording the beliefs of the Persians about India), and of a history of Assyria and Persia in 23 books, called *Persica*, written in opposition to Herodotus in the Ionic dialect, and professedly founded on the Persian royal archives. The first six books treated of the history of Assyria and Babylon to the foundation of the Persian empire; the remaining seventeen went down to the year 398. Of the two histories we possess abridgments by Photius, and fragments are preserved in Athenaeus, Plutarch and especially Diodorus Siculus, whose second book is mainly from Ctesias. As to the worth of the *Persica* there has been much controversy, both in ancient and modern times. Being based upon Persian authorities, it was naturally looked upon with suspicion by the Greeks and censured as untrustworthy.

For an estimate of Ctesias as a historian see G. Rawlinson's *Herodotus*, i. 71-74; also the edition of the fragments of the *Persica* by J. Giimore (1888, with introduction and notes and list of authorities).

CTESIPHON

A large village on the left bank of the Tigris, opposite to Seleucia, of which it formed a suburb, about 25 m. below Bagdad. It is first mentioned in the year 220 by Polybius v. 45. 4. When the Parthian Arsacids had conquered the lands east of the Euphrates in 129 B.C., they established their winter residence in Ctesiphon. They dared not stay in Seleucia, as this city, the most populous town of western Asia, always maintained her Greek self-government and a strong feeling of independence, which made her incline to the west whenever a Roman army attacked the Parthians. The Arsacids also were afraid of destroying the wealth and commerce of Seleucia, if they entered it with their large retinue of barbarian officials and soldiers (Strabo xvi. 743, Plin. vi. 122, cf. Joseph. *Ant.* xviii. 9, 2). From this time Ctesiphon increased in size, and many splendid buildings rose; it had the outward appearance of a large town, although it was by its constitution only a village. From A.D. 36-43 Seleucia was in rebellion against the Parthians till at last it was forced by King Vardanes to yield. It is very probable that Vardanes now tried to put Ctesiphon in its place; therefore he is called founder of Ctesiphon by Ammianus Marcellinus (xxiii. 6. 23), where King Pacorus (78-110) is said to have increased its inhabitants and built its walls. Seleucia was destroyed by the Romans in A.D. 164. When Ardashir I. founded the Sassanian empire (226), and fixed his residence at Ctesiphon, he built up Seleucia again under the name of Veh-Ardashir. Later kings added other suburbs; Chosroes I. in 540 established the inhabitants of Antiochia in Syria, whom he had led into captivity, in a new city, "Chosrau-Antioch" (or "the Roman city") near his residence. Therefore the Arabs designate the whole complex of towns which lay together around Seleucia and Ctesiphon and formed the residence of the Sassanids by the name Madāin, "the cities,"—their number is often given as seven. In the wars between the Roman and

Persian empires, Ctesiphon was more than once besieged and plundered, thus by Odaenathus in 261, and by Canis in 283; Julian in 363 advanced to Ctesiphon, but was not able to take it (Ammianus xxiv. 7). After the battle of Kadisiya (Qādisīya) Ctesiphon and the neighbouring towns were taken and plundered by the Arabs in 637, who brought home an immense amount of booty (see CALIPHATE). From then, these towns decayed before the increasing prosperity of the new Arab capitals Basra and Bagdad. The site is marked only by the ruins of one gigantic building of brick-work, called Takhti Khesra, "throne of Khosrau" (*i.e.* Chosroes). It is a great vaulted hall ornamented with pilasters, the remainder of the palace and the most splendid example of Sassanian architecture (see ARCHITECTURE, vol. ii. p. 558, for further details and illustration). (Ed. M.)

CUBA

(the aboriginal name), a republic, the largest and most populous of the West India Islands, included between the meridians of 74° 7′ and 84° 57′ W. longitude and (roughly) the parallels of 19° 48′ and 23° 13′ N. latitude. It divides the entrance to the Gulf of Mexico into two passages of nearly equal width,—the Strait of Florida, about 110 m. wide between Capes Hicacos in Cuba and Arenas in Florida (Key West being a little over 100 m. from Havana); and the Yucatan Channel, about 130 m. wide between Capes San Antonio and Catoche. On the N.E., E. and S.E., narrower channels separate it from the Bahamas, Haiti (50 m.) and Jamaica (85 m.). In 1908, by the opening of a railway along the Florida Keys, the time of passage by water between Cuba and the United States was reduced to a few hours.

The island is long and narrow, somewhat in the form of an irregular crescent, convex toward the N. It has a decided pitch to the S. Its length from Cape Maisí to Cape San Antonio along a medial line is about 730 m.; its breadth, which averages about 50 m., ranges from a maximum of 160 m. to a minimum of about 22 m. The total area is estimated at 41,634 sq. m. without the surrounding keys and the Isle of Pines (area about 1180 sq. m.), and including these is approximately 44,164. The geography of the island is still very imperfectly known, and all figures are approximate only. The coast line, including larger bays, but excluding reefs, islets, keys and all minute sinuosities, is about 2500 m. in length. The N. littoral is characterized by bluffs, which grow higher and higher toward the east, rising to 600 ft. at Cape Maisí. They are marked by distinct terraces. The southern coast near Cape Maisí is low and sandy. From Guantánamo to Santiago it rises in high escarpments, and W. of Santiago, where the Sierra Maestra runs close to the sea, there is a very high abrupt shore. To the W. of Manzanillo it sinks again, and throughout most of the remaining distance to Cape San Antonio is low, with a sandy or marshy littoral; at places sand hills fringe the shore; near Trinidad there are hills of considerable height; and the coast becomes high and rugged W. of Point Fisga, in the province of Pinar del Rio. On both the N. and the S. side of the island there are long chains of islets and reefs and coral keys (of which it is estimated there are 1300), which limit access to probably half of the coast, and on the N. render navigation difficult and dangerous. On the S. they are covered with mangroves. A large part of the southern littoral is subject to overflow, and much more of it is permanently marshy. The Zapata Swamp near Cienfuegos is 600 sq. m. in area; other large swamps are the Majaguillar, E. of Cárdenas, and the Ciénaga del Buey, S. of the Cauto river. The Isle of Pines in its northern part is hilly and wooded; in its southern part, very low, level and rather barren; a tidal swamp almost cuts the island in two. A remarkable feature of the Cuban coast is the number of excellent anchorages, roadsteads and harbours. On the N. shore, beginning at the W., Bahía Honda, Havana, Matanzas, Cardenas, Nuevitas and Nipe; and on the S. shore running westward Guantánamo, Santiago and Cienfuegos, are harbours of the first class, several of them among the best of the world. Mariel, Cabañas, Banes, Sagua la Grande and Baracoa on the N., and Manzanillo, Santa Cruz, Batabanó and Trinidad on the S. are also excellent ports or anchorages. The peculiar pouch-shape of almost all the harbours named (Matanzas being a marked exception) greatly increases their security and defensibility. These pouch harbours are probably "drowned" drainage basins. The number of small bays that can be utilized for coast trade traffic is extraordinary.

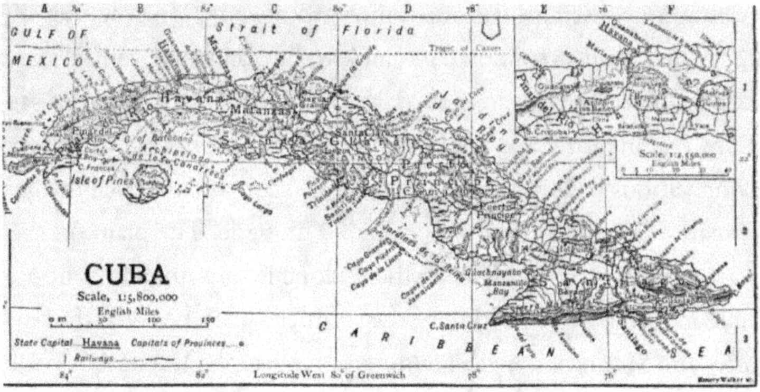

(Click to enlarge.)

In popular language the different portions of the island are distinguished as the Vuelta Abajo ("lower turn"), W. of Havana; the Vuelta Arriba ("upper turn"), E. of Havana to Cienfuegos—Vuelta Abajo and Vuelta Arriba are also used colloquially at any point in the island to mean "east" and "west"—Las Cinco Villas—*i.e.* Villa Clara, Trinidad, Remedios, Cienfuegos and Sancti Spiritus—between Cienfuegos and Sancti Spiritus; and Tierra Adentro, referring to the region between Cienfuegos and Bayamo. These names are extremely common. The province and city of Puerto Príncipe are officially known as Camagüey, their original Indian name, which has practically supplanted the Spanish name in local usage.

Five topographic divisions of the island are fairly marked. Santiago (now Oriente) province is high and mountainous. Camagüey is characterized by rolling, open plains, slightly broken, especially in the W., by low mountains. The E. part of Santa Clara province is decidedly rough and broken. The W. part, with the provinces of Matanzas and Havana, is flat and rolling, with occasional hills a few hundred feet high. Finally, Pinar del Rio is dominated by a prominent mountain range and by outlying piedmont hills and mesas. There are mountains in Cuba from one end of the island to the other, but they are not derived from any central mass and are not continuous. As just indicated there are three distinctively mountainous districts, various minor groups lying outside these. The three main systems are known in Cuba as the occidental, central and oriental. The first, the Organ mountains, in Pinar del Rio, rises in a sandy, marshy region near Cape San Antonio. The crest runs near the N. shore, leaving various flanking spurs and foothills, and a coastal plain which at its greatest breadth on the S. is some 20 m. wide. The plain on the N. is narrower and higher. The southern slope is smooth, and abounds in creeks and rivers. The portion of the southern plain between the bays of Cortés and Majana is the most famous portion of the Vuelta Abajo tobacco region. The mountain range is capriciously broken at points, especially near Bejucal. The highest part is the Pan de Guajaibón, near Bahía Honda, at the W. end of the chain; its altitude has been variously estimated from 2500 to 1950 ft. The central system has two wings, one approaching the N. coast, the other covering the island between Sancti Spiritus and Santa Clara. It comprehends a number of independent groups. The highest point, the Pico Potrerillo, is about 2900 ft. in altitude. The summits are generally well rounded, while the lower slopes are often steep. Frequent broad intervals of low upland or low level plain extend from sea to sea between and around the mountains. Near the coast runs a continuous belt of plantations, while grazing, tobacco and general farm lands cover the lower slopes of the hills, and virgin forests much of the uplands and mountains.

The oriental mountain region includes the province of Oriente and a portion of Camagüey. In extent, in altitude, in mass, in complexity and in geological interest, it is much the most important of the three systems. Almost all the mountains are very bold. They are imperfectly known. There are two main ranges, the Sierra Maestra, and a line of various groups along the N. shore. The former runs from Cape Santa Cruz eastward along the coast some 125 m. to beyond the river Baconao. The Sierra de Cobre, a part of the system in the vicinity of Santiago, has a general elevation of about 3000 ft. Monte Turquino, 7700-8320 ft. in altitude, is the highest peak of the island. Gran Piedra rises more than 5200 ft., the Ojo del Toro more than 3300, the Anvil de Baracoa is somewhat lower, and Pan de Matanzas is about 1267 ft. The western portions of the range rise abruptly from the ocean, forming a bold and beautiful coast. A multitude of ravines and gullies, filled with torrential streams or dry, according to the season of the year, and characterized by many beautiful cascades, seam the narrow coastal plain and the flanks of the mountains. The spurs of the central range are a highly intricate complex, covered with dense forests of superb woods. Many points are inaccessible, and the scenery is wild in the extreme. The mountains beyond Guantánamo are locally known by a variety of names, though topographically a continuation of the Sierra Maestra. The same is true of the chains that coalesce with these near Cape Maisí and diverge northwesterly along the N. coast of the island. The general character of this northern marginal system is much the same as that of the southern, save that the range is much less continuous. A dozen or more groups from Nipe in the E. to the coast N. of Camagüey in the W. are known only by individual names. The range near Baracoa is extremely wild and broken. The region between the lines of the two coastal systems is a much dissected plateau, imperfectly explored. The Cauto river, the only one flowing E. or W. and the largest of Cuba, flows through it westward to the southern coast near Manzanillo. The scenery in the oriental portion of the island is very beautiful, with wild mountains and tropical forests. In the central part there are extensive prairies. In the west there are swelling hills and gentle valleys, with the royal palm the dominating tree. The valley of the Yumurí, near Matanzas, a small circular basin crossed by a river that issues through a glen to the sea, is perhaps the most beautiful in Cuba.

A very peculiar feature of Cuba is the abundance of caverns in the limestone deposits that underlie much of the island's surface. The caves of Cotilla near Havana, of Bellamar near Matanzas, of Monte Libano near Guantánamo, and those of San Juan de los Remedios, are the best known, but there are scores of others. Many streams are "disappearing," part of their course being through underground tunnels. Thus the Rio San Antonio suddenly disappears near San Antonio de los Baños; the cascades of the Jatibónico del Norte disappear and reappear in a surprising manner; the Moa cascade (near Guantánamo) drops 300 ft. into a cavern and its waters later reissue from the earth; the Jojo river disappears in a great "sink" and later issues with violent current at the edge of the sea. The springs of fresh water that bubble up among the keys of the S. coast are also supposedly the outlets of underground streams.

The number of rivers is very great, but almost without exception their courses are normal to the coast, and they are so short as to be of but slight importance. The Cauto river in Oriente province is exceptional; it is 250 m. long, and navigable by small vessels for about 75 m. Inside the bar at its mouth (formed by a storm in 1616) ships of 200 tons can still ascend to Cauto. In Camagüey province the Jatibónico del Sur; in Oriente the Salado, a branch of the Cauto; in Santa Clara the Sagua la Grande (which is navigable for some 20 m. and has an important traffic), and the Damuji; in Matanzas, the Canimar; and in Pinar del Rio the Cuyaguateje, are important streams. The water-parting in the four central provinces is very indefinite. There are few river valleys that are noteworthy—those of the Yumurí, the Trinidad and the Güines. At Guantánamo and Trinidad are other valleys, and between Mariel and Havana is the fine valley of Ariguanabo. Of lakes, there are a few on the coast, and a very few in the mountains. The finest is Lake Ariguanabo, near Havana, 6 sq. m. in area. Of the almost innumerable river cascades, those of the Sierra Maestra Mountains, and in particular the Moa cascade, have already been mentioned. The Guamá cascade in Oriente province and the Hanabanilla Fall near Cienfuegos (each more than 300 ft. high), the Rosario Fall in Pinar del Rio, and the Almendares cascade near Havana, may also be mentioned.

Geology.—The foundation of the island is formed of metamorphic and igneous rocks, which appear in the Sierra Maestra and are exposed in other parts of the island wherever the comparatively thin covering of later beds has been worn away. A more or less continuous band of serpentine belonging to this series forms the principal watershed, although it nowhere rises to any great height. It is in this band that the greater part of the mineral wealth of Cuba is situated. These ancient rocks have hitherto yielded no fossils and their age is therefore uncertain, but they are probably pre-Cretaceous at least. Fossiliferous Cretaceous limestones containing *Rudistes* have been found in several parts of the island (Santiago de los Baños, Santa Clara province, &c.). At the base there is often an arkose, composed largely of fragments of serpentine and granite derived from the ancient floor. At Esperanza and other places in the Santa Clara province, bituminous plant-bearing beds occur beneath the Tertiary limestones, and at Baracoa a Radiolarian earth occupies a similar position. The latter, like the similar deposits in other West Indian islands, is probably of Oligocene age. It is the Tertiary limestones which form the predominant feature in the geology of Cuba. Although they do not exceed 1000 ft. in thickness, they probably at one time covered the whole island except the summits of the Sierra Maestra, where they have been observed, resting upon the older rocks, up to a height of 2300 ft. They contain corals, but are not coral reefs. The shells which have been found in them indicate that they belong for the most part to the Oligocene period. They are frequently very much disturbed and often strongly folded. Around the coast there is a raised shelf of limestone which was undoubtedly a coral reef. But it is of recent date and does not attain an elevation of more than 40 or 50 ft.

Minerals are fairly abundant in number, but few are present in sufficient quantity to be industrially important. Traditions of gold and silver, dating from the time of the Spanish conquest, still endure, but these metals are in fact extremely rare. Oriente province is distinctively the mineral province of the island. Large copper deposits of peculiar richness occur here in the Sierra de Cobre, near the city of Santiago; and both iron and manganese are abundant. Besides the deposits in Oriente province, iron is known to exist in considerable amount in Camagüey and Santa Clara, and copper in Camagüey and Pinar del Rio provinces. The iron ores mined at Daiquiri near Santiago are mainly rich hematites running above 60% of iron, with very little sulphur or phosphorus admixture. The copper deposits are mainly in well-marked fracture planes in serpentine; the ore is pyrrhotite, with or without chalcopyrite. Manganese occurs especially along the coast between Santiago and Manzanillo; the best ores run above 50%. Chromium and a number of other rare minerals are known to exist, but probably not in commercially available quantities. Bituminous products of every grade, from clear translucent oils resembling petroleum and refined naphtha, to lignite-like substances, occur in all parts of the island. Much of the bituminous deposits is on the dividing line between asphalt and coal. There is an endless amount of stone, very little of which is hard enough to be good for building material, the greatest part being a soft coralline limestone. The best buildings in Havana are constructed of a very rich white limestone, soft and readily worked when fresh, but hardening and slightly darkening with age. There are extensive and valuable deposits of beautiful marbles in the Isle of Pines, and lesser ones near Santiago. The Organ Mountains contain a hard blue limestone; and sandstones occur on the N. coast of Pinar del Rio province. Clays of all qualities and colours abound. Mineral waters, though not yet important in trade, are extremely abundant, and a score of places in Cuba and the Isle of Pines are already known as health resorts. Those near San Diego, Guanabacoa and Santa Maria del Rosario (near Havana) and Madruga (near Güines) are the best known.

The soil of the island is almost wholly of modern formation, mainly alluvial, with superficial limestones as another prominent feature. In the original formation of the island volcanic disturbances and coral growth played some part; but there are only very slight superficial evidences in the island of former volcanic activity. Noteworthy earthquakes are rare. They have been most common in Oriente province. Those of 1776, 1842 and 1852 were particularly destructive, and of earlier ones those of 1551 and 1624 at Bayamo and of 1578 and 1678 at Santiago. Every year there are seismic disturbances, and though Santiago is the point of most frequent visitation, they occur in all parts of the island, in 1880 affecting the entire western end. Notable seismic disturbances in Cuba have coincided with similar activity in Central America so often as to make some connexion apparent.

Flora.—The tropical heat and humidity of Cuba make possible a flora of splendid richness. All the characteristic species of the West Indies, the Central American and Mexican and southern Florida seaboard, and nearly all the large trees of the Mexican tropic belt, are embraced in it. As many as 3350 native flowering species were catalogued in 1876. The total number of species of the island flora was estimated in 1892 by a writer in the *Revista Cubana* (vol. xv. pp. 5-16) to be between 5000 and 6000, but hardly one-third of this number had then been gathered into a herbarium, and all parts of the island had not then been explored. It was estimated officially in 1904 that the wooded lands of the island comprised 3,628,434 acres, of which one-third were in Oriente province, another third in Camagüey, and hardly any in Havana province. Much of this area is of primeval forest; somewhat more than a third of the total, belonging to the government, was opened to sale (and speculative exspoliation) in 1904. The woods are so dense over large districts as to be impenetrable, except by cutting a path foot by foot through the close network of vines and undergrowth. The jagüey (*Ficus* sp.), which stifles in its giant coils the greatest trees of the forest, and the copei (*Clusia rosea*) are remarkable parasitic lianas. Of the palm there are more than thirty species. The royal palm is the most characteristic tree of Cuba. It attains a height of from 50 to 75 ft., and sometimes of more than 100 ft. Alone, or in groups, or in long aisles, towering above the plantations or its fellow trees of the forest, its beautiful crest dominates every landscape. Every portion, from its roots to its leaves, serves some useful purpose. From it the native draws lumber for his hut, utensils for his kitchen, thatch for his roof, medicines, preserved delicacies, and a long list of other articles. The corojo palm (*Cocos crispa*) rivals the royal palm in beauty and utility; oil, sugar, drink and wood are derived from it. The coco palm (*Cocos nucifera*) is also put to varied uses. The mango is planted with the royal palm along the avenues of the plantations. The beautiful ceiba (*Bombax ceiba* L., *Ceiba pentandra*) or silk cotton tree is the giant of the Cuban forests; it often grows to a height of 100 to 150 ft. with enormous girth. The royal piñon (*Erythrina velatina*) is remarkable for the magnificent purple flowers that cover it. The tamarind and banyan are also noteworthy. Utilitarian trees and plants are legion. There are at least forty choice cabinet and building woods. Of these, ebonies, mahogany (for the bird's-eye variety such enormous prices are paid as $1200 to $1800 per thousand board-feet), cullá (or cuyá, *Bumelia retusa*), cocullo (cocuyo, *Bumelia nigra*), ocuje (*Callophyllum viticifolia*, *Ornitrophis occidentalis*, *O. cominia*), jigüe (jique, *Lysiloma sabicu*), mahagua (*Hibiscus tiliaceus*), granadillo (*Brya ebenus*), icaquillo (*Licania incania*) and agua-baría (*Cordia gerascanthes*) are perhaps the most beautiful. Other woods, beautiful and precious, include guayacan (Guaiacum sanctum), baría (varía, *Cordia gerascanthoides*)—the fragrant, hard-wood Spanish elm—the quiebra-hacha (*Copaifera hymenofolia*), which three are of wonderful lasting qualities; the jiquí (*Malpighia obovata*), acana (*Achras disecta*, *Bassia albescens*), caigarán (or caguairan, *Hymenaea floribunda*), and the dagame (*Calicophyllum candidissimum*), which four, like the cullá, are all wonderfully resistant to humidity; the caimatillo (*Chrysophyllum oliviforme*), the yaya (or yayajabico, yayabito: *Erythalis fructicosa*, *Bocagea virgata*, *Guateria virgata*, *Asimina Blaini*), a magnificent construction wood; the maboa (*Cameraria latifolia*) and the jocuma (jocum: *Sideroxylon mastichodendron*, *Bumelia saticifolia*), all of individual beauties and qualities. Many species are rich in gums and resins; the calambac, mastic, copal, cedar, &c. Many others are oleaginous, among them, peanuts, sun-flowers, the bene seed (sesame), corozo, almond and palmachristi. Others (in addition to some already mentioned) are medicinal; as the palms, calabash, manchineel, pepper, fustic and a long list of cathartics, caustics, emetics, astringents, febrifuges, vermifuges, diuretics and tonics. Then, too, there are various dyewoods; rosewood, logwood (or campeachy wood), indigo, manajú (*Garcinia Morella*), Brazil-wood and saffron. Textile plants are extremely common. The majagua tree grows as high as 40 ft.; from its bark is made cordage of the finest quality, which is scarcely affected by the atmosphere. Strong, fine, glossy fibres are yielded by the exotic ramie (*Boehmeria nivea*), whose fibre, like that of the majagua, is almost incorruptible; by the maya or rat-pineapple (*Bromelia Pinguin*), and by the daquilla (or daiguiya—*Lagetta lintearia*, *L. valenzuelana*), which like the maya yields a brilliant, flexible product like silk; stronger cordage by the corojo palms, and various henequén plants, native and exotic (especially *Agave americana*, *A. Cubensis*); and various plantains, the exotic *Sansevieria guineensis*, okra, jute, *Laportea*, various lianas, and a great variety of reeds, supply varied textile materials of the best quality. The yucca is a source of starch. For building and miscellaneous purposes, in addition to the rare woods above named, there are cedars (used in great quantities for cigar boxes); the pine, found only in the W., where it gives its name to the Isle of Pines and the province of Pinar del Rio; various palms; oaks of varying hardness and colour, &c. The number of alimentary plants is extremely great. Among economic plants should be mentioned the coffee, cacao, citron, cinnamon, cocoa-nut and rubber tree. Wheat, Indian corn and many vegetables, especially tuberous, are particularly important. Plantain occurs in several varieties; it is in part a cheap and healthful substitute for bread, which is also made from the bitter cassava, after the poison is extracted. The sweet cassava yields tapioca. Bread-trees are fairly common, but are little cared for. White and sweet potatoes, yams, sweet and bitter yuccas, sago and okra, may also be mentioned.

Fruits are varied and delicious. The pineapple is the most favoured by Cubans. Four or five annual crops grow from one plant, but not more than three can be marketed, unless locally, as the product deteriorates. The better ("purple") varieties are mainly consumed in the island, and the smaller and less juicy "white" varieties exported. The tamarind is everywhere. Bananas are grown particularly in the region about Nipe, Gibara and Baracoa, whence they are exported in large quantities, though there is a tendency to lessen their culture in these parts in favour of sugar. Mangoes, though exotic, are extremely common, and in the E. grow wild in the forests. They are the favourite fruit of the negroes. Oranges are little cultivated, although they offer apparently almost unlimited possibilities; their culture decreased steadily after 1880, but after about 1900 was again greatly extended. Lemons yield continuously through the year, but like oranges, not much has yet been done with them commercially. Pomegranates are as universally used in Cuba as apples in the United. States. Figs and grapes degenerate in Cuba. Dates grow better, but nothing has been done with them. The coco-nut palm is most abundant in the vicinity of Baracoa. Among the common fruits are various anonas—the custard apple (*Anona cherimolia*), sweet-sop (*A. squamosa*), sour-sop (*A. muricata*), mamón (*A. reticulata*), and others,—the star-apple (*Chrysophyllum cainito, C. pomiferum*), rose-apple (*Eugenia jambos*), pawpaw, the sapodilla (*Sapota achras*), the caniste (*Sapota Elongata*), jagua (*Genipa americana*), alligator pear (*Persea gratissima*), the yellow mammee (*Mammea americana*) and so-called "red mammee" (*Lucuma mammosa*) and limes.

Fauna.—The fauna of Cuba, like the flora, is still imperfectly known. Collectively it shows long isolation from the other Antilles. Only two land mammals are known to be indigenous. One is the hutía (agouti) or Cuban rat, of which three species are known (*Capromys Fournieri*, *C. melanurus* and *C. Poey*). It lives in the most solitary woods, especially in the eastern hills. The other is a peculiar insectivore (*Solenodon paradoxus*), the only other representatives of whose family are found in Madagascar. Various animals, apparently indigenous, that are described by the early historians of the conquest, have disappeared. An Antillean rabbit is very abundant. Bats in prodigious numbers, and some of them of extraordinary size, inhabit the many caves of the island; more than twenty species are known. Rats and mice, especially the guayabita (*Mus musculus*), an extremely destructive rodent, are very abundant. The manatee, or sea-cow, frequents the mouths of rivers, the sargasso drifts, and the regions of submarine fresh-water springs off the coast. Horses, asses, cows, deer, sheep, goats, swine, cats and dogs were introduced by the early Spaniards. The last three are common in a wild state. Deer are not native, and are very rare; a few live in the swamps.

Of birds there are more than 200 indigenous species, it is said, and migratory species are also numerous. Waders are represented by more than fifty species. Vultures are represented by only one species, the turkey buzzard, which is the universal scavenger of the fields, and until recent years even of the cities, and has always been protected by custom and the Laws of the Indies. Falcons are represented by a score of species, at least, several of them nocturnal. Kestrels are common. The gallinaceous order is rich in *Columbidae*. Trumpeters are notably represented, and climbers still more so. Among the latter are species of curious habits and remarkable colouring. Woodpeckers (*Coloptes auratus*), macaws, parrakeets and other small parrots, and trogons, these last of beautifully resplendent plumage, deserve particular mention. The Cuban mocking-bird is a wonderful songster. Of humming-birds there are said to be sixty species, probably only one indigenous. Of the other birds mere mention may be made of the wild pigeon, raven, indigo-bird, English lady-bird and linnet.

Reptiles are numerous. Many tortoises are notable. The crocodile and cayman occur in the swampy littoral of the south. Of lizards the iguana (*Cyclura caudata*) is noteworthy. Chameleons are common. Snakes are not numerous, and it is said that none is poisonous or vicious. There is one enormous boa, the maja (*Epicrates angulifer*), which feeds on pigs, goats and the like, but does not molest man.

Fishes are present in even greater variety than birds. Felipa Poey, in his *Ictiologia Cubana*, listed 782 species of fish and crustaceans, of which 105 were doubtful; but more than one-half of the remainder were first described by Poey. The fish of Cuban waters are remarkable for their metallic colourings. The largest species are found off the northern coast. Food fishes are relatively not abundant, presumably because the deep sea escarpments of the N. are unfavourable to their life. Shell fish are unimportant. Two species of blind fish, of extreme scientific interest, are found in the caves of the island. Of the "percoideos" there are many genera. Among the most important are the robalo (*Labrax*), an exquisite food fish, the tunny, eel, Spanish sardine and mangua. Of the sharks the genus *Squalus* is represented by individuals that grow to a length of 26 to 30 ft. The hammer-head attains a weight at times of 600 □. The saw-fish is common. Of fresh-water fish the lisa, dogro, guayácón and viajocos (*Chromis fuscomaculatus*) are possibly the most noteworthy.

Molluscs are extraordinarily numerous; and many, both of water and land, are rarities among their kind for size and richness of colour. Of crustaceans, land-crabs are remarkable for size and number. Arachnids are prodigiously numerous. Insect life is abundant and beautiful. The bite of the scorpion and of the numerous spiders produces no serious effects. The nigua, the Cuban jigger, is a pest of serious consequence, and the mal de nigua (jigger sickness) sometimes causes the death of lower animals and men. Sand-flies and biting gnats are lesser nuisances. Lepidoptera are very brilliant in colouring. The cucujo or Cuban firefly (*Pyrophorus noctilucus*) gives out so strong a light that a few of them serve effectively as a lantern. The *Stegomyia* mosquito is the agent of yellow fever inoculation. Sponges grow in great variety.

Climate.—The climate of Cuba is tropical and distinctively insular in characteristics of humidity, equability and high mean temperature. There are two distinct seasons: a "dry" season from November to April, and a hotter, "wet" season. About two-thirds of the total precipitation falls in the latter. Droughts, extensive in area and in duration, are by no means uncommon. At Havana the mean temperature is about 76° F., with extreme monthly oscillations ranging on the average from 6° to 12° F. for different months, and with a range between the means of the coldest and warmest months of 10° (70° to 80°); temperatures below 50° or above 90° being rare. The mean rainfall at Havana is about 40.6 in. (sometimes over 80), and the mean absolute humidity of different months ranges from 70 to 80%. These figures represent fairly well the conditions of much of the northern coast. In the N.E. the rainfall is much greater. The equability of heat throughout the day is masked and relieved by the afternoon sea breezes. The trades are steady through the year, and in the dry season the western part of the island enjoys cool "northers." Despite this the interior is somewhat cooler than the coast, and in the uplands frost is not uncommon. The southern littoral is also (except in sheltered points such as Santiago, which is one of the hottest cities of the island) somewhat cooler than the northern.

More than eight or ten years rarely pass without tornadoes or hurricanes of local severity at least. Notably destructive ones occurred in 1768, 1774, 1842, 1844, 1846, 1865, 1870, 1876, 1885 and 1894. Those of 1842 and 1844 caused extreme distress in the island. In 1846, 300 vessels and 2000 houses were destroyed at Havana; in 1896 the banana groves of the N.E. coast were ruined and the banana industry prostrated; and in 1906 Havana suffered damage. The autumn months, particularly October and November, are those in which such storms most frequently occur.

Health.—Convincing evidence is offered by the qualities of the Spanish race in Cuba that white men of temperate lands can be perfectly acclimatized in this tropical island. As for diseases, some common to Cuba and Europe are more frequent or severe in the island, others rarer or milder. There are the usual malarial, bilious and intermittent fevers, and liver, stomach and intestinal complaints prevalent in tropical countries; but unhygienic living is, in Cuba as elsewhere, mainly responsible for their existence. Yellow fever (which first appeared in Cuba in 1647) was long the only epidemic disease, Havana being an endemic focus. Aside from the recurrent loss of life, the pecuniary loss from such epidemics was enormous, and the interference with commerce and social intercourse with other countries extremely vexatious. The Cuban coast was uninterruptedly full of infection, and the danger of an outbreak in each year was never absent, until the work of the United States army in 1901-1902 conclusively proved that this disease, though ineradicable by the most extreme sanitary measures, based on the accepted theory of its origin as a filth-disease, could be eradicated entirely by removing the possibility of inoculation by the *Stegomyia* mosquito. Since then yellow fever has ceased to be a scourge in Cuba. Small-pox was the cause of a greater mortality than yellow fever even before the means of combating the latter had been ascertained. The remarkable sanitary work begun during the American occupation and continued by the republic of Cuba, has shown that the ravages of this and other diseases can be greatly diminished. Leprosy is rather common, but seemingly only slightly contagious. Consumption is very prevalent.

Agriculture.—Soils are of four classes: calcareous-ferruginous, alluvial, argillous and silicious. Calcareous lands are predominant, especially in the uplands. Deep residual clay soils derived from underlying limestones, and coloured red or black according to the predominance of oxides of iron or vegetable detritus, characterize the plains. A red-black soil known as "mulatto" or tawny is perhaps the best fitted for general cultivation. Tobacco is most generally cultivated on loose red soils, which are rich in clays and silicates; and sugar-cane preferably on the black and mulatto soils; but in general, contrary to prevalent suppositions, colour is no test of quality and not a very valuable guide in the setting of crops. Almost without exception the lands throughout the island are of extreme fertility. The lowlands about Cienfuegos, Trinidad, Mariel and Matanzas are noted for their richness. The census of 1899 showed that farm lands occupied three-tenths of the total area; the cultivated area being one-tenth of the farms or 3% of the whole. At the end of 1905 it was officially estimated that 16% was in cultivation. In 1902 it was officially estimated that the public land available for permanent agrarian cultivation, including forest lands, was only 186,967 hectares (416,995 acres), almost wholly in the province of Oriente. The average size of a farm in 1899 was 143 acres. More than 85% of all cultivated lands were then occupied by whites; and somewhat more than one-half (56.6%) of all occupiers were renters. Holdings of more than 32 acres constituted only 7% of the total. As regards crops, 47% of the cultivated area was given over to sugar, 11% to sweet potatoes, 9% to tobacco and almost 9% to bananas. But owing to the disturbed conditions created by the war it is probable that these figures by no means represent normal conditions. The actual sugar crop of 1899-1900, for example, was not a quarter of that of 1894. With the establishment of peace in 1898 and the influx of American and other capital and of a heavy immigration, great changes took place in agriculture as in other industrial conditions.

Sugar has been the dominant crop since the end of the 18th century. Before the Civil War of 1895-1898 the capital invested in sugar estates was greater by half than that represented by tobacco and coffee plantations, live-stock ***Sugar.*** ranches and other farms. Since that time fruit and live-stock interests have increased. The dependence of the island on one crop has been an artificial economic condition often of grave momentary danger to prosperity; but generally speaking, the progress of the industry has been steady. The competition of the sugar-beet has been felt severely. During and after the war of 1868-1878, when many Cuban estates were confiscated, many families emigrated, and many others were ruined, the ownership of plantations largely passed from the hands of Cubans to Spaniards. Under the conditions of free labour, the development of railways abroad, the improvement of machinery both in cane and beet producing countries, the general competition of the beet, and the fall of prices, it was impossible for the Cuban industry to survive without radical betterment of methods. About 1885 began an immense development of centralization (the tendency having been evident many years before this). Plantations have increased greatly in size (and also diminished in number), greater capital is involved, bagasse furnaces have been introduced, double grinding mills have increased by more than a half the yield of juice from a given weight of cane, and extractive operations instead of being carried on on all plantations have been (since 1880) concentrated in comparatively few "centrals" (168 in Feb. 1908). Three-fourths of all are in the jurisdictions of Cienfuegos, Cárdenas, Havana, Matanzas and Sagua la Grande, which are the great sugar centres of the island (three-fourths of the crop coming from Matanzas and Santa Clara provinces). Caibarién, Guantánamo and Manzanillo are next in importance. A comparatively low cost of labour, the fact that labour is not, as in the days of slavery, that of unintelligent blacks but of intelligent free labourers, the centralized organization and modern methods that prevail on the plantations, the remarkable fertility of the soil (which yields 5 or 6 crops on good soil and with good management, without replanting), and the proximity of the United States, in whose markets Cuba disposes of almost all her crop, have long enabled her to distance her smaller West Indian rivals and to compete with the bounty-fed beet. The methods of cultivation, however, are still distinctly extensive, and the returns are much less than they would be (and in some other cane countries are) under more intensive and scientific methods of cultivation. Indeed, conditions were relatively primitive so late as 1880, if compared with those of other sugar-producing countries. More than four-fifths of the total area sown to cane in the island is in the three provinces of Santa Clara, Matanzas and Oriente (formerly Santiago), the former two representing two-thirds of the area and three-fourths of the crop. The majority of the sugar estates are of an area less than 3000 acres, and the most common area is between 1500 and 2000 acres; but the extremes range from a very small size to 60,000 acres. Only a part of the great estates is ever planted in any one season. The most profitable unit is calculated to be a daily consumption of 1500 tons of cane, or 150,000 in a grinding season of 100 days, which implies a feeding area not above 6000 acres. In the season of 1904-1905, which may be taken as typical, 179 estates, with a planted area of 431,056 acres, produced 11,576,137 tons of cane, and yielded—in addition to alcohol, brandy and molasses—1,089,814 tons of sugar. Of this amount 416,862 tons were produced by 24 estates yielding more than 11,000 tons each, including one (planting 28,050 acres) that yielded 33,609, and 4 others more than 22,000 tons each. The production of the island from 1850 to 1868 averaged 469,934 tons yearly, rising from 223,145 to 749,000; from 1869 to 1886 (continuing high during the period of the Ten Years' War), 632,003 tons; from 1887 to 1907—omitting the five years 1896-1900 when the industry was prostrated by war,—909,827 tons (and including the war period, 758,066); and in the six harvests of 1901-1906, 1,016,899 tons. Prior to 1902 the million mark, was reached only twice—in 1894 and 1895. Following the resuscitation of the industry after the last war, the island's crop rose steadily from one-sixth to a full quarter of the total cane sugar output of the world, its share in the world's product of sugar of all kinds ranging from a tenth to an eighth. Of this enormous output, from 98.3% upward went to the United States;[1] of whose total importation of all sugars and of cane sugar the proportion of Cuban cane—steadily rising—was respectively 49.8 and 53.7% in the seasons of 1900-1901 and 1904-1905.

If sugar is the island's greatest crop, tobacco is her most renowned in the markets of the world. Three-fourths of the tobacco of Cuba comes from Pinar del Rio province; the rest mainly from the provinces of Havana and ***Tobacco.*** Santa Clara,—the description *de partido* being applied to the leaf not produced in Havana and Pinar del Rio provinces, and sometimes to all produced outside the *vuelta abajo*. This district, including the finest land, is on the southern slope of the Organ Mountains between the Honda river and Mantua; bananas are cultivated with the tobacco. "Vegas" (tobacco fields) of especially good repute are also found near Trinidad, Remedios, Yara, Mayarí and Vicana. The tobacco industry has been uniformly prosperous, except when crippled by the destruction of war in 1868-1878 and 1895-1898. Even in the time of slavery tobacco was generally a white-man's crop; for it requires intelligent labour and intensive care. In recent years the growth of the leaf under cloth tents has greatly increased, as it has been abundantly proved that the product thus secured is much more valuable—lighter in colour and weight, finer in texture, with an increased proportion of wrapper leaves, and more uniform qualities, and with lesser amounts of cellulose, nicotine, gums and resins. In these respects the finest Cuban tobacco crops, produced in the sun, hardly rival the finest Sumatra product; but produced under cheese-cloth they do. "Cuban tobacco" does not mean to-day, as a commercial fact, what the words imply; for the original *Nicotiana Tabacum*, variety *havanensis*, can probably be found pure to-day only in out-of-the-way corners of Pinar del Rio. After the Ten Year's War seed of Mexican and United States tobaccos was in great demand to re-seed the ruined vegas, and was introduced in great quantities; and although by a later law the destruction of these exotic species was ordered, that destruction was in fact quite impossible. "Lusty growers and coarser than the genuine old-time Cuban ... Mexican tobaccos (*Nicotiana Tabacum*, variety *macrophyllum*) are to-day predominant in a large part of Cuban vegas.... Ordinary commercial Cuban seed of to-day is largely, and often altogether, Mexican tobacco." Though improved in the Cuban environment, the foreign tobaccos introduced after the Ten Years' War did not lose their exotic character, but prevailed over the indigenous forms: "Tobaccos with exactly the character of the introduced types are now the prevalent forms" (quotation from Bulletin of the *Estación Central Agronómica*, Feb. 1908). In the markets of the world Cuban tobacco has always suffered less competition than Cuban sugar, and still less has been done than in the case of sugar cane in the study of methods of cultivation, which in several respects are far behind those of other tobacco-growing countries. The crop of 1907 was 201,512 bales (109,562,400 □ Sp.).

Coffee-raising was once a flourishing and very promising industry. It first attained prominence with the settlement in eastern Cuba, late in the 18th century, of French refugee immigrants from San Domingo. Some "cafetales" ***Coffee.*** were established by the newcomers near Havana, but the industry has always been almost exclusively one of Oriente province; with Santa Clara as a much smaller producer. Before the war of 1868-1878 the production amounted to about 25,000,000 □ yearly. The war of 1895-1898 still further diminished the vitality of the industry. In 1907 the crop was 6,595,700 □. The berries are of fine quality, and despite the competition of Brazil there is no (agricultural) reason why the home market at least should not be supplied from Cuban estates.

Of other agricultural crops those of fruits are of greatest importance—bananas (which are planted about once in three years), pine-apples (planted about once in five years), coco-nuts, oranges, &c. The coco-nut industry has long been largely confined to the region about Baracoa, owing to the ruin of the trees elsewhere by a disease not yet thoroughly understood, which, appearing finally near Baracoa, threatened by 1908 to destroy the industry there as well. Yams and sweet-potatoes, yuccas, malangas, cacao, rice—which is one of the most important foods of the people, but which is not yet widely cultivated on a profitable basis—and Indian corn, which grows everywhere and yields two crops yearly, may be mentioned also. In very recent years gardening has become an interest of importance, particularly in the province of Pinar del Rio. Save on the coffee, tobacco and sugar plantations, where competition in large markets has compelled the adoption of adequate modern methods, agriculture in Cuba is still very primitive. The wooden ploughstick, for instance—taking the country as a whole—has never been displaced. A central agricultural experiment station (founded 1904) is maintained by the government at Santiago de las Vegas; but there is no agricultural college, nor any special school for the scientific teaching and improvement of sugar and tobacco farming or manufacture.

Stock-breeding is a highly important interest. It was the all-important one in the early history of the island, down to about the latter part of the 18th century. Grasses grow luxuriantly, and the savannahs of central Cuba are, in this respect, excellent cattle ranges. The droughts to which the island is recurrently subject are, however, a not unimportant drawback to the industry; and though the best ranges, under favourable conditions, are luxuriant, nevertheless the pastures of the island are in general mediocre. Practically nothing has yet been done in the study of native grasses and the introduction of exotic species. The possibilities of the stock interest have as yet by no means been realized. The civil wars were probably more disastrous to it than to any other agricultural interest of the island. It has been authoritatively estimated, for example, that from 90 to 95% of all horses, neat cattle and hogs in the entire island were lost in the war years of 1895-1898. In the decade after 1898 particularly great progress was made in the raising of live-stock. The fishing and sponge industries are important. Batabanó and Caibarién are centres of the sponge fisheries.

Manufactures.—The manufacturing industries of Cuba have never been more than insignificant as compared with what they might be. In

1907 48.5% of all wage-earners were engaged in agriculture, fishing and mining, 16.3 in manufactures, and 17.7 in trade and transportation. Such manufactures as are of any consequence are mostly connected with the sugar and tobacco industries. Forest resources have been but slightly touched (more so since the end of Spanish rule) except mahogany, which goes to the United States, and cedar, which is used to box the tobacco products of the island, much going also to the United States. The value of forest products in 1901-1902 amounted to $320,528. There are some tanneries, some preparation of preserves and other fruit products, and some old handicraft industries like the making of hats; but these have been of comparatively scant importance. Despite natural advantages for all meat industries, canned meats have generally been imported. The leading manufactures are cigars and cigarettes, sugar, rum and whisky. The tobacco industries are very largely concentrated in Havana, and there are factories in Santiago de las Vegas and Bejucal. The yearly output of cigars was locally estimated in 1908 at about 500,000,000, but this is probably too high an estimate. In 1904-1906 the yearly average sent to the United States was 234,063,652 cigars, 29,776,429 □ of leaf and 14,203,571 packages of cigarettes. The sugar industry is not similarly centralized. With the improvement of methods the old partially refined grades (moscobados) have disappeared.

Mining.—Mining is of very considerable importance. The Cobre copper mines near Santiago were once the greatest producers of the world. They were worked from 1524 until about 1730, when they were abandoned for almost a century, after which they were reopened and greatly developed. In 1828-1840 about two million dollars' worth of ore was shipped yearly to the United States alone. After 1868 the mines were again abandoned and flooded, the mining property being ruined during the civil war. Finally, after 1900 they again became prosperous producers. The "Cobre" mine is only the most famous and productive of various copper properties. The copper output has not greatly increased since 1890, and is of slight importance in mineral exports. Iron and manganese have, on the contrary, been greatly developed in the same period. Iron is now the most important mineral product. The iron ores are even more accessible than the famous ones of the Lake Superior region in the United States. No shafts or tunnels are necessary except for exploration; the mining consists entirely in open-cut and terrace work. The cost of exploitation is accordingly slight. Daiquiri, near Santiago, and mines near Nipe, on the north coast, are the chief centres of production. Nearly the entire product goes to the United States. The first exports from the Daiquiri district were made by an American company in 1884; the Nipe (Cagimaya) mines became prominent in promise in 1906. The shipments from Oriente province from 1884 to 1901 aggregated 5,053,847 long tons, almost all going to the United States (which is true of other mineral products also). After 1900 production was greatly increased and by 1906 had come to exceed half a million tons annually. There are small mines in Santa Clara and Camagüey provinces. Manganese is mined mainly near La Maya and El Cristo in Oriente. The traditions as to gold and silver have already been referred to. Evidences of ancient workings remain near Holguin and Gibara, and it is possible that some of these workings are still exploitable. Mining for the precious metals ceased at a very early date, after rich discoveries were made on the continent. Bituminous products, though, as already stated, widely distributed, are not as yet much developed. The most promising deposits and the most important workings are in Matanzas and Santa Clara provinces. Petroleum has been used to some extent both as a fuel and as an illuminant. Small amounts of asphalt have been sent to the United States. Locally, asphalts are used as gas enrichers. Grahamite and glance-pitch are common, and are exported for use in varnish and paint manufactures. The commercial product of stones, brick and cement is of rapidly increasing importance. The foundation of the island is in many places almost pure carbonate of lime, and there are numerous small limekilns. The product is used to bleach sugar, as well as for construction and disinfection purposes. The number of small brick plants is legion, almost all very primitive.

Commerce.—Commerce (resting largely upon specialized agriculture) is vastly more prominent as yet than manufacturing and mining in the island's economy. The leading articles of export are sugar, tobacco and fruit products; of import, textiles, foodstuffs, lumber and wood products, and machinery. Sugar and tobacco products together represent seven-eighths (in 1904-1907 respectively 60.3 and 27.3%) of the normal annual exports. In the quinquennial period 1890-1894 (immediately preceding the War of Independence) the average yearly commerce of the island in and out was $86,875,663 with the United States; and $28,161,726 with Spain.[2] During the American military occupation of the island in 1899-1902, of the total imports 45.9% were from the United States, 14 from other American countries, 15 from Spain, 14 from the United Kingdom, 6 from France and 4 from Germany; of the exports the corresponding percentages for the same countries were 70.7, 2, 3, 10, 4 and 7. No special favours were enjoyed by the United States in this period, and about the same percentages prevailed in the years following. The total commercial movement of the island in the five calendar years 1902-1906 averaged $177,882,640 (for the five fiscal years 1902-1903 to 1906-1907, $185,987,020) annually, and of this the share of the United States was $108,431,000 yearly, representing 45.8% of all imports and 81.9% of all exports. The proportion of imports taken from the United States is greatest in foodstuffs, metals and metal manufactures, timber and furniture, mineral oils and lard. The trade of the United States with the island was as great in 1900-1907 as with Mexico and all the other West Indies combined; as great as its trade with Spain, Portugal and Italy combined; and almost as great as its trade with China and Japan.

Communications.—Poor means of communication have always been a great handicap to the industries of the island. The first railroad in Cuba (and the first in Spanish lands) was opened from Havana to Güines in 1837. In succeeding years a fairly ample system was built up between the cities of Pinar del Rio and Santa Clara, with a number of short spurs from the chief ports farther eastward into the interior. After the first American occupation a private company built a line from Santa Clara to Santiago, more than half the length of the island, finally connecting its two ends (1902). The policy of the railways was always one rather of extortion than of fairness or of any interest in the development of the country, but better conditions have begun. There was ostensible government regulation of rates after 1877, but the roads were guaranteed outright against any loss of revenue, and in fact practically nothing was ever done in the way of reform in the Spanish period. In 1900 the total length of railways was 2097 m., of which 1226 were of 17 public roads and 871 m. of 107 private roads. In August 1908 the mileage of all railways (including electric) in Cuba was 2329.8 m. The telegraph and telephone systems are owned by the government. Cables connect the island with Florida, Jamaica, Haiti and San Domingo, Porto Rico, the lesser Antilles, Panama, Venezuela and Brazil. Havana, Santiago and Cienfuegos are cable ports. Wagon roads are still of small extent and primitive character save in a very few localities. The peculiar two-wheeled carts of the country, carrying enormous loads of 4 to 6 tons, destroy even the finest road. Similar carts, slightly lighter, used in the cities, quickly destroy any paving but stone block. The only good highways of any considerable length in 1908 were in the two western provinces and in the vicinity of Santiago. During the second American occupation work was begun on a network of good rural highways.

Population.—Various censuses were taken in Cuba beginning in 1774; but the results of those preceding the abolition of slavery, at least, are probably without exception extremely untrustworthy. The census of 1887 showed a population of 1,631,687, that of 1899 a population of 1,572,792 (the decrease of 3.6% is explained by the intervening war); and by the census of 1907 there were 2,048,980 inhabitants, 30.3% more than in 1899. The average of settlement per square mile varied from 169.7 in Havana province to 11.8 in Camagüey, and was 46.4 for all of Cuba; the percentage of urban population (in cities, that is, with more than 1000 inhabitants) in the different provinces varied from 18.2 in Pinar del Rio to 74.7 in Havana, and was 43.9 for the entire island. There were five cities having populations above 25,000—Havana, 297,159; Santiago, 45,470; Matanzas, 36,009; Cienfuegos, 30,100; Puerto Príncipe (or Camagüey), 29,616; and fourteen more above 8000—Cardenas, Manzanillo, Guanabacoa, Santa Clara, Sagua la Grande, Sancti Spiritus, Guantánamo, Trinidad, Pinar del Rio, San Antonio de los Baños, Jovellanos, Marianao, Caibarién and Güines. The proportion of the total population which in 1907 was in cities of 8000 or more was only 30.3%; and the proportion in cities of 25,000 or more was 21.4%. Mainly owing to the large element of transient foreign whites without families (long characteristic of Cuba), males outnumber females—in 1907 as 21 to 19. Native whites, almost everywhere in the majority, constituted 59.8% of all inhabitants; persons of negro and mixed blood, 29.7%; foreign-born whites, 9.9%; Chinese less than 0.6%. Foreigners constituted 25.6% of the population in the city of Havana; only 7% in Pinar del Rio province. Native blood is most predominant in the provinces of Oriente and Pinar del Rio. After the end of the war of 1895-1898 a large immigration from Spain began; the inflow from the United States was very small in comparison. The Republic strongly encourages immigration. In 1900-1906 there were 143,122 immigrants, of whom 124,863 were Spaniards, 4557 were from the United States, 2561 were Spanish Americans, and a few were Italian, Syrian, Chinese, French, English, &c. The Chinese element is a remnant of a former coolie population; their numbers in 1907 (11,217) were less than a fourth the number in 1887. Their introduction began in 1847 and ended in 1871. Conjugal conditions in Cuba are peculiar. In 1907 only 20.7% of the total population were legally married; an additional 8.6% were living in more or less permanent consensual unions, these being particularly common among the negroes. Including all unions the total is below the European proportion, but above that of Porto Rico or Jamaica in 1899.

The negro element is strongest in the province of Oriente and weakest in Camagüey; in the former it constituted 43.1% of the population, in the latter 18.3%, and in Havana City 25.5%. In Guantánamo, in Santiago de Cuba, and in seven other towns they exceeded the whites in number.

Caibarién and San Antonio de los Baños had the largest proportion of white population. The position of the negroes in Cuba is exceptional. Despite the long period of slavery they are decidedly below the whites in number. The Spanish slave laws (although in practice often frightfully abused) were always comparatively generous to the slave, making relatively easy, among other things, the purchase of his freedom, the number of free blacks being always great. Since the abolition of slavery the status of the black has been made more definite, and his rights naturally much greater. The wars of 1868-1878 and 1895-1898 and the threatened war of 1906 all helped to give to the negro element its high position. There is no antagonism between the divisions of the coloured race. All hold their own with the white in industrial usefulness to the community, and though the blacks are more backward in education and various other tests of social advancement, still their outlook is full of promise. There is practically no colour caste in Cuba; politically the negro is the white man's equal; socially there is very little ostensible inequality and almost perfect toleration. The negro in Cuba shows promising though undeveloped traits of landlordship. Women labour habitually in the fields. Miscegenation of blacks and whites was extremely common before emancipation. It is sometimes said that since then there has been a counter-tendency, but it is impossible to prove such a statement conclusively except with the aid of future censuses. Few of the negroes are black; some of the blackest have the regular features of the Caucasian; and racial mixtures are everywhere evidenced by colour of skin and by physiognomy. Its seems certain that the African element has been holding its own in the population totals since emancipation.

Cuba is overwhelmingly Roman Catholic in religion, but under the new Republic there is a complete separation of church and state, and liberalism and indifference are increasing. Illiteracy is extremely widespread. In 1907 the census showed 56.6% (43.3 in 1899) of persons above ten years who could read. Of the voting population 53.2% of native white, and 37.3% of coloured Cuban citizens, and 71.6% of Spanish citizens could read. A revolution in education was begun the first year of the United States military occupation and continued under the Republic.

Constitution.—The constitution upon which the government of Cuba rests was framed during the period of the United States military government; it was adopted the 21st of February 1901, and certain amendments or conditions required by the United States were accepted on the 12th of June 1901. The constitution is republican and modelled on the Constitution of the United States, with some marked differences of greater centralization, due to colonial experience under the rule of Spain, notably as regards federalism; the provinces of the island being less important than the states of the American Union. The president of the Republic, who is elected for four years by an electoral college, and cannot hold office for more than two successive terms, has a cabinet whose members he may appoint and remove freely, their number being determined by law. He sanctions, promulgates and executes the laws, and supplements them (partly co-ordinately with congress) by administrative regulations in harmony with their ends; holds a veto power and pardoning power; controls with the senate political appointments and removals; and conducts foreign relations, submitting treaties to the senate for ratification. Congress consists of two houses. The senate contains four members from each province, chosen for eight years by a provincial electoral board, which consists of the provincial councilmen plus a double number of electors (half of them paying high taxes) who are selected at a special election by their fellow citizens. Half of the senators retire every four years. The senate is the court of trial for the president, officers of the cabinet, and provincial governors when accused of political offences. It also acts jointly with the president in political appointments and treaty making. The house of representatives, whose members are chosen directly by the citizens for four years, one-half retiring every two years, has the special power of impeaching the president and cabinet officers. Congress meets twice annually, in April and November. Its powers are extensive, including, in addition to ordinary legislative powers, control of financial affairs, foreign affairs, the power to declare war and approve treaties of peace, amnesties, electoral legislation for the provinces and municipalities, control of the electoral vote for president and vice-president, and designation of an acting president in case of the death or incapacity of these officers. The subjects of legislative power are very similar to those of the United States congress; but control of railroads, canals and public roads is explicitly given to the federal government. Justice is administered by courts of various grades, with a supreme court at Havana as the head; the members of this being appointed by the president and senate. This court passes on the constitutionality of all laws, decrees and regulations.

There are six provinces—Pinar del Rio, Havana, Matanzas, Santa Clara, Camagüey or Puerto Príncipe, and Oriente. Each has a provincial governor and assembly chosen directly by the people, generally charged with independent control of matters affecting the province; but the president may interfere against an abuse of power by either the governor or the assembly. Municipalities are administered by mayors (alcaldes) and assemblies elected by the people, and control strictly municipal affairs. The "termino municipal" is the chief political and administrative civil division. It is an urban district together with contiguous rural territory. Its divisions are "barrios." The president may interfere if necessary in the municipality as in the province; and so may the governor of the province. But all interference is subject to review of claims by the courts. Both provinces and municipalities are forbidden by the constitution to contract debts without a coincident provision of permanent revenue for their settlement.

The franchise is granted to every male Cuban twenty-one years of age, not mentally incapacitated, nor previously a convict of crime, nor serving in the army or navy of the state. Foreigners may become citizens in five years by naturalization. Church and state are completely separated, toleration being guaranteed for the profession and practice of all religious beliefs, and the government may not subsidize any religion.

Primary education is declared by the constitution to be free and compulsory; and its expenses are paid by the central government so far as it may be beyond the power of the province or municipality to bear them. Secondary ***Education.*** and advanced education is controlled by the state. In the last days of Spanish rule (1894), there were 904 public and 704 private schools, and not more than 60,000 pupils enrolled; in 1000 there were 3550 public schools with an enrolment of 172,273 and an average attendance of 123,362. In the four school years from 1903-1904 to 1906-1907 the figures of enrolment and average attendance were: 201,824 and 110,531; 194,657 and 105,706; 186,571 and 98,329; and 189,289 and 93,865. In 1906-1907 the percentage (31.6) of attendants to children of school age was twice as large as in 1898-1899. Private schools, some of very high grade, draw many pupils. Almost all schools are primary. The university of Havana (founded 1728) was given greatly improved facilities, especially of material equipment, by the American military government, and seems to have begun an ambitious progress. In 1907 the number of students was 554. Below the university there are six provincial institutes, one in each province, in each of which there is a preparatory department, a department of secondary education, and (this due to peculiar local conditions) a school of surveying; and in that of Havana commercial departments in addition. In Havana, also, there is a school of painting and sculpture, a school of arts and trades, and a national library, all of which are supported or subventioned by the national government, as are also a public library in Matanzas, and the Agricultural Experiment Station at Santiago de las Vegas. In connexion with the university is a botanical garden; with the national sanitary service, a biological laboratory, and special services for small-pox, glanders and yellow fever. Independent of the government are various schools and learned societies in Havana (*q.v.*). A school was established by the government in Key West, Florida (U.S.A.), in 1905, for the benefit of the Cuban colony there. Finally, the government sustains about two score of penal establishments, reform schools, hospitals, dispensaries and asylums, which are scattered all over the island,—every town of any considerable size having one or more of these charities.

Under the colonial rule of Spain the head of government was a supreme civil-military officer, the governor and captain-general. His control of the entire administrative life of the island was practically absolute. Originally ***Former government.*** residents at Santiago de Cuba, the captains-general resided after 1589 at Havana. Because of the isolation of the eastern part of the island, the dangers from pirates, and the important considerations which had caused Santiago de Cuba (*q.v.*) to be the first capital of the island, Cuba was divided in 1607 into two departments, and a governor, subordinate in military matters to the captain-general at Havana, was appointed to rule the territory east of Puerto Príncipe. In 1801, when the audiencia—of which the captain-general was *ex officio* president—began its functions at that point, the governor of Santiago became subordinated in political matters as much as in military. Two chief courts of justice (audiencias) sat at Havana (after 1832) and Puerto Príncipe (1800-1853); appeals could go to Spain; below the audiencias were "alcaldes mayores" or district judges and ordinary "alcaldes" or local judges. The audiencias also held important political powers under the Laws of the Indies. The captaincy-general of Cuba was not originally, however, by any means so broad in powers as the viceroyalties of Mexico and Peru; and by the creation in 1765 of the office of intendant—the delegate of the national treasury—his faculties were very greatly curtailed. The great powers of the intendant were, however, merged in those of the governor-general in 1853; and the captain-general having been given by royal order in 1825 (several times later explicitly confirmed, and not revoked until 1870) the absolute powers (to be assumed at his initiative and discretion) of the governor of a besieged city, and by a royal order of 1834 the power to banish at will persons supposed to be inimical to the public peace; and being by virtue of his office the president and dominator of all the important administrative boards of the government, held the government of the island, and in any emergency the liberty and property of its inhabitants, in his hand. The royal orders following 1825 developed a system of extraordinary and extreme repression. In 1878, as the result of the Ten Years' War, various administrative reforms, of a decentralizing tendency, were introduced. The six provinces were created, and had governors and assemblies ("diputaciones"); and a municipal law was provided that in many ways was a sound basis for local government. But centralization remained very great. In the municipality the alcalde (mayor) was appointed by the governor-general, and the ayuntamiento (council) was controlled by the veto of the provincial governor and by the assembly of the province. The deputation was subject in turn to the same veto of the provincial governor, and he controlled by the governor-general. There was besides a provincial commission of five lawyers named by the governor-general from the members of the deputation, who settled election questions, and questions of eligibility in this body, gave advice as to laws, acted for the deputation when it was not sitting, and in general facilitated centralized control of the administrative system. The character of this body was altered in 1890, and in 1898, in which latter year its functions were reduced to the essentially judicial. Despite superficial decentralization after 1878 any real growth of local self-government was rendered impossible. Moreover, no great reforms were made in the abuses naturally incident to the old personal system. Exile and imprisonment at the will of the government and without trial were common. Personal liberty, liberty of conscience, speech, assembly, petition, association, press, liberty of movement and security of home, were without real guarantee even within the extremely small limits in which they nominally existed. Under the constitution of the Republic the sphere of individual liberty is large and constitutionally protected against the government.

Finance.—There has been a great change in the budget of Cuba since the advent of the Republic. In 1891-1896 the average annual income was $20,738,930, the annual average expenditure $25,967,139. More than half of the revenue was derived from customs duties (two-thirds of the total being collected at Havana). Of the expenditure more than ten million dollars annually went for the public debt, 5.5 to 6 millions for the army and navy, as much more for civil administration (including more than two millions for purely Peninsular services with which the colony was burdened); and on an average probably one million more went for sinecures. Every Cuban paid about twice as heavy taxes as a Spaniard of the Peninsula. Very little was spent on sanitation, roads, other public works and education. The revenue receipts under the Republic have increased especially over those of the old régime in the item of customs duties; and the expenditure is very differently distributed. Lotteries which were an important source of revenue under Spain were abolished under the Republic. The debt resting on the colony in 1895 (a large part of it as a result of the war of 1868-1878, the entire cost of which was laid upon the island, but a part as the result of Spain's war adventures in Mexico and San Domingo, home loans, &c.) was officially stated at $168,500,000. The attainment of independence freed the island from this debt, and from enormous contemplated additions to cover the expense incurred by Spain during the last insurrection. The debt of the Republic in April 1908 was $48,146,585, including twenty-seven millions which were assumed in 1902 for the payment of the army of independence, four for agriculture, and four for the payment of revolutionary debts, and $2,196,585, representing obligations assumed by the revolution's representative in the United States during the War of Independence. United States and British investments, always important in the agriculture and manufactures of the island, greatly increased following 1898, and by 1908 those of each nation were supposed to exceed considerably $100,000,000.

Archaeology.—Archaeological study in Cuba has been limited, and has not produced results of great importance. Almost nothing is actually known of prehistoric Cuba; and a few skulls and implements are the only basis existing for conjecture. Very little also is known as to the natives who inhabited the island at the time of the discovery. They were a tall race of copper hue; fairly intelligent, mild in temperament, who lived in poor huts and practised a limited and primitive agriculture. How numerous they were when the Spaniards first came among them cannot be said; undoubtedly tradition has greatly exaggerated their number. They are supposed to have been practically extinct by 1550. Even in the 19th century reports were spread of communities in which Indian blood was supposedly still plainly dominant; but the conclusion of the competent scientists who have investigated such rumours has been that at least absolutely nothing of the language and traditions of the aborigines has survived.

History.—Cuba was discovered by Columbus in the course of his first voyage, on the 27th of October 1492. He died believing Cuba was part of a continent. In 1508 Sebastian de Ocampo circumnavigated it. In 1511 Diego Velazquez began the conquest of the island. Baracoa (the landing point), Bayamo, Santiago de Cuba, Puerto Príncipe, Sancti Spiritus, Trinidad and the original Havana were all founded by 1515. Velazquez's reputation and legends of wealth drew many immigrants to the island. From Cuba went the expeditions that discovered Yucatan (1517), and explored the shores of Mexico, Hernando Cortés's expedition for the invasion of Mexico, and de Soto's for the exploration of Florida. The last two had a pernicious effect on Cuba, draining it of horses, money and of men. At least as early as 1523 the African slave trade was begun. In 1544 the Indians, so far as they had not succumbed to the labour of the mines and fields to which they were put by the Spaniards, were proclaimed emancipated. The administration in the 16th century was loose and violent. The local authorities were divided among themselves by bitter feuds—the ecclesiastical against the civil, the *ayuntamiento* against the governors, the administrative officers among themselves; brigandage, mutinies and intestinal struggles disturbed the peace. As a result of the transfer of Jamaica to England, the population of Cuba was greatly augmented by Jamaican immigrants to about 30,000 in the middle of the 17th century.

The activity of English and French pirates began in the 16th century, and reached its climax in the middle of the 17th century. So early also began dissatisfaction with the economic regulations of the colonial system, even grave resistance to their enforcement; and illicit trade with privateers and foreign colonies had begun long before, and in the 17th and 18th centuries was the basis of the island's wealth. In 1762 Havana was captured after a long resistance by a British force under Admiral Sir George Pocock and the earl of Albemarle, with heavy loss to the besiegers. It was returned to Spain the next year in exchange for the Floridas. From this date begins the modern history of the island. The British opened the port to commerce and the slave trade and revealed its possibilities. The government of Spain, beginning in 1764, made notable breaches in the old monopolistic system of colonial trade throughout America; and Cuba received special privileges, also, that were a basis for real prosperity. Spain paid increasing attention to the island, and in harmony with the policy of the Laws of the Indies many decrees intended to stimulate agriculture and commerce were issued by the crown, first in the form of monopolies, then with increased freedom and with bounties. Various colonial products and the slave trade were favoured in this way. After the cession of the Spanish portion of San Domingo to France hundreds of Spanish families emigrated to Cuba, and many thousand more immigrants, mainly French, followed them from the entire island during the revolution of the blacks. Most of them settled in Oriente province, where their names and blood are still apparent, and with their cafetales and sugar plantations converted that region from neglect and poverty to high prosperity.

Under a succession of liberal governors (especially Luis de las Casas, 1790-1796, and the marqués de Someruelos, 1799-1813), at the end of the 18th century and the first part of the 19th, when the wars in Europe cut off Spain almost entirely from the colony, Cuba was practically independent. Trade was comparatively free, and worked a revolution in culture and material conditions. General Las Casas, in particular, left behind him in Cuba an undying memory of good efforts. Free commerce with foreigners—a fact after 1809—was definitely legalized in 1818 (confirmed in 1824). The state tobacco monopoly was abolished in 1817. The reported populations by the (untrustworthy) censuses of 1774, 1792 and 1817 were 161,670, 273,301 and 553,033. Something of political freedom was enjoyed during the two terms of Spanish constitutional government under the constitution of 1812. The sharp division between creoles and peninsulars (*i.e.* between those born in Cuba and those born in Spain), the question of annexation to the United States or possibly to some other power, the plotting for independence, all go back to the early years of the century.

Partly because of political and social divisions thus revealed, conspiracies being rife in the decade 1820-1830, and partly as preparation for the defence against Mexico and Colombia, who throughout these same years were threatening the island with invasion, the captains-general, in 1825, received the powers above referred to; which became, as time passed, monstrously in disaccord with the general tendencies of colonial government and with increasing liberties in Spain, but continued to be the spiritual basis of Spanish rule in the island. Among the governors of the 19th century Miguel Tacon, governor in 1834-1839, a forceful and high-handed soldier, deserves mention, especially in the annals of Havana; he ruled as a tyrant, made many reforms as regarded law and order, and left Havana, in particular, full of municipal improvements. The good he did was limited to the spheres of public works and police; in other respects his rule was a pernicious influence for Cuba. Politically his rule was marked by the proclamation at Santiago in 1836, without his consent, of the Spanish constitution of 1834; he repressed the movement, and in 1837 the deputies of Cuba to the Cortes of Spain (to which they were admitted in the two earlier constitutional periods) were excluded from that body, and it was declared in the national constitution that Cuba (and Porto Rico) should be governed by "special laws." The inapplicability of many laws passed for the Peninsula—all of which under a constitutional system would apply to Cuba as to any other province, unless that system be modified—was indeed notorious; and Cuban opinion had repeatedly, through official bodies, protested against laws thus imposed that worked injustice, and had pleaded for special consideration of colonial conditions. The promise of "special laws" based upon such consideration was therefore not, in itself, unjust, nor unwelcome. But as the colony had no voice in the Cortes, while the "special laws" were never passed (Cuba expected special fundamental laws, reforming her government, and the government regarded the old Laws of the Indies as satisfying the obligation of the constitution) the arbitrary rule of the captains-general remained quite supreme, under the will of the crown, and colonial discontent became stronger and stronger. The rule of Leopoldo O'Donnell was marked in 1844 by a cruel and bloody persecution of negroes for a supposed plot of servile war; O'Donnell's actions being partly due to the inquietude that had prevailed for some years over the supposed machinations of English abolitionists and even of English official residents in the island, and also over the mutual jealousies and supposed annexation ambitions of Great Britain and the United States.

A Cuban international question had arisen before 1820. Spain, the United States, England, France, Colombia and Mexico were all involved in it, the first four continually. In the eighteen-fifties a strong pro-slavery interest in the United States advocated the acquisition of the island. One feature of this was the "Ostend Manifesto" (see Buchanan, James), in which the ministers of the United States at London, Paris and Madrid declared that if Spain refused a money offer for the colony the United States should seize it. Their government gave this document publicity. The Cuban policy of Presidents Pierce and Buchanan (during 1853-1861) was vainly directed to acquiring the island. From 1849 to 1851 there were three abortive filibustering expeditions from the United States, two being under a Spanish general, Narciso Lopez (1798-1851). The domestic problem, the problem of discontent in the island, had become acute by 1850, and from this time on to 1868 the years were full of conflict between liberal and reactionary sentiment in the colony, centreing about the asserted connivance of the captains-general in the illegal slave trade (declared illegal after 1820 by the treaties of 1817 and 1835 between Great Britain and Spain), the notorious immorality and prodigal wastefulness of the government, and the selfish exploitation of the colony by Spaniards and the Spanish government. From early in the 19th century there had always been separatists, reformists and repressionists in the island, but they were individuals rather than groups. The last were peninsulars, the others mainly creoles, and among the wealthy classes of the latter the separatists gradually gained increasing support.

An ineffective and extremely corrupt administration, a grave economic condition, new and heavy taxes, military repression, recurring heavy deficits in the budget, adding to a debt (about $150,000,000 in 1868) already very large and burdensome, and the complete fiasco of the *junta* of inquiry of Cuban and Porto Rican representatives which met in Madrid in 1866-1867—all were important influences favouring the outbreak of the Ten Years' War. Among those who waged the war were men who fought to compel reforms, others who fought for annexation to the United States, others who fought for independence. The reformists demanded, besides the correction of the above evils, action against slavery, assimilation of rights between peninsulars and creoles and the practical recognition of equality, *e.g.* in the matter of office-holding, a grievance centuries old in Cuba as in other Spanish colonies, and guarantees of personal liberties. The separatists, headed by Carlos Manuel de Céspedes (1819-1874), a wealthy planter who proclaimed the revolution at Yara on the 10th of October, demanded the same reforms, including gradual emancipation of the slaves with indemnity to owners, and the grant of free and universal suffrage. War was confined throughout the ten years almost wholly to the E. provinces. The policy of successive captains-general was alternately uncompromisingly repressive and conciliatory. The Spanish volunteers committed horrible excesses in Havana and other places; the rebels also burned and killed indiscriminatingly, and the war became increasingly cruel and sanguinary. Intervention by the United States seemed probable, but did not come, and after alternations in the fortunes of war, Martinez Campos in January 1878 secured the acceptance by the rebels of the convention (pacto) of Zanjón, which promised amnesty for the war, liberty to slaves in the rebel ranks, the abolition of slavery, reforms in government, and colonial autonomy. A small rising after peace (the "Little War" of 1879-1880) was easily repressed. Gradual abolition of slavery was declared by a law of the 13th of February 1880; definitive abolition in 1886; and in 1893 the equal civil status of blacks and whites in all respects was proclaimed by General Calleja. There is no more evidence to warrant the wholly erroneous statement sometimes made that emancipation was an economic set-back to Cuba than could be gathered to support a similar statement regarding the United States. Coolie importation from China had been stopped in 1871.

As for autonomy and political reforms it has already been remarked that the change from the old régime was only superficial. The Spanish constitution of 1876 was proclaimed in Cuba in 1881. In 1878-1895 political parties had a complex development. The Liberal party was of growing radicalism, the Union Constitutional party of growing conservatism; and after 1893 a Reformist party was launched that drew the compromisers and the waverers. The demands of the Liberals were as in 1868; those for personal and property rights were much more definitely stated, and among explicit reforms demanded were the separation of civil and military power, general recognition of administrative responsibility under a colonial autonomous constitutional régime; also among economic matters, customs reforms and reciprocity with the United States were demanded. As for the representation accorded Cuba in the Spanish Cortes, as a rule about a quarter of her deputies were Cuban-born, and the choice of only a few autonomists was allowed by those who controlled the elections. Reciprocity with the United States was in force from 1891 to 1894 and was extremely beneficial to Cuba. Its cessation greatly increased disaffection.

Discontent grew, and another war was prepared for. On the 23rd of February 1895 General Calleja suspended the constitutional guarantees. The leading chiefs of the Ten Years' War took the field again—Máximo Gómez, Antonio Macéo, Jose Martí, Calixto García and others. Unlike that war, this was carried to the western provinces, and indeed was fiercest there. Among the military means adopted by the Spaniards to isolate their foe were "trochas" (*i.e.* entrenchments, barbwire fences, and lines of block-houses) across the narrow parts of the island, and "reconcentracion" of non-combatants in camps guarded by the Spanish forces. The latter measure produced extreme suffering and much starvation (as the reconcentrados were largely thrown upon the charity of the beggared communities in which they were huddled). In October 1897 the Spanish premier, P. M. Sagasta, announced the policy of autonomy, and the new dispensation was proclaimed in Cuba in December. But again all final authority was reserved to the captain-general. The system was never to have a practical trial, although a full government was quickly organized under it. The American people had sent food to the reconcentrados; President McKinley, while opposing recognition of the rebels, affirmed the possibility of intervention; Spain resented this attitude; and finally, in February 1898, the United States battleship "Maine" was blown up—by whom will probably never be known—in the harbour of Havana.

On the 20th of April the United States demanded the withdrawal of Spanish troops from the island. War followed immediately. A fine Spanish squadron seeking to escape from Santiago harbour was utterly destroyed by the American blockading force on the 3rd of July; Santiago was invested by land forces, and on the 15th of July the city surrendered. Other operations in Cuba were slight. By the treaty of Paris, signed on the 10th of December, Spain "relinquished" the island to the United States in trust for its inhabitants; the temporary character of American occupation being recognized throughout the treaty, in accord with the terms of the American declaration of war, in which the United States disclaimed any intention to control the island except for its pacification, and expressed the determination to leave the island thereupon to the control of its people. Spanish authority ceased on the 1st of January 1899, and was followed by American "military" rule (January 1, 1899-May 20, 1902). During these three years the great majority of offices were filled by Cubans, and the government was made as different as possible from the military control to which the colony had been accustomed. Very much was done for public works, sanitation, the reform of administration, civil service and education. Most notable of all, yellow fever was eradicated where it had been endemic for centuries. A constitutional convention sat at Havana from the 5th of November 1900 to the 21st of February 1901. The provisions of the document thus formed have already been referred to. In the determination of the relations that should subsist between the new republic and the United States certain definite conditions known as the Platt Amendment were finally imposed by the United States, and accepted by Cuba (12th of June 1901) as a part of her constitution. By these Cuba was bound not to incur debts her current revenues will not bear; to continue the sanitary administration undertaken by the military government of intervention; to lease naval stations (since located at Bahía Honda and Guantánamo) to the United States; and finally, the right of the United States to intervene, if necessary, in the affairs of the island was explicitly affirmed in the provision, "That the government of Cuba consents that the United States may exercise the right to intervene for the protection of Cuban independence, the maintenance of a government adequate for the protection of life, property and individual liberty, and for discharging the obligations with respect to Cuba imposed by the treaty of Paris on the United States, now to be assumed and undertaken by the government of Cuba." The status thus created is very exceptional in the history of international relations. The status of the Isle of Pines was left an open question by the treaty of Paris, but a decision of the Supreme Court of the United States has declared it (in a question of customs duties) to be a part of Cuba, and though a treaty to the same end did not secure ratification (1908) by the United States Senate, repeated efforts by American residents thereon to secure annexation to the United States were ignored by the United States government.

The first Cuban congress met on the 5th of May 1902, prepared to take over the government from the American military authorities, which it did on the 20th of May. Tomas Estrada Palma (1835-1908) became the first president of the Republic. In material prosperity the progress of the island from 1902 to 1906 was very great; but in its politics, various social and economic elements, and political habits and examples of Spanish provenience that ill befit a democracy, led once more to revolution. Congress neglected to pass certain laws which were required by the constitution, and which, as regards municipal autonomy, independence of the judiciary, and congressional representation of minority parties, were intended to make impossible the abuses of centralized government that had characterized Spanish administration. Political parties were forming without very evident basis for differences outside questions of political patronage and the good or ill use of power; and, in the absence of the laws just mentioned, the Moderates, being in power, used every instrument of government to strengthen their hold on office. The preliminaries of the elections of December 1905 and March 1906 being marked by frauds and injustice, the Liberals deserted the polls at those elections, and instead of appealing to judicial tribunals controlled by the Moderates, issued a manifesto of revolution on the 28th of July 1906.[3] This insurrection rapidly assumed large proportions. The government was weak and lacked moral support in the whole island. After repeated petitions from President Palma for intervention by the United States, commissioners (William H. Taft, Secretary of War, and Robert Bacon, Acting Secretary of State) were sent from Washington to act as peace mediators.

All possible efforts to secure a compromise that would preserve the Republic failed. The president resigned (on the 28th of September), Congress dispersed without choosing a successor, and as an alternative to anarchy the United States was compelled to proclaim on the 29th of September 1906 a provisional government,—to last "long enough to restore order and peace and public confidence," and hold new elections. The insurrectionists promptly disbanded. Government was maintained under the Cuban flag,—the diplomatic and consular relations with even the United States remaining in outward forms unchanged; and the regular forms of the constitution were scrupulously maintained so far as possible. No use was made of American military force save as a passive background to the government. The government of intervention at first directed its main effort simply to holding the country together, without undertaking much that could divide public opinion or seem of unpalatably foreign impulse; and later to the establishment of a few fundamental laws which, when intervention ceased, should give greater simplicity, strength and stability to a new native government. These laws strictly defined the powers of the president; more clearly separated the executive departments, so as to lessen friction and jealousies; reformed the courts; reformed administrative routine; and increased the strength of the provinces at the expense of the municipalities. On the 28th of January 1909 the American administration ceased, and the Republic was a second time inaugurated, with General José Miguel Gomez (b. 1856), the leader of the Miguelista faction of the Liberal party, as president, and Alfredo Zayas, the leader of the Zayista faction of the same party, as vice-president. The last American troops were withdrawn from the island on the 1st of April 1909.

AUTHORITIES.—**General Description.**—There is no trustworthy recent description. The best books are E. Pechardo, *Geografía de la isla de Cuba* (4 tom., Havana, 1854); M. Rodriguez-Ferrer, *Naturaleza y civilización de ... Cuba*, vol. i. (Madrid, 1876). See also *United States Geological Survey, Bulletin 192* (1902), H. Gannett, "A Gazetteer of Cuba." Of general descriptions in English, in addition to travels cited below, may be cited R. T. Hill, *Cuba and Porto Rico with the other West Indies* (New York, 1898).

Fauna and Flora.—A. H. R. Grisebach, *Catalogus plantarum Cubensium* (Leipzig, 1866), and F. A. Sauvalle, *Flora Cubana: revisio catalogi Grisebachiani* (Havana, 1868); and *Flora Cubana: enumeratio nova plantarum Cubensium* (Havana, 1873); F. Poey et al., *Repertorio fisico-natural de la isla de Cuba* (2 vols., Havana, 1865-1868), and F. Poey, *Memorias sobre la historia natural de ... Cuba* (3 tom., Havana, 1851-1860); Ramon de la Sagra, with many collaborators, *Historia física, política y natural de ... Cuba* (Paris, 1842-1851, 12 vols.; issued also in French; vols. 3-12 being the "Historia Natural"); *Anales* of the Academia de Ciencias (Havana, 1863- , annual); M.

Gomez de la Maza, *Flora Habanera* (Havana, 1897); S. A. de Morales, *Flora arborícola de Cuba aplicada* (Havana, 1887, only part published); D. H. Seguí, *Ojeado sobre la flora médica y tóxica de Cuba* (Havana, 1900); J. Gundlach, *Contribucion à la entomologia Cubana* (Havana, 1881); J. M. Fernandez y Jimenez, *Tratado de la arboricultura Cubana* (Havana, 1867).

Geology and Minerals.—M. F. de Castro, "Pruebas paleontologicas de que la isla de Cuba ha estado unida al continento americano y breve idea de su constitucion geologica," *Bol. Com. Mapa Geol. de Esp.* vol. viii. (1881), pp. 357-372; M. F. de Castro and P. Salterain y Legarra, "Croquis geologico de la isla de Cuba," *ibid.* vol. viii. pl. vi. (published with vol. xi., 1884). Many articles in *Anales* of the Academy; also, R. T. Hill in *Harvard College Museum of Comparative Zöology, Bulletin*, vol. 16, pp. 243-288 (1895); *United States Geological Survey*, 22nd Annual Report, 1901, C. W. Hayes et al., "Geological Reconnaissance of Cuba"; *Civil Report of General Leonard Wood*, governor of Cuba (1902), vol. v., H. C. Brown, "Report on Mineral Resources of Cuba."

Climate.—See the *Boletin Oficial de la Secretaria de Agricultura*, and publications of the observatory of Havana. **Sanitation.**—For conditions 1899-1902, see *Civil Reports* of American military governors. For conditions since 1902 consult the *Informe Mensual* (1903-) of the Junta Superior de Sanidad.

Agriculture.—Consult the *Boletin* above mentioned, publications of the Estación Central Agronómica, and current statistical serial reports of the treasury department (Hacienda) on natural resources, live-stock interests, the sugar industry (annual), &c.

Industries, Commerce, Communications.—See the works of Sagra and Pezuela. For conditions about 1899 consult R. P. Porter (Special Commissioner of the United States government), *Industrial Cuba* (New York, 1899); W. J. Clark, *Commercial Cuba* (New York, 1898); reports of foreign consular agents in Cuba; and the statistical annuals of the Hacienda on foreign commerce and railways.

Population.—The early censuses were extremely unreliable. Illuminating discussions of them can be found in Humboldt's *Essay*, Saco's *Papeles* and Pezuela's *Diccionario.* See *United States Department of War, Report on the Census of Cuba 1899* (Washington, 1899); *U.S. Bureau of the Census, Cuba: Population, History and Resources, 1907* (1909).

Education.—See *Civil Reports* of the American military government, 1899-1902; United States commissioner of education, *Report, 1897-1898*; current reports in *Informe del superintendente de escuelas de Cuba* ... (Havana, 1903-). On Letters and Culture.—E. Pechardo y Tapia, *Diccionario ... de voces Cubanas* (Havana, 1836, 4th ed., 1875; all editions with many errors); Antonio Bachiller y Morales, *Apuntes para la historia de las letras y de la instrucción pública de Cuba* (3 tom., Havana, 1859-1861); J. M. Mestre, *De la filosofía en la Habana* (Havana, 1862); A. Mitjans, *Estudio sobre el movimiento científico y literario de Cuba* (Havana, 1890); biographies of Varela and Luz Caballero by Rodriguez (see below); files of *La Revista de Cuba* (16 vols., Havana, 1877-1884) and *La Revista Cubana* (21 vols., Havana, 1885-1895). The literature of Travel is rich. It suffices to mention *Letters from the Havannah*, by the English consul (London, 1821); E. M. Masse, *L'Île de Cuba* (Paris, 1825); D. Turnbull, *Travels in the West* (London, 1840), and R. R. Madden, *The Island of Cuba* (London, 1853)—two very important books regarding slavery; J. B. Rosemond de Beauvallon, *L'Île de Cuba* (Paris, 1844); J. G. Taylor, *The United States and Cuba* (London, 1851); F. Bremer, *The Homes of the New World* (2 vols., New York, 1853); M. M. Ballou, *History of Cuba, or Notes of a Traveller* (Boston, 1854); R. H. Dana, *To Cuba and Back* (Boston, 1859); J. von Sivers, *Die Perle der Antillen* (Leipzig, 1861); A. C. N. Gallenga, *The Pearl of the Antilles* (London, 1873); S. Hazard, *Cuba with Pen and Pencil* (Hartford, Conn., 1873); H. Piron, *L'Île de Cuba* (Paris, 1876). Of later books, F. Matthews, *The New-Born Cuba* (New York, 1899); R. Davey, *Cuba Past and Present* (London, 1898). Among the writers who have left short impressions are A. Granier de Cassagnac (1844), J. J. A. Ampère (1855), A. Trollope (1860), J. A. Froude (1888).

Administration.—Consult the literature of history and colonial reform given below. Also: Leandro Garcia y Gragitena, *Guia del empleado de hacienda* (Havana, 1860), with very valuable historical data; Carlos de Sedano y Cruzat, *Cuba desde 1850 à 1873. Coleccion de informes, memorias, proyectos y antecedentes sobre el gobierno de la isla de Cuba* (Madrid, 1875); Vicente Vasquez Queipo, *Informe fiscal sobre fomento de la poblacion blanca* (Madrid, 1845); *Informacion sobre reformas en Cuba y Puerto Rico celebrada en Madrid en 1866 y 67 por los representantes de ambas islas* (2 tom., New York, 1867; 2nd ed., New York, 1877); and the *Diccionario* of Pezuela. These, with the works of Saco, Sagra, Arango and Alexander von Humboldt's work, *Essai politique sur l'île de Cuba* (2 vols., Paris 1826; Spanish editions, 1 vol., Paris, 1827 and 1840; English translation by J. S. Thrasher, with interpolations, New York, 1856), are indispensable. For conditions at the end of the 18th century, Fran. de Arango y Parreño, *Obras* (2 tom., Havana, 1888). For later conditions, E. Valdes Dominguez, *Los Antiguos Diputados de Cuba* (Havana, 1879); B. Huber, *Aperçu statistique de l'île de Cuba* (Paris, 1826); Humboldt; Sagra, vols. 1-2 of the book cited above, being the *Historia física y política*, and also the earlier work on which they are based, *Historia económica-política y estadística de ... Cuba* (Havana, 1831); treatises on administrative law in Cuba by J. M. Morilla (Havana, 1847; 2nd ed., 1865, 2 vols.) and A. Govin (3 vols., Havana, 1882-1883); A. S. Rowan and M. M. Ramsay, *The Island of Cuba* (New York, 1896); *Coleccion de reales ordenes, decretos y disposiciones* (Havana, serial, 1857-1898); *Spanish Rule in Cuba. Laws Governing the Island. Reviews Published by the Colonial Office in Madrid ...* (New York, for the Spanish legation, 1896); and compilations of Spanish colonial laws listed under article Indies, Laws of the. On the new Republican régime: *Gaceta Oficial* (Havana, 1903-); reports of departments of government; M. Romero Palafox, *Agenda de la republica de Cuba* (Havana, 1905).

See also the *Civil Reports* of the United States military governors, J. R. Brooke (2 vols., 1899; Havana and Washington, 1900), L. Wood (33 vols., 1900-1902; Washington, 1901-1902).

History.—The works (see above) of Sagra, Humboldt and Arango are indispensable; also those of Francisco Calcagno, *Diccionario biográfico Cubano* (ostensibly, New York, 1878); Vidal Morales y Morales, *Iniciadores y primeros mártires de la revolución Cubana* (Havana, 1901); José Ahumada y Centurión, *Memoria histórica política de ... Cuba* (Havana, 1874); Jacobo de la Pezuela, *Diccionario geográfico-estadístico-histórico de ... Cuba* (4 tom., Madrid, 1863-1866); *Historia de ... Cuba*, (4 tom., Madrid, 1868-1878; supplanting his *Ensayo histórico de ... Cuba*, Madrid and New York, 1842); and José Antonio Saco, *Obras* (2 vols., New York, 1853), *Papeles* (3 tom., Paris, 1858-1859), and *Coleccion postuma de Papeles* (Havana, 1881). Also: Rodriguez Ferrer, *op. cit.* above, vol. 2 (Madrid, 1888); P. G. Guitéras, *Historia de ... Cuba* (2 vols., New York, 1865-1866). Of great value is J. Zaragoza, *Las Insurrecciones en Cuba. Apuntes para la historia política* (2 tom., Madrid, 1872-1873); also J. I. Rodriguez, *Vida de ... Félix Varela* (New York, 1878), and *Vida de D. José de la Luz* (New York, 1874; 2nd ed., 1879). On early history see *Coleccion de documentos inéditos relativos al descubrimiento ... de ultramar* (series 2, vols. 1, 4, 6, Madrid, 1885-1890). On archaeology, N. Fort y Roldan, *Cuba indigena* (Madrid, 1881); M. Rodriguez Ferrer (see above); and especially A. Bachiller y Morales, *Cuba primitiva* (Havana, 1883). For the history of the Cuban international problem consult José Ignacio Rodriguez, *Idea de la anexion de la isla de Cuba à los Estados Unidos de America* (Havana, 1900), and J. M. Callahan, Cuba and International Relations (Johns Hopkins University, Baltimore, 1898), which supplement each other. On the domestic reform problem there is an enormous literature, from which may be selected (see general histories above and works cited under § Administration of this bibliography): M. Torrente, *Bosquejo económico-político* (2 tom., Madrid-Havana, 1852-1853); D. A. Galiano, *Cuba en 1858*

(Madrid, 1859); José de la Concha, twice Captain-General of Cuba, *Memorias sobre el estado político, gobierno y administración de ... Cuba* (Madrid, 1853); A. Lopez de Letona, *Isla de Cuba, reflexiones* (Madrid, 1856); F. A. Conte, *Aspiraciones del partido liberal de Cuba* (Havana, 1892); P. Valiente, *Réformes dans les îles de Cuba et de Porto Rico* (Paris, 1869); C. de Sedano, *Cuba: Estudios políticos* (Madrid, 1872); H. H. S. Aimes, *History of Slavery in Cuba, 1511-1868* (New York, 1907); F. Armas y Cèspedes, *De la esclavitud en Cuba* (Madrid, 1866), and *Régimen político de las Antillas Españolas* (Palma, 1882); R. Cabrera, *Cuba y sus Jueces* (Havana, 1887; 9th ed., Philadelphia, 1895; 8th ed., in English, *Cuba and the Cubans*, Philadelphia, 1896); P. de Alzola y Minondo, *El Problema Cubano* (Bilbao, 1898); various works by R. M. de Labra, including *La Cuestion social en las Antillas Españolas* (Madrid, 1874), *Sistemas coloniales* (Madrid, 1874), &c.; R. Montoro, *Discursos ... 1878-1893* (Philadelphia, 1894); Labra *et al.*, *El Problema colonial contemporánea* (2 vols., Madrid, 1894); articles by Em. Castelar *et al.*, in Spanish reviews (1895-1898). On the period since 1899 the best two books in English are C. M. Pepper, *To-morrow in Cuba* (New York, 1899); A. G. Robinson, *Cuba and the Intervention* (New York, 1905).

(F. S. P.)

---

1 Other countries taking only 27,462 long tons out of a total of 5,719,777 in the seven fiscal years 1899-1900 to 1905-1906.

2 In these same years the trade of the United States with Cuba and Porto Rico was: importations from the islands, $59,221,444 annually; exportations to the islands, $20,017,156. The corresponding figures for Spain were $7,265,142 and $20,035,183; and for the United Kingdom, $714,837 and $11,971,129, the trade with other countries being of much less amount.

3 In the preliminary registration by Moderate officials a total electorate was registered of 432,313,—about 30% of the supposed population of the island.

www.ingramcontent.com/pod-product-compliance
Lightning Source LLC
LaVergne TN
LVHW050300210926
848481LV00014BA/63